Pair Correlations in Many-Fermion Systems

Pair Correlations in Many-Fermion Systems

Edited by

Vladimir Z. Kresin
Lawrence Berkeley Laboratory
University of California
Berkeley, California

Plenum Press • New York and London

Library of Congress Cataloging-in-Publication Data

On file

Proceedings of an Advanced Study Institute School on Pair Correlations in
Many-Fermion Systems, held June 5 – 15, 1997, in Erice, Sicily, Italy

ISBN 0-306-45823-3

© 1998 Plenum Press, New York
A Division of Plenum Publishing Corporation
233 Spring Street, New York, N.Y. 10013

http://www.plenum.com

10 9 8 7 6 5 4 3 2 1

All rights reserved

No part of this book may be reproduced, stored in a retrieval system, or transmitted in any form
or by any means, electronic, mechanical, photocopying, microfilming, recording, or otherwise,
without written permission from the Publisher

Printed in the United States of America

PREFACE

At first glance, the articles in this book may appear to have nothing in common. They cover such seemingly disparate subjects as the properties of small metallic clusters and the behavior of superfluid He3, nuclear physics and organic materials, copper oxides and magnetic resonance. Why have they been brought together, particularly in our time of narrow specialization?

In fact, the properties and effects described in this book touch upon one and the same fundamental phenomenon: pair correlation. Introduced in the theory of superconductivity by J. Bardeen, L. Cooper, and J. Schrieffer (BCS), this effect plays a key role in various Fermi systems.

The book consists of several sections. The first chapter is concerned with conventional and high Tc superconductors. The second chapter describes two relatively young families of superconductors: organics and fullerenes. Chapter III addresses the superfluidity of He^3. The discovery of this phenomenon in 1971 was a big event in physics and last year was acknowledged by a Nobel prize. This book contains the text of the Nobel lecture. Chapters IV and V are devoted to correlations in finite Fermi systems such as small metallic clusters, C_{60} anions, and atomic nuclei.

The book thus covers a broad range of problems, illuminating the close ties between various areas of physics.

Almost all of the included papers were presented as Lectures at the International School on Pair Correlation in Many-Fermion Systems (Erice, Sicily, June 1997). The idea of the School was to bring together leading scientists working in different fields and demonstrate to the students (and now to the readers) the mutual influence of various branches of science. This multidisciplinary scenario found a welcoming home at the superb Majorana Scientific Centre in Erice.

I would like to express special thanks to Prof. G. Benedek (University di Milane), Director of the School of Solid State Physics, and to Prof. A. Bussman-Holder (Max-Planck Institut, Stuttgart) for their help in organizing the School.

I am also grateful to A. Bill and to E. Dynin for their help in the preparation of this book.

Vladimir Z. Kresin

CONTENTS

I. Superconducting State in Conventional and High Tc Oxides

Foundations of Superconductivy and Extension to Less-Common Systems 3
L. Gor'kov

The Isotope Effect in Superconductors 25
A. Bill, V. Z. Kresin, and S. Wolf

Hyperfine Interactions in Metals 57
W. D. Knight

Perovskite Oxides: A Rick and Fascinating Crystal Class Family 63
A. Bussmann-Holder

Josephson Effect: Low-Tc vs. High-Tc Superconductors 75
A. Barone

The Spectrum of Thermodynamic Fluctuations in Short Coherence Length
Superconductors 89
A. Gauzzi

Microwave Response of High-Tc Superconductors 111
A. Agliolo Gallitto, I. Ciccarello, M. Guccione, and M. Li Vigni,
D. Persano Adorno

II. Organic Superconductivity. Fullerenes

Structure and Phase Diagram of Organic Superconductors 135
T. Ishiguro and H. Ito

Pairing and Its Mechanism in Organic Superconductors 147
T. Ishiguro

Superconductivity in Doped C_{60} Compounds 155
O. Gunnarsson, E. Koch, and R. Martin

On the Pair Correlations between Electrosolitions 173
L. Brizhik and A. Eremko

III. Superfluid He^3

The Pomeranchuk Effect 187
   R. Richardson

Pair Correlations in Superfluid Helium 3 205
   D. Vollhardt

IV. Finite Systems. Clusters

Non-Adiabacity and Pairing in the Finite Systems 223
   V. Z. Kresin

Fullerene Anions and Pairing in Finite Systems 235
   Th. Jolicoeur

Quantized Electronic States in Metal Microclusters: Electronic Shells, Structural
   Effects, and Correlations 245
   V. V. Kresin and W. Knight

V. Finite Systems. Nuclei

Role of Pairing Correlations in Cluster Decay Processes 265
   D. S. Delion, A. Insolia, and R. J. Liotta

Index 295

I. SUPERCONDUCTING STATE IN CONVENTIONAL AND HIGH Tc OXIDES.

FOUNDATIONS OF SUPERCONDUCTIVITY AND EXTENSION TO LESS-COMMON SYSTEMS

Lev P. Gor'kov[1,2]

[1]National High Magnetic Field Laboratory, Florida State University
[2]L.D.Landau Institute for Theoretical Physics
Russian Academy of Sciences

Abstract

The review of fundamentals of the microscopic theory of superconductivity, as a whole, serves to the purpose to elicit a knowledge about theoretical results and qualitative statements which would be robust enough to allow a judgment regarding the nature of superconductivity in such uncommon superconductors, as cuprates or superconductors of the Heavy Fermion family.

INTRODUCTION

Phenomenon of superconductivity long ago is an integral part of our education and our everyday life. It is especially so now, after the discovery by Bednorz and Müller of the so-called "High Temperature Superconductivity" in cuprates, for which critical temperatures, T_c's, amount to the range of $\sim 100K$, far exceeding the nitrogen's liquefaction temperature (of 77K).

Needless to say, we are now witnessing enormous experimental and theoretical struggle in attempts to explain unusual cuprates' properties and disclose the mechanism for so high a transition temperature. However, whenever superconductivity itself is discussed, even in cuprates more often than not, one applies views and concepts borrowed from the experience accumulated in about four-decades-long studies of more "common" (low T_c) superconductors. During that time a few scientific generations have changed, and, as today, it seems very timely to outline the basic concepts and methods by which superconductivity can be approached.

In the classical Bardeen, Cooper, and Schrieffer's (BCS) theory[1] of superconductivity, one is to distinguish between its two main constitutions: the phonon-mediated attraction between electrons, as a mechanism for superconductivity, and possibility of the Fermi liquid's instability caused by a weak interaction, commonly known as the "Cooper phenomenon". This latter one has proved to be of a broader significance, while giving rise to more or less straightforward generalization of the original BCS theoretical scheme. The well-known realization of the views are properties of superfluid

^{3}He, superconductivity in the so-called "Heavy Fermions" (HF) materials and, possibly, superconductivity in high-T_c cuprates.

In what follows we outline the main results of the theory of superconductivity by adopting the view that transition into superconducting state, as induced by a Cooper instability[2], is phase transition: it changes the electrons' symmetry and, hence, must be characterized by an order parameter[3]. Starting with discussion of the Cooper instability, we develop symmetry approach to superconductivity. Competing interactions, weak coupling limit vs. strong coupling (in the frameworks of the phonon model) etc., are also topics of our interest. The model is then generalized to anisotropic metals and arbitrary shape of the Fermi surface, with a more extended discussion of broken symmetries of the superconducting order parameter. The effects of impurities are also considered in some detail. At the end we briefly summarize our review and comment on the current situation in superconducting cuprates.

The discussion, as a whole, serves to the purpose to elicit a knowledge about theoretical results and qualitative statements which would be robust enough to allow a judgment regarding the nature of superconductivity in such uncommon superconductors, as cuprates or superconductors of the Heavy Fermion family.

Throughout the text the basic knowledge of diagrammatic methods (Green functions, Feynman diagrams, etc.) is supposed.[4,5]

COOPER'S INSTABILITY

Although the original paper by Cooper[2] was based on an oversimplified model and some unjustified assumptions, its conclusions, nevertheless, have removed the main hurdle on the way to understanding the phenomenon of superconductivity. Namely, although superconductivity was known to be a widespread property of metals, it appeared to be a low temperature phenomenon, as characterized by critical temperatures in the range of $1 \div 10K$ for common superconductors –an energy scale exceedingly small compared with typical electronic energies ($\sim 1eV$), or even with characteristic phonon frequencies, $\bar{\omega} \sim 10^2 K$. (The lattice's participation in the phenomenon has been known since the discovery of the isotope effect in 1950[6]). Smallness of T_c, hence, would have implied a weakness of the effective interactions, while the quantum mechanics teaches us that for the three-dimensional processes even attractive but arbitrary weak interactions would produce no singularities. According to [2], the situation drastically changes for electrons in the presence of the Fermi sea. A more rigorous proof sketched below, shows that the normal ground state of a metal may become unstable when e-e-interactions are taken into account (see [4] for more details).

Let $u(\mathbf{p}'_1 - \mathbf{p}_1)$ be the Fourier component of some short range interaction, $U(\mathbf{r} - \mathbf{r}')$, between electrons (shown in Figure 1a as the dashed line), which gives rise to the matrix element for scattering of two electrons, $(\sigma_1, \mathbf{p}_1)$ and $(\sigma_2, \mathbf{p}_2)$, with the momentum transfer $(\mathbf{p}'_1 - \mathbf{p}_1)$. Consider now the perturbation series in the form of the ladder diagrams of Figure 1b, which may be summed up by writing it as the diagrammatic equation for the vertex $\Gamma_{\sigma,-\sigma;\sigma',-\sigma'}(p, q - p; p', q - p')$ in Figure 1c. (Here we use the 4-dimensional notations, $p = (\mathbf{p}, \omega)$, where ω is the discrete thermodynamic frequency, $\omega \equiv i\omega_n; \omega_n = (2n + 1)\pi T$ for the fermionic lines). The kernel of this integral equation, according to Figure 1b, is given by

$$-T \sum_{\varepsilon} \int \frac{u(\mathbf{p} - \bar{\mathbf{p}})}{(2\pi)^3} \mathcal{G}(\bar{p}) \mathcal{G}(q - \bar{p}) d^3\bar{\mathbf{p}} \tag{1}$$

where $\mathcal{G}(\bar{p})$ are the thermodynamic Green functions of free electrons:

$$\mathcal{G}(\bar{p}) = \frac{1}{i\varepsilon_n - \varepsilon(\bar{\mathbf{p}}) + \mu} \tag{2}$$

The reason for selecting the ladder diagrams in Figure 1b becomes clear if $q = 0$ (or is small). Two electronic states $(\sigma, \mathbf{p})$ and $(-\sigma, -\mathbf{p})$ are then related to each other by time-reversal symmetry; this remains to be also true in each intermediate state along the ladder, as represented by the product of two Green functions in eq.(1); in fact, $\varepsilon(\bar{p}) \equiv \varepsilon(-\bar{p})$, independently on the specific choice of electronic spectrum in a metal. Summation over $\varepsilon_n = (2n+1)\pi T$ and integration over momentum $\bar{p}$ provide a singular contribution in the vicinity of the Fermi surface:

$$-T \sum_{\varepsilon_n} \int \mathcal{G}(\bar{p})\mathcal{G}(-\bar{p}) \frac{d^3\bar{p}}{(2\pi)^3} = -T \sum_{\varepsilon_n} \int \frac{d\xi}{(2\pi)^3} \frac{v_F^{-1}(\bar{p})dS_{\bar{p}}}{\varepsilon_n^2 + \xi^2} \quad (3)$$

Here

$$\xi = v_F(\bar{p})(p - p_F(\bar{p})) \equiv \varepsilon(\bar{p}) - \mu \quad (3')$$

is the energy dependence on the perpendicular to the Fermi surface momentum component, while integration over $S_{\bar{p}}$ would average any other dependence on position of momentum $\bar{p}$ at the Fermi surface. (The integral $\int \frac{dS_{\bar{p}}}{v_F(\bar{p})(2\pi)^3}$ is the normal density of states, $\nu(\varepsilon_F)$, per one spin direction. As for T itself, it is supposed to be of order of T_c, the critical temperature for superconductivity onset).

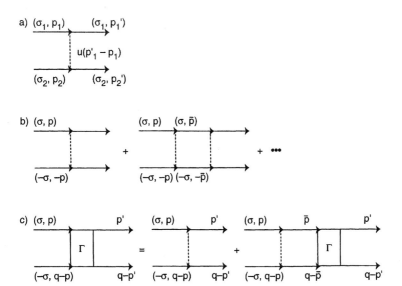

Figure 1. a) Matrix element of e-e-interaction; b) Ladder diagrams; c) Integral equation for the vertex part.

Summation over ε_n at large n can be approximated by integration over ε: $T \sum_{\varepsilon_n} \cong \frac{1}{2\pi} \int d\varepsilon$, and one obtains in (3) the following integral:

$$\int \int_{-\infty}^{+\infty} \frac{d\varepsilon d\xi}{\xi^2 + \varepsilon^2} \quad (3'')$$

which logarithmically diverges both at large and small (ξ, ε). The upper-limit-divergence may be removed by details of interactions, say, by a cut-off at some large energy, $\tilde{\omega}$, while the low limit is made finite solely by the fact that frequency ε_n is actually the discrete variable, $\varepsilon_n = (2n+1)\pi T$. Therefore, eq.(3) contains the logarithmic contribution of a general form:

$$\nu(\varepsilon_F) \ln \frac{\tilde{\omega}}{T} \qquad (4)$$

which becomes arbitrarily large when T tends to zero ($T \ll \tilde{\omega}$). At low enough T large value of (4) can compensate any smallness of interaction $u(\mathbf{p}-\bar{\mathbf{p}})$ in eq.(1), resulting in non-perturbative solutions for the vertex part, Γ, in Figure 1c.

Assuming that this is the case, one may omit in $u(\mathbf{p}-\bar{\mathbf{p}})$ the dependence on perpendicular (to the Fermi surface) components both of $\mathbf{p}$ and $\bar{\mathbf{p}}$ leaving only the angular dependence between $(\mathbf{p},\bar{\mathbf{p}})$. As for the vertex in Figure 1c, $\Gamma(\mathbf{p},\mathbf{p}')$ now depends on $\mathbf{p},\mathbf{p}'$, both momenta are at the Fermi surface. The equation now reads (for the isotropic spectrum):

$$\Gamma(\mathbf{p}-\mathbf{p}') = u(\mathbf{p}-\mathbf{p}') - (\ln \frac{\tilde{\omega}}{T}) \frac{mp_0}{(2\pi)^3} \int u(\mathbf{p}-\bar{\mathbf{p}}) \Gamma(\bar{\mathbf{p}}-\mathbf{p}') d\Omega \qquad (5)$$

For $u(\mathbf{p}-\mathbf{p}') \equiv u_0$, as in the isotropic BCS-theory, one obtains

$$\Gamma = \frac{u_0}{1 + \frac{u_0 m p_0}{2\pi^2} \ln \frac{\tilde{\omega}}{T}} \qquad (6)$$

At negative $u_0 < 0$ expression (6) as the function of temperature, has the pole at

$$T_c = \tilde{\omega} \exp\left(-\frac{2\pi^2}{mp_0 |u_0|}\right) \qquad (6')$$

so that solution of (5) is not possible below T_c. The singularity at T_c signals that the normal phase can not exist anymore. The Bethe-Solpiter's form of equation for Γ in Figure 1c implies, indeed, that it is the tendency of two electronic states connected near the Fermi surface to each other by the time-reversal transformation, to form a bound pair, which is responsible for the superconducting instability.

Eq.(5) allows an easy generalization to the case of an anisotropic metal; there are also some other points with regard to (5) which press for a more detailed discussion. We will come back to it later. Before to conclude this section, let us write down the dependence of the vertex part Γ in Figure 1b,c as the function of ω_0 and $\mathbf{q}$, at small $\omega_0, v_F q \ll T_c$ (above T_c):

$$\Gamma = -\frac{\nu^{-1}(\varepsilon_F)}{-i\frac{\omega_0}{8\pi T_c} + \frac{T-T_c}{T_c} + \frac{(v_F q)^2 7\xi(3)}{6(\pi T_c)^2}} \qquad (7)$$

Two conclusions can be drawn from (7); the first one is that an attractive weak interaction, $|u_0| \nu(\varepsilon_F) \ll 1$, renormalizes itself (near T_c) to the atomic strength, although in a rather narrow frequency and momentum interval:

$$T \sim \omega_0 \sim v_F |\mathbf{q}| \sim T_c \qquad (8)$$

Secondly, the vertex $\Gamma(\omega_0)$, if continued from the thermodynamical axis onto the upper half-plane of the physical frequencies, $z = i\omega_0$, is analytical there only at T above T_c. To restore the proper analytical behavior below T_c, transition into new (superconducting) phase is to be explicitly taken into account.

EQUATIONS FOR SUPERCONDUCTING PHASE.

Notion of the bound electronic pairs with the total momentum $\mathbf{q} = 0$ brings into consideration a new, bosonic degree of freedom. Thus, for the free Bose gas below the

Bose-condensation temperature, a finite fraction, n_0, of the *total* number of particles, N, in the whole volume V, occupies the single quantum state with $\mathbf{k} \equiv 0 : N_0 = <\hat{b}_0^+ \hat{b}_0> = n_0 N \gg 1$. The matrix elements of the corresponding creation, $\hat{b}_0^+$, and annihilation, $\hat{b}_0$, operators are then so large that the operators themselves play a role of the c-numbers. There is no direct correspondence, of course, between Bose particles and the Cooper pairs: strength of the interaction (7) rapidly decreases with increase of the total pair momentum $\mathbf{q}$, outside the interval (8). (It also means that the center of mass of a Cooper pair is never defined with an accuracy better than $\xi_0 \sim v_F/T_c$). However, at *small* $\mathbf{q}$ the interaction gets so strong that one may assume that, indeed, many electronic pairs are created very close to T_c, and they all go then in a coherent quantum state with $\mathbf{q} = 0$ below T_c. The role of the Bose operators, $\hat{b}_0^+, \hat{b}_0$, would be taken over by fermionic operators products of the type[3]:

$$\sum_p \hat{a}_{p\sigma}^+ \hat{a}_{-p-\sigma}^+; \quad \sum_p \hat{a}_{p\sigma} \hat{a}_{-p-\sigma}$$

With the above in mind, consider the equations of motion for the fermion field operators, $\hat{\psi}(x), \hat{\psi}^+(x)$. In what follows, we will mostly discuss thermodynamic properties of superconductors. Correspondingly, all equations below are written in the thermodynamic technic and, hence, $x = (\tau, \mathbf{r})$, where τ is the "time" variable ($0 < \tau < 1/T$) for the Matsubara operators in the Heisenberg representation[4,5]:

$$\hat{\psi}_\alpha(x) = e^{\tau(\hat{H}-\mu\hat{N})} \hat{\psi}_\alpha(\mathbf{r}) e^{-\tau(H-\mu\hat{N})} \quad (9)$$

(with the same connection between the Heisenberg- and Schroedinger representations for the operator $\hat{\psi}_\alpha^+(x)$).

The Hamiltonian, $\hat{H}$, for simplicity sake, is chosen in the form:

$$\hat{H} = \hat{H}_0 + \frac{1}{2}\int\int \hat{\psi}_\lambda^+(\mathbf{y})\hat{\psi}_\rho^+(\mathbf{y}')U(\mathbf{y}-\mathbf{y}')\hat{\psi}_\rho(\mathbf{y}')\hat{\psi}_\lambda(\mathbf{y})d^3\mathbf{y}d^3\mathbf{y}' \quad (10)$$

Here $\hat{H}_0$ is the Hamiltonian of free electrons (for an anisotropic metal its spectrum in the momentum space has an arbitrary dispersion, $\varepsilon(\mathbf{p})$). The second term corresponds to an interaction in the real space. To match the results of the previous section, we assume again that $U(\mathbf{r}-\mathbf{r}')$ is a short range interaction, so that $|\mathbf{r}-\mathbf{r}'| \sim a$, where $a_0 \sim p_F^{-1}$ is an atomic scale. Taking the time derivative of $\hat{\psi}_\alpha(x)$ in eq.(9), one obtains:

$$\frac{\partial \hat{\psi}_\alpha(x)}{\partial \tau} = [\hat{H} - \mu\hat{N}, \hat{\psi}_\alpha(x)] = -(\varepsilon_0(\hat{p}) - \mu)\hat{\psi}_\alpha(x)$$
$$- \int U(\mathbf{r}-\mathbf{y})\hat{\psi}_\rho^+(\mathbf{y})\hat{\psi}_\rho(\mathbf{y})\hat{\psi}_\alpha(x)d^3\mathbf{y} \quad (11)$$

Since $U(\mathbf{r})$ is otherwise small enough to neglect by more usual Fermi liquid corrections due to the e-e-interaction, the only distinction in the superconducting phase comes from the "Bose condensation" of the Cooper pairs, which manifests itself in new non-zero averages of the product in (11) for two $\hat{\psi}$-operators:

$$\hat{\psi}_\rho(y)\hat{\psi}_\alpha(x) \simeq <\hat{\psi}_\rho(y)\hat{\psi}_\alpha(x)> \quad (12)$$

where $<...>$ is taken over the Gibbs' thermodynamical ensemble. It is more convenient to introduce the notation

$$\Delta_{\rho\alpha}(x-y) = -U(\mathbf{x}-\mathbf{y})<\hat{\psi}_\rho(y)\hat{\psi}_\alpha(x)> \quad (13)$$

and re-write eq.(11) in the form:

$$-\frac{\partial \hat{\psi}_\alpha(x)}{\partial \tau} = (\varepsilon_0(\hat{p}) - \mu)\hat{\psi}_\alpha^+(x) - \int \hat{\psi}_\rho^+(y)\Delta_{\rho\alpha}(x-y)d^3\mathbf{y} \quad (11')$$

If the above derivation were repeated for the operator $\hat{\psi}_\beta^+(x)$, the result is

$$\frac{\partial \hat{\psi}_\alpha^+(x)}{\partial \tau} = (\varepsilon_0(\hat{\mathbf{p}}) - \mu)\hat{\psi}_\alpha^+(x) - \int \Delta_{\alpha\lambda}^+(x-y)\hat{\psi}_\lambda(y)d^3\mathbf{y} \tag{14}$$

where

$$\Delta_{\alpha\lambda}^+(x-y) = -U(\mathbf{x}-\mathbf{y}) < \hat{\psi}_\alpha^+(x)\hat{\psi}_\lambda^+(y) > \tag{13'}$$

The order parameters $\hat{\Delta}, \hat{\Delta}^+$, indeed, couple together the electronic and the hole states.

The above equations (11-14) suggest that the well-known diagrammatic formalism in the normal phase may be restored below T_c in the superconducting phase[3]. Consider, in addition to the Green functions $\mathcal{G}_{\alpha\rho}(x,x') = - < T_\tau(\hat{\psi}_\alpha(x)\hat{\psi}_\beta^+(x')) >$ (the average $< \ldots >$ is taken over the full Gibb's ensemble, and T_τ is the time ordering with respect to the Matsubara "time", τ), two "anomalous" (Gor'kov's) Green functions:

$$\begin{aligned}\mathcal{F}_{\alpha\beta}(x,x') &=< T_\tau(\hat{\psi}_\alpha(x)\hat{\psi}_\beta(x')) > \\ \mathcal{F}_{\alpha\beta}^+(x,x') &=< T_\tau(\hat{\psi}_\alpha^+(x)\hat{\psi}_\beta^+(x')) > \end{aligned} \tag{15}$$

Generalization of the Wick's theorem in the form:

$$\begin{aligned} < T_\tau(\hat{\psi}_\alpha(x_1)\hat{\psi}_\beta(x_2)\hat{\psi}_\gamma^+(x_3)\hat{\psi}_\delta^+(x_4)) > = \\ = -\mathcal{G}_{\alpha\gamma}(x_1,x_3)\mathcal{G}_{\beta\delta}(x_2,x_4) + \mathcal{G}_{\alpha\delta}(x_1,x_4)\mathcal{G}_{\beta\gamma}(x_2,x_3) + \\ + \mathcal{F}_{\alpha\beta}(x_1,x_2)\mathcal{F}_{\gamma\delta}^+(x_3,x_4)\end{aligned} \tag{16}$$

when combined with equations of motion for the operators $\hat{\psi}_\alpha(x), \hat{\psi}_\alpha^+(x)$ in their form (11), immediately makes it possible to decouple products of the four $\hat{\psi}$-operators arising due to the interaction term in (11, 14) (as before, the first two terms in (16) containing products of $\mathcal{G}$-functions, may be omitted as insignificant Fermi-liquid corrections). Before writing the resulting equations for functions $\mathcal{G}_{\alpha\beta}$ and $\mathcal{F}_{\alpha\beta}$ and $\mathcal{F}_{\alpha\beta}^+$, it is worth our while to discuss the symmetry properties of the latter in some more details[3,7].

The functions $\mathcal{F}_{\alpha\beta}(x,x')$ and $\mathcal{F}_{\alpha\beta}^+(x,x')$ in eq.(15) are the order parameters of the new phase. Since at any gradient transformation

$$\mathbf{A}(\mathbf{r}) \Rightarrow \mathbf{A}(\mathbf{r}) + \nabla\chi(\mathbf{r}) \tag{17}$$

the wave functions, (i.e., operators $\hat{\psi}(x), \hat{\psi}^+(x)$), would transform as

$$\hat{\psi}(x) \Rightarrow \hat{\psi}(x)\exp\left[\frac{ie\chi(\mathbf{r})}{\hbar c}\right]; \quad \hat{\psi}^+(x) \Rightarrow \hat{\psi}^+(x)\exp\left[\frac{-ie}{\hbar c}\chi(\mathbf{r})\right] \tag{17'}$$

the order parameters of eq.(15) are not invariant under transformation (17), (17'). In other words, at the superconducting transition the gauge invariance is the symmetry that gets broken. (The Abelian group of transformations (17, 17') is usually denoted as $U(1)$). In absence of external fields the system is homogeneous, and at an appropriate gauge choice all Green functions would depend on the coordinates difference.

Note that although for the the Gibbs ensemble the total number of particles N, is not fixed, both $\mathcal{F}$ and $\mathcal{F}^+$ are comprised of non-diagonal matrix elements:

$$\begin{aligned}\mathcal{F} &\Rightarrow < N-2 \mid \hat{\psi}\hat{\psi} \mid N > \\ \mathcal{F}^+ &\Rightarrow < N \mid \hat{\psi}^+\hat{\psi}^+ \mid N-2 > \end{aligned} \tag{18}$$

Therefore, if a change in the total particle number becomes a matter of importance (for instance, at tunneling phenomena in the Josephson junctions), the appropriate thermodynamical variable is N-the number of particles, not the chemical potential, μ. In these variables the functions (15) would always include an additional "time" dependences[3]:

$$e^{-2\mu\tau}\mathcal{F}_{\alpha\beta}(x-x'), \quad e^{2\mu\tau}\mathcal{F}^+_{\alpha\beta}(x-x') \tag{15'}$$

(with $\exp(2\mu\tau) \Rightarrow \exp(i2\mu t)$ on the physical time axis).

Consider now averages (15) at coinciding $\tau_1 = \tau_2 + 0$:

$$\mathcal{F}_{\alpha\beta}(\mathbf{r},\mathbf{r}';0_+) = <\hat{\psi}_\alpha(\mathbf{r})\hat{\psi}_\beta(\mathbf{r}')>$$
$$\mathcal{F}^+_{\alpha\beta}(\mathbf{r},\mathbf{r}';0_+) = <\hat{\psi}^+_\alpha(\mathbf{r})\hat{\psi}^+_\beta(\mathbf{r}')> \tag{19}$$

The fermion operators in (19) anti-commute at equal times. As a result, one obtains from (19):

$$\mathcal{F}_{\alpha\beta}(\mathbf{r},\mathbf{r}';0_+) = -\mathcal{F}_{\beta\alpha}(\mathbf{r}',\mathbf{r};0_+)$$
$$\mathcal{F}^+_{\alpha\beta}(\mathbf{r},\mathbf{r}';0_+) = -\mathcal{F}^+_{\beta\alpha}(\mathbf{r}',\mathbf{r};0_+)$$
$$(\mathcal{F}_{\alpha\beta}(\mathbf{r},\mathbf{r}';0_+))^* = \mathcal{F}^+_{\beta\alpha}(\mathbf{r}',\mathbf{r};0_+) \tag{20}$$

Let's emphasize again that eqs.(20) are nothing but the direct consequence of the Fermi-statistics for electronic operators. If spin is a good variable in the metallic system, the above sign change at the permutation of operators may be attached either to spin or to space variables. Considering dependence on $\mathbf{r} - \mathbf{r}'$ as a dependence on the relative coordinate of two electrons constituting the pair, one arrives at the conclusion that if spin S of the pair is zero, in

$$\mathcal{F}_{\alpha\beta}(\mathbf{r}-\mathbf{r}';0_+) = -i(\sigma_2)_{\alpha\beta}\mathcal{F}(\mathbf{r}-\mathbf{r}') \tag{21}$$

the new function $\mathcal{F}(\mathbf{r}-\mathbf{r}')$ is to be even under the inversion (i.e., it may consist only of even spherical harmonics), and vice versa, at $S=1$. The antisymmetric matrix

$$\hat{I} = -i\hat{\sigma}_2; \quad I^2 = -\hat{1} \tag{22}$$

substitutes for the singlet spin wave function. For the triplet case, $S=1$, instead (21), one may write

$$\mathcal{F}_{\alpha\beta}(\mathbf{r}-\mathbf{r}';0_+) = [(\hat{\sigma}\cdot\mathbf{d})\hat{I}]_{\alpha\beta}\mathcal{F}(\mathbf{r}-\mathbf{z}') \tag{21'}$$

where a vector $\mathbf{d}$ provides the axis of quantization for the total spin, $S=1$, of the pair. The spin part in (20') is now symmetric under permutation of the spin indices.

The above symmetry properties for functions $\mathcal{F}, \mathcal{F}^+$ at coinciding time variables, $\tau = \tau' + 0$, as it follows from equations below, give rise to exactly the same behavior at permutations of the 4-component variables, x and x', in the definitions (15). On the other hand, one may introduce such a symmetry as a *definition*. It opens a new possibility that functions $\mathcal{F}_{\alpha\beta}(x-x')$ may be odd at permutation of the *time* variables, τ and τ'. This would lead to the so-called "odd frequency" superconductivity[8,9].

Returning to eqs.(13), (13'), one may re-write Δ's in the new notations ($S=0$):

$$\Delta_{\rho\alpha}(x-y) = \hat{I}_{\alpha\rho}U(\mathbf{x}-\mathbf{y})\mathcal{F}(\mathbf{x}-\mathbf{y})$$
$$\Delta^+_{\alpha\lambda}(x-y) = -\hat{I}_{\alpha\lambda}U(\mathbf{x}-\mathbf{y})\mathcal{F}^+(\mathbf{x}-\mathbf{y}) \tag{23}$$

and in the Fourier components:

$$\Delta(\mathbf{p}) = -\frac{1}{(2\pi)^3}\int u(\mathbf{p}-\mathbf{p}')\mathcal{F}(\mathbf{p}';0_+)d^3\mathbf{p}'$$
$$\Delta^+(\mathbf{p}) = -\frac{1}{(2\pi)^3}\int u(\mathbf{p}-\mathbf{p}')\mathcal{F}^+(\mathbf{p}';0_+)d^3\mathbf{p}' \tag{22'}$$

(Spin indices were omitted in (22'): in absence of magnetic phenomena the "diagonal" Green function $\mathcal{G}_{\alpha\beta}(1,2) \equiv \delta_{\alpha\beta}\mathcal{G}(1,2)$, and all equations may be written in terms of scalar functions. This is not exactly so generally, in case, say, of a more exotic superconductivity, which we hope to discuss later).

With all preliminary work already done above, it is straightforward from definitions of all the Green functions and with use of equations of motion for the operators $\hat{\psi}(x), \hat{\psi}^+(x)$, to obtain a complete set of equations which embraces four Green functions. They can be unified into the following 2x2-matrix form:

$$\hat{G}^{-1}\hat{G} = 1 \tag{24}$$

or, in explicit form:

$$\begin{pmatrix} \{-\frac{\partial}{\partial \tau} - H_0(\hat{\mathbf{p}}) + \mu\} & \Delta \\ -\Delta^* & \{\frac{\partial}{\partial \tau} - \bar{H}_0(\hat{\mathbf{p}}) + \mu\} \end{pmatrix} \cdot$$
$$\begin{pmatrix} \mathcal{G}(x,x') & \mathcal{F}(x,x') \\ \mathcal{F}^+(x;x') & -\mathcal{G}(x',x) \end{pmatrix} = \hat{1} \tag{23'}$$

(the matrix multiplication in principle, means integration in the real space over intermediate coordinate: compare with eqs.(11'), (14)).

Operators $H_0(\hat{\mathbf{p}})$ and $\bar{H}_0(\hat{\mathbf{p}})$ differ in presence of an external magnetic field because they have the effect, as one may see from (23'), on operators $\hat{\psi}$ and $\hat{\psi}^+$, correspondingly. The standard procedure of including magnetic field effects into consideration, is to introduce the vector potential of the field, $\mathbf{A}(\mathbf{r})$, in the bare Hamiltonians, $H_0(\hat{\mathbf{p}})$ and $\bar{H}_0(\hat{\mathbf{p}})$, substituting correspondingly $\hat{\mathbf{p}} \Rightarrow \hat{\mathbf{p}} - \frac{e}{c}\mathbf{A}$ and $\hat{\mathbf{p}} \Rightarrow \hat{\mathbf{p}} + \frac{e}{c}\mathbf{A}$ into them. The system of equations (24) and (23') preserves the gauge invariance of any physical quantities. The Green functions themselves are, of course, not gauge-invariant, and their transformation behavior[7] follows from eqs.(17), (17') and their definitions (15).

Needless to say that in static fields the 2x2-Green function $\hat{G}(x,x')$ may be written in the form:

$$G(\mathbf{r},\mathbf{r}';i\omega_n) = \sum_\lambda \frac{1}{i\omega_n - E_\lambda} \begin{pmatrix} u_\lambda(\mathbf{r}) \\ v_\lambda(\mathbf{r}) \end{pmatrix} \otimes (u_\lambda^*(\mathbf{r}'), v_\lambda^*(\mathbf{r}')) \tag{24'}$$

where E_λ and $(u_\lambda(\mathbf{r}), v_\lambda(\mathbf{r}))$ are correspondingly, the eigenvalues and the eigenfunctions of the homogeneous equation in (24):

$$\begin{pmatrix} E_\lambda - (H_0(\hat{\mathbf{p}}) + \mu) & \Delta(\mathbf{r}) \\ \Delta^*(\mathbf{r}) & E_\lambda + \bar{H}_0(\hat{\mathbf{p}}) \end{pmatrix} \begin{pmatrix} u_\lambda(\mathbf{r}) \\ v_\lambda(\mathbf{r}) \end{pmatrix} = 0 \tag{23''}$$

In eqs. (24') and (23") we adjust (u,v) to the BCS- and the Bogolubov-notations.

In absence of a magnetic field, all functions depend on the coordinates difference. Therefore in the momentum space eqs.(23') acquire the algebraic character. In accordance with [4], we define:

$$\mathcal{G}(x) = T\sum_n \int e^{-i\omega_n \tau + i\mathbf{p}\mathbf{r}} \mathcal{G}(p) \frac{d^3\mathbf{p}}{(2\pi)^3} \tag{25}$$

(similarly, for $\mathcal{F}(p)$ and $\mathcal{F}^+(p)$; $\omega_n = (2n+1)\pi T$)). One obtains from (23'):

$$(i\omega_n - \xi(\mathbf{p}))\mathcal{G}(p) + \Delta(\mathbf{p})\mathcal{F}^+(p) = 1$$
$$(i\omega_n + \xi(\mathbf{p}))\mathcal{F}^+(p) + \Delta^+(\mathbf{p})\mathcal{G}(p) = 0 \tag{26}$$

Actually, in zero magnetic field, $\mathcal{F} = \mathcal{F}^+, \Delta = \Delta^+$. The solution for $\mathcal{G}(p)$, $\mathcal{F}^+(p)$ is

$$\mathcal{G}(p) = -\frac{i\omega_n + \xi(\mathbf{p})}{\omega_n^2 + \xi^2(\mathbf{p}) + |\Delta(\mathbf{p})|^2}; \quad \mathcal{F}^+ = \frac{\Delta(\mathbf{p})}{\omega_n^2 + \xi^2(\mathbf{p}) + |\Delta(\mathbf{p})|^2} \tag{27}$$

The continuation of the frequency $z = i\omega_n$ from the thermodynamical axis determines the gapped BCS spectrum as the real frequency poles in the Green functions:

$$E(\mathbf{p}) = \pm\sqrt{\xi^2(\mathbf{p}) + |\Delta(\mathbf{p})|^2} \tag{28}$$

In (22') we preserved the p-dependence because of the intention to discuss below some further implications, as far as an anisotropy in metals is concerned. For the isotropic model $\Delta \equiv$ const. It is well-known that the BCS-theory not only perfectly agrees with the vast experimental data (within an accuracy $\sim 10 \div 15\%$ over the whole temperature range below T_c) for most common superconductors– the explanation of the Hebel-Slichter peak[10] in the NMR-relaxation rate, $T_1^{-1}(T)$, at T_c can be considered as a *qualitative* proof of the basic BCS ideas. It is more subtle to address the issue whether or not the basic ideas, as we have discussed them, want some essential modification in superconductors of the Heavy Fermion family, or in cuprates. A more detailed analysis below is aimed to establish some robust features in theoretical predictions.

The system (23'), or (26) becomes complete when definition for $\Delta(\mathbf{p})$, eq.(22') is added to eqs.(26):

$$\Delta(\mathbf{p}) = -T \sum_{\omega_n} \int d^3p' \frac{u(\mathbf{p}-\mathbf{p}')}{(2\pi)^3} \frac{\Delta(\mathbf{p}')}{\omega_n^2 + \xi^2(\mathbf{p}') + |\Delta(\mathbf{p}')|^2} \tag{29}$$

At the temperature of transition, T_c, $\Delta(\mathbf{p}) \to 0$. In the remaining integral of eq.(29) one immediately recognizes the homogeneous part of eq.(5) for the vertex $\Gamma(\mathbf{p},\mathbf{p}')$: temperature of transition is the eigenvalue for the homogeneous solution of (5) to exist. To start with, in the isotropic case $u(\mathbf{p}) \equiv u_0$, and one gets again result (6'). On the other hand, at $T < T_c$, the gap, $\Delta(T)$, depends on temperature and is to be determined from the relation:

$$1 = \frac{|u_0|mp_F}{2\pi^2}\left[T\sum_{\omega_n=(2n+1)\pi T}\int d\xi \left(\frac{1}{\omega_n^2 + \xi^2 + \Delta^2(T)}\right)\right] \tag{29'}$$

It is exactly the same integral as in (3")(at large (ω,ξ)), but its low energy cut-off is now also due to presence of $\Delta^2(T)$ in (29'). Therefore, if definition of T_c (6') includes a high-energy cut-off, $\tilde{\omega}$, the same cut-off in (29') being expressed in terms of T_c, eq.(29') establishes relation between $\Delta(T)$ and T, T_c. The difference

$$\int d\xi \left[T_c \sum_{\omega_n=(2n+1)\pi T_c}\left(\frac{1}{\omega_n^2+\xi^2}\right) - T\sum_{\omega_n=(2n+1)\pi T}\left(\frac{1}{\omega_n^2+\xi^2+\Delta^2(T)}\right)\right] = 0 \tag{30}$$

now converges at high (ξ,ω_n) independently on any cut-off parameter. Once T_c is known, $\Delta(0)$ is also known. The BCS result from (30):

$$\Delta(0) = T_c\frac{\pi}{\gamma} \simeq 1.75 T_c$$

In other words, the ratio $\Delta(T)/\Delta(0)$ is a function of T/T_c only:

$$\frac{\Delta(T)}{\Delta(0)} = f(T/T_c) \tag{31}$$

In the BCS theory eq.(31) is true for any other physical quantity, while the corresponding functions, $f(T/T_c)$, are to be calculated microscopically. T_c is the basic parameter of the theory: if T_c is determined experimentally, the thermodynamics of a superconductor becomes known. The scale for the thermodynamical magnetic field, H_{CT}, is also set by the thermodynamical relation:

$$F_n(T) - F_s(T) = \frac{H_{CT}^2}{8\pi} \tag{32}$$

(F_n and F_s are the free energies in the normal and superconducting states, respectively). The second experimentally independent parameter, related to the fact that superconducting currents can flow below T_c to screen the penetration of a weak magnetic field, is the London's penetration depth, δ_L:

$$\delta_L^{-2}(0) = \frac{4\pi n e^2}{mc^2} \tag{33}$$

It is expressible in terms of the so-called "plasma-frequency", ω_{0p}.

To summarize, once T_c (and ω_{0p}) are known experimentally, the behavior of superconductor below T_c in the isotropic model gets pre-determined. As was already mentioned above, accuracy of the isotropic BCS theory is very good. As for expression (6') for T_c in terms of a cut-off parameter, $\tilde{\omega}$, it serves only to the purpose of removing the logarithmic divergence in (3"): it does not matter whether the cut-off applies to ω_n, or ξ, or both, for all other results of the BCS theory do not contain these divergences. However, for calculations of the T_c-value a microscopic picture of interactions is necessary. In the phonon model the cut-off, $\tilde{\omega}$, is of order of characteristic phonon frequencies, $\bar{\omega}$, and from the above point of view, is applied to summation over frequencies, ω_n.

Eqs.(26), or, more generally, eqs.(24, 23') can be presented in the diagrammatic form[7] of Figure 2: to distinguish between $\mathcal{G}$ and $\mathcal{F}, \mathcal{F}^+$ we provide any operator $\hat{\psi}(x)$ with the arrow directed out of the point x; correspondingly, an operator $\hat{\psi}^+(x')$ is represented by the arrow directed to x'. If the wiggly lines were identified with the interaction, $U(\mathbf{x}-\mathbf{y})$, in eq.(23), one may consider (22') as the new, anomalous self-energy parts, $\Sigma_2(\mathbf{p};\omega), \Sigma_2^+(\mathbf{p};\omega)$ (the regular self-energy corrections, $\Sigma_1(\mathbf{p};\omega)$, were omitted in Figure 2 assuming the weak coupling limit).

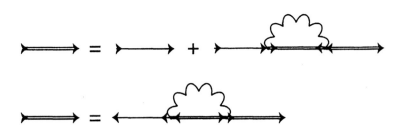

Figure 2. Diagrammatic equations for the Green Function, $\mathcal{G}(x,x')$ (two arrows in the same direction), and for the Gor'kov's functions, $\mathcal{F}(x,x')$ and $\mathcal{F}^+(x,x')$ (arrows in opposite directions, as explained in the text).

It is now straightforward to take explicitly into account the phonon mechanism (or any other mechanism in which electron-electron interaction is mediated by exchange of excitations of a bosonic character). The "wiggly" line becomes the phonon Green function, $g_{ph}^2 \mathcal{D}(x-x')$, which in the momentum representation is[4]:

$$g_{ph}^2 \mathcal{D}(\mathbf{q}, \varepsilon_n) = -\frac{g_{ph}^2 \Omega^2(\mathbf{q})}{\varepsilon_n^2 + \omega_0^2(\mathbf{q})}; \quad (\varepsilon_n = 2n\pi T) \tag{34}$$

(Here g_{ph} stands for the electron-phonon coupling constant, $\omega_0(\mathbf{q})$ is a phonon frequency mode. As for $\Omega^2(\mathbf{q})$, it is of the order of a typical phonon frequency, $\bar{\omega}$, but has no direct physical meaning - its momentum dependence also includes the momentum dependence of interaction between electrons and a given phonon mode). One would have then:

$$\Delta(\mathbf{p},\omega) = g_{ph}^2 T \sum_{\omega_n'} \int \frac{d^3\mathbf{p}'}{(2\pi)^3} \frac{\Omega^2(\mathbf{p}-\mathbf{p}')\mathcal{F}(\mathbf{p}';\omega_n')}{(\omega_n - \omega_n')^2 + \omega_0^2(\mathbf{p}-\mathbf{p}')} \qquad (35)$$

and the high energy cut-off, unlike eq.(29), is imposed by the very fact of the frequency dependence in the phonon Green function (34): integration in (35) over ξ and summation over ω_n' converge at $\xi \sim \omega_n' \sim \omega_0$. Of the utmost importance, however, is the fact that the phonon model can be solved at the arbitrary strength of interaction $(g_{ph}^2 \sim 1)^{11}$. This results from the Migdal's theorem (see in[4]), according to which any further corrections of the form shown in Figure 3, are small due to the adiabaticity parameter $(m/M) \ll 1$.

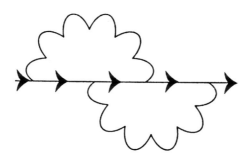

Figure 3. Strong-coupling corrections into the self-energy part for the Green function.

Eq.(35) provides a base for realistic calculations of the transition temperature, T_c, in a metal. As a matter of fact, in most common superconductors $T_c \ll \bar{\omega}$, indicating some weakness of the effective electron-phonon interaction. Note the exponential dependence of T_c on interaction's strength in the BCS expression (6')! In any event, in the phonon mechanism T_c is much smaller than electronic energy scale either because the interaction is weak, or merely due to the fact that phonons' frequencies are much below ~ 1 eV. For any other hypothetical boson exchange mechanism there is not the adiabatical parameter, and one could not neglect corrections of the type shown in Figure 3, except in the weak coupling limit.

Another problem is the role of the Coulomb repulsion. The physical reasons for "reducing" its strength are well-known: two electrons of the pair remain in distances of order of $\hbar v_F/\bar{\omega}$ in the real space. This diminishes Coulomb repulsion for the phonon model. On the other hand, the repulsive interaction is not a principle obstacle for superconductivity to occur, if it is short ranged, but has a structure in the momentum space. To demonstrate this point, consider an oversimplified model. In Figure 4 the isotropic Fermi sphere is artificially subdivided into two parts, as shown in Figure 4a. Let us take two electrons, ($\uparrow$ p) and ($\downarrow$ -p) with positions on its two semispheres, correspondingly, and consider diagrams of Figure 1b with the only difference; namely, assume that interaction in Figure 1a depends on whether a given electron, say ($\uparrow$ p), after scattering remains on its (right, R) part of the Fermi surface, or goes to the opposite (left, L) side; $u_{LL} = u_{RR}$ and $u_{LR} = u_{RL}$, correspondingly. The homogeneous part in the equation for Γ in Figure 1c consists of two components, Γ_1 and Γ_2, whose meanings are shown in Figure 4b.

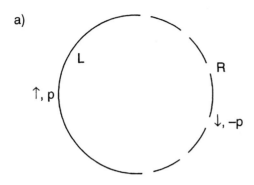

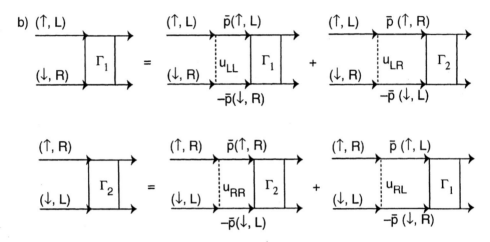

Figure 4. Simplified model for anisotropic pairing: a) two parts of spherical Fermi surface; b) Diagrammatic equations for scattering processes between the left (L) and right (R) pieces of the Fermi surface.

Assuming same energy cut-off, $\tilde{\omega}$, for both u_{LL} and u_{LR}, and making use of the analysis which had led us to eq.(4), the system of equations in Figure 4b becomes of the form:

$$\Gamma_1 = -\frac{A}{2}\nu(\varepsilon_F)u_{LL}\Gamma_1 - \frac{A}{2}\nu(\varepsilon_F)u_{LR}\Gamma_2$$
$$\Gamma_2 = -\frac{A}{2}\nu(\varepsilon_F)u_{LL}\Gamma_2 - \frac{A}{2}\nu(\varepsilon_F)u_{LR}\Gamma_1 \qquad (36)$$

($A = \ln \tilde{\omega}/T$), or in terms of $\Gamma_\pm = \Gamma_1 \pm \Gamma_2$

$$\Gamma_+[1 + \frac{A}{2}\nu(\varepsilon_F)(u_{LL} + u_{LR})] = 0$$
$$\Gamma_-[1 + \frac{A}{2}\nu(\varepsilon_f)(u_{LL} - u_{LR})] = 0 \qquad (36')$$

The asymmetrical part (with respect to momenta, at fixed spin directions), Γ_- would have a solution if $u_{LR} > u_{LL}$, even though interactions are repulsive! The diagonal part, u_{LL}, includes in this model the Coulomb scattering with small momenta transfers. In metals, in presence of the screening effects such processes are not singled out, so that a solution in (36), (36') may exist. For a rigorous BCS-like treatment interactions are to be weak enough to disregard corrections in Figure 3.

ANISOTROPY

At derivation of eq.(3) it has been mentioned that anisotropy in real metals does not change much the mathematics of the weak coupling BCS-scheme, for T_c is usually rather small, and superconductivity affects only a narrow vicinity of the Fermi surface. Nevertheless, simplicity of the isotropic model, such as dependencies of all characteristics on T/T_c only (eq.(31)), strictly speaking, disappears in a more general case. This is particularly true in the case of a "broken symmetry" of the order parameter, i.e. then the gap, $\Delta(\mathbf{p})$ has a lower symmetry than the metal initially.

To simplify the discussion below, in eq.(29) we first sum over the frequency variable[4]:

$$T \sum_{\omega_n} \frac{1}{\omega_n^2 + E^2(\mathbf{p}')} = \left(\frac{1}{2E(\mathbf{p}')}\right) th \frac{E(\mathbf{p}')}{2T}, \qquad (37)$$

write $u(\mathbf{p} - \mathbf{p}')$ in a more general form

$$u(\mathbf{p} - \mathbf{p}') \Rightarrow u(\mathbf{p}; \mathbf{p}') \qquad (38)$$

(both $\mathbf{p}$ and $\mathbf{p}'$ are at the Fermi surface of an arbitrary shape), and finally, transform

$$d^3\mathbf{p} \Rightarrow d\xi \frac{dS_\mathbf{p}}{v_F(\mathbf{p})} \qquad (38')$$

as in eq.(3). After it is done, eq.(29) reduces to:

$$\Delta(\mathbf{p}) = \frac{1}{(2\pi)^3} \int \frac{u(\mathbf{p};\mathbf{p}')dS'_{\mathbf{p}'}}{v_F(\mathbf{p}')} \Delta(\mathbf{p}') \int_0^{\tilde{\omega}} \left[\frac{1}{E(\mathbf{p}')} th\left(\frac{E(\mathbf{p}')}{2T}\right)\right] d\xi \qquad (39)$$

The logarithmic divergence is seen more explicitly and is easier for treatment. Re-write (39) in the form

$$\Delta(\mathbf{p}) = \frac{1}{(2\pi)^3} \int \frac{u(\mathbf{p};\mathbf{p}')}{v_F(\mathbf{p}')} \Delta(\mathbf{p}') \{\ln\left(\frac{\tilde{\omega}C}{T}\right) +$$
$$+ \int_0^\infty \left[\frac{1}{E(\mathbf{p}')} th\left(\frac{E(\mathbf{p}')}{2T}\right) - \frac{1}{\xi} th\left(\frac{\xi}{2T}\right)\right] d\xi\} dS_{\mathbf{p}'} \qquad (39')$$

i.e. we added and subtracted the term with $\Delta(\mathbf{p}') \equiv 0$ in $E(\mathbf{p}')$. The difference converges at $\xi \to \infty$, while the logarithmic term can be calculated explicitly (C is a constant, which we do not specify). Let us also introduce the notations:

$$y(\mathbf{p}) = \frac{1}{\sqrt{v_F(\mathbf{p})}} \Delta(\mathbf{p}); \quad \lambda(T) = \ln\left(\frac{\tilde{\omega}C}{T}\right);$$

$$K(\mathbf{p},\mathbf{p}') = \frac{1}{(2\pi)^3} \frac{u(\mathbf{p},\mathbf{p}')}{\sqrt{v_F(\mathbf{p})v_F(\mathbf{p}')}} \qquad (40)$$

In terms of (40), eq.(39') acquires a form of the integral equation:

$$y(\mathbf{p}) = \lambda \int K(\mathbf{p},\mathbf{p}')y(\mathbf{p}')dS_{\mathbf{p}'} +$$
$$+ \int K(\mathbf{p},\mathbf{p}')y(\mathbf{p}')F(T,\Delta(\mathbf{p}'))dS_{\mathbf{p}'} \qquad (41)$$

Here

$$F(T,\Delta(\mathbf{p}')) = \int_0^\infty \left[\frac{1}{E(\mathbf{p}')} th\left(\frac{E(\mathbf{p}')}{2T}\right) - \frac{1}{\xi} th\left(\frac{\xi}{2T}\right)\right] d\xi \qquad (41')$$

The linear part is the Fredholm integral equation with a symmetric kernal, $K(\mathbf{p},\mathbf{p}')$ in (40). Its highest eigenvalue, λ_c, defines the critical temperature, T_c, and the eigen function of the solution, which we write as:

$$y_c(\mathbf{p}) = \frac{\psi(\mathbf{p})}{\sqrt{v_F(\mathbf{p})}} \qquad (40')$$

Suppose now that $T < T_c$ and transform (39') to the form:

$$y(\mathbf{p}) - \lambda_c \int K(\mathbf{p},\mathbf{p}')y(\mathbf{p}')dS_{\mathbf{p}'} =$$
$$= \int K(\mathbf{p},\mathbf{p}')y(\mathbf{p}')\{\ln\frac{T_c}{T} + F(T,\Delta(\mathbf{p}'))\}dS_{\mathbf{p}'} \qquad (42)$$

Because λ_c is large, $(\sim 1/u \gg 1)$ two terms in the left hand side are each of order of unity in the weak coupling limit, while R.H.S. of (42) is of order of $u \ll 1$ and may be considered as small perturbation with respect to L.H.S. For (42) to have the meaning, the "orthogonality condition" of R.H.S. to the eigenfunction $\psi(\mathbf{p})$ gives rise to the full description of superconducting phase below T_c[12]:

$$\Delta(\mathbf{p}) = \Delta(T)\psi(\mathbf{p}) \qquad (43)$$

$$\int \frac{|\psi(\mathbf{p})|^2}{v_F(\mathbf{p})} \{\ln\frac{T_c}{T} + F(T,\Delta(T)\psi(\mathbf{p}))\}dS_{\mathbf{p}} = 0$$

Therefore the energy gap, $|\Delta(\mathbf{p})|$ is not a constant along the Fermi surface anymore. Hence, thermodynamical properties cease to be universal functions of $\frac{T}{T_c}$ only and may essentially depend on the specific form of $\psi(\mathbf{p})$ in (40'). (Thus, at low temperature the excitation's gap is defined by the minimal value of $\Delta(\mathbf{p})$, Δ_{min}). Although isotropic model describes properties of most superconductors rather well, as mentioned above, it cannot be so in case of the so-called "non-s-wave" superconductivity. Such a superconductivity type is, probably, characteristic for cuprates ("d-wave"). It is worthwhile to say a few words in this connection.

The normal metallic phase is characterized by its lattice symmetry. One usually means by that the transformation group, with respect to which the lattice is invariant; the latter consist of a point rotation group, G, and inversion, I. If no magnetism is involved in the normal phase, all properties are also invariant with respect to the time-reversal symmetry, R. We have also mentioned above the gradient (gauge) invariance, $U(1)$. Correspondingly, the kernal $K(\mathbf{p},\mathbf{p}')$ in eqs.(40), (41) possesses the lattice symmetry. On the other hand, its solutions $\psi(\mathbf{p})$ are not obliged to remain invariant under all transformations of the lattice group. As it is well known, the kernal can be written down in the form:

$$K(\mathbf{p};\mathbf{p}') = \sum_\lambda u_\lambda \psi_\lambda(\mathbf{p})\psi_\lambda^*(\mathbf{p}') \qquad (44)$$

where summation index λ runs over all eigenvalues of the operator K, and $\psi_\lambda(\mathbf{p})$ are the corresponding eigenfunctions and u_λ is a set of parameters of any sign, as is specific to the interaction.

Although the total number of eigenvalues in (44) is infinite, the eigenfunctions, $\psi_\lambda(\mathbf{p})$, may be classified by their symmetry behavior. For the isotropic case, for instance,

$$K(\mathbf{p};\mathbf{p}') \equiv K(\mathbf{p}-\mathbf{p}') = \sum_l u_l P_l(\cos\theta)P_l(\cos\theta') \qquad (44')$$

where $P_l(\cos\theta)$ are the Legendre polynomials, l is the angular momentum value, and each term in (44') realizes the corresponding representation of the three-dimensional rotation group, $O(3)$. For the point crystal group G the total number of different representations is finite, many different $\psi_\lambda(p)$ would possess the same transformation properties. For our example case, cuprates, one usually presumes the tetragonal group, $\mathcal{D}_4$. With respect to rotations the solutions (40') may transform as: $x^2 + y^2$, or z^2; xy; $x^2 - y^2$; x, y. The first two remain invariant under all tetragonal transformation (the "identical" representation, called nowadays the "s-wave" type, because of its being a straightforward generalization of the isotropic case). The last representation is two-dimensional and would need a further study[13]. The second and the third ones have zeros along appropriate directions of the tetragonal lattice. They both can be called as a "d-wave" type, but the existing experimental data are in favor of the $(x^2 - y^2)$ combination. It is worthwhile to stress that a form, such as

$$\mathbf{p}_x^2 - \mathbf{p}_y^2 \Rightarrow \frac{1}{\sqrt{v_F}}(\cos^2\varphi - \sin^2\varphi) \qquad (45)$$

is not the solution for the kernal $K(\mathbf{p}, \mathbf{p}')$ (44). Eq.(45) adequately reflects the fact that such a solution $\psi(\mathbf{p})$ changes sign along the diagonal tetragonal directions, so that the gap will pass through zero at these points (lines) at the Fermi surface. However, this is all we know: the actual $\psi(\mathbf{p})$ may significantly differ from (45) away from the zeros. At best eq.(45) can be chosen only to perform the calculations in eq.(43) to see how good an agreement with experiment would follow from it. On the other hand, the statement that gap has zeros at these points or lines is of the *qualitative* character and would result for instance, in the T^2- behavior of the specific heat at lowest temperatures.

The symmetry approach, i.e. enumeration of all the possibilities for different lattice symmetries, has been first developed for superconductors in [13] in connection with superconductivity in the so-called "Heavy Fermions" (UBe$_{13}$, UPt$_3$, etc.). The analysis is modified by the presence of strong spin-orbital interactions. Both $S=0$ and $S=1$ for the total spin of the Cooper pair in eqs.(20, 20') can be treated on the equal footing, although physical conclusions may differ for the two cases. The main difference stems from the fact that for $S=1$ the gap zeroes may exist only at a few symmetry *points* (nodes) on the Fermi surface resulting in a T^3-law for the low temperature specific heat. Among these phases some would posses magnetic properties (for more details see [13] and two reviews [14]).

From the above it is clear that anisotropy, especially anisotropy of the gap parameters, inevitable at a broken gap symmetry, introduces numerous complications into the microscopic theory due to the need of knowing of the solution $\psi(\mathbf{p})$ in eqs.(43). Detailed measurements of the temperature behavior for various physical quantities may prove to be helpful in restoring $\Delta(\mathbf{p}) = \Delta(T)\psi(\mathbf{p})$, although it has not been a matter of the utmost importance.

As for most of the physical phenomena (except the low temperature features due to the gap zeroes), they can be well understood in terms of the phenomenological Ginsburg-Landau theory. The corresponding equations for both isotropic and anisotropic cases have been derived in [7,15] on the microscopic basis at $|T - T_c| \ll T$. One word of caution in this regard: not all the possibilities enumerated in [13] would follow from the mean field BCS-like approach.

Finally, consider the self-consistency of the above scheme. The transition into superconducting phase has been accounted for by introducing the corresponding order parameters of eqs.(23). Meanwhile, above T_c the Cooper effect manifests itself in growth of the effective interaction between electrons, eq.(7), in the narrow interval (8) of the phase space. To estimate corrections due to this, consider (at $T > T_c$) the self-energy diagram for the Green function in Figure 5a with $\Gamma(p, q-p; q-p, p)$ from (7). The vertex Γ is of order of $(\nu(\varepsilon_F)t)^{-1}$ at $(\varepsilon_{qn} = 0)$ in momenta volume $\sim |\mathbf{q}|^3 \sim t^{3/2}(T_c^3/v_F^3) \sim t^{3/2}\xi_0^{-3}$. (Here $t = |T - T_c|/T_c$). The Green function $\mathcal{G}(q-p) \sim 1/T_c$. As a result, the self-energy in Figure 5 is of the order of

$$\nu^{-1}(\varepsilon_F)t^{1/2}\xi_0^{-3} \qquad (46)$$

This is to be compared with the "self-energy" due to the superconductivity onset (second order effect in Δ, see in Figure 5b):

$$|\Delta|^2 \mathcal{G}(-p) \sim T_c t \tag{46'}$$

One immediately sees that (46) is small compared to (46') everywhere outside an exceedingly small vicinity of T_c:

$$|t|^{1/2} \equiv \left(\frac{\Delta T}{T_c}\right)^{1/2} \sim \frac{T_c^2}{\varepsilon_F^2} \ll 1 \tag{47}$$

In other words, outside (47) fluctuations are negligibly small in common superconductors.

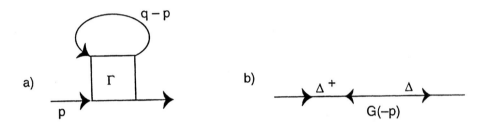

Figure 5. Two self-energy corrections: a) Contribution from the Cooper channel; b) Effective self-energy part due to onset of superconductivity (terms in the Green function quadratic in Δ).

The condition for applicability of the BCS-theory may be also formulated as

$$(\xi_0 p_F)^2 \gg 1 \tag{47'}$$

i.e. the coherence length, $\xi_0 \sim \hbar v_F / T_c$ is to be much larger than atomic distances.

IMPURITY EFFECTS

We will discuss below the influence of impurities only on thermodynamic properties of superconductors. As for the transport phenomena, such as the field penetration into superconductors, scattering of electrons by defects of any kind results in a behavior which phenomenologically is rather similar for any type of pairing.

For the isotropic BCS-superconductor defects produce a minor effect on T_c and other thermodynamical characteristics which depend on the gap value, in other words, the superconducting gap itself, $\Delta(T)$, is not sensitive to the impurities' presence (at least, at low enough concentrations, x, of order of a few percent). This result is known in the literature as "the Anderson's Theorem"[16], which states that at such a concentration, changes in the density of states at the Fermi level would be very small, of order of $(1/\tau \varepsilon_F) \ll 1$ (τ- a scattering time); since superconducting pairing involves two electronic states related by the time-reversal symmetry, and this symmetry being still preserved, the corresponding densities of states and a short-ranged attractive interaction remain unchanged in presence of any usual (non-magnetic) defects, leading to the same energetics of pairing. Actually, this result has been first explicitly derived in [17] by using the so-called "cross-technic", invented [17,18] to treat scattering by impurities diagrammatically.

We will not dwell upon the full treatment of impurities in superconductor at arbitrary temperatures, limiting ourselves to the behavior of the critical temperature, T_c, as a function of concentration, and its dependence on the pairing type or the impurity properties.

a)

b)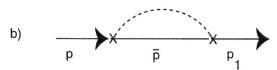

Figure 6. Cross-technic for impurities: a) perturbation expansion in the impurities potential (crosses); b) Self-energy part after averaging over positions of impurities.

As for the "cross-technic", mentioned above, its basic rules are rather simple: let $u(\mathbf{r}-\mathbf{r}_a)$ be the potential of an impurity centered at the point $\mathbf{r}_a$. The potential produces in momentum space the Born correction to Green function shown in Figure 6a. The cross here stands for $\sum_{\mathbf{r}_a} u(\mathbf{p}'-\mathbf{p})\exp\{i(\mathbf{p}-\mathbf{p}')\mathbf{r}_a\}$ -the Fourier component. Averaging Green function over impurity positions in this term gives some constant ($nu(0)$, n is the volume concentration of impurities) which is of no interest since it is equivalent to some shift in the chemical potential. The second diagram in Figure 6a has two crosses, and summation runs independently over all impurities. While averaging over $\mathbf{r}_a$ and $\mathbf{r}_b$, terms with $\mathbf{r}_a \neq \mathbf{r}_b$ reproduce the above result. The new contribution arises when $\mathbf{r}_a = \mathbf{r}_b$ in both crosses. It is shown diagrammatically in Figure 6b: the dashed line emphasizes that two crosses stand for the same impurity atom. The effective self-energy part is of the form:

$$n \int |u(\mathbf{p}-\mathbf{p}')|^2 \mathcal{G}(p') \frac{d^3\mathbf{p}'}{(2\pi)^3} \tag{48}$$

Substitution of the Green function (2),(3') into (48) shows that (48) has contributions both from ξ close to the Fermi surface, and far away from it. The latter contribution is real and, again, of no importance (actually, these terms are nothing but corrections to the Born amplitude, $u(0)$). As for the vicinity of the Fermi surface, it gives after integration over ξ:

$$(sign\omega_n)\pi i \frac{n}{(2\pi)^3}\int |u(\mathbf{p}-\mathbf{p}')|^2 \frac{dS_{\mathbf{p}'}}{v_F(\mathbf{p}')} \equiv \frac{i}{2\tau_\mathbf{p}} sign\omega_n \tag{49}$$

Eq.(49) is the definition of the scattering time, $\tau_\mathbf{p}$, which is $\mathbf{p}$-dependent, of course, for anisotropic metals (Consider substitution $u(\mathbf{p}-\mathbf{p}') \Rightarrow u(\mathbf{p};\mathbf{p}')$ in eq.(48)). As far as the Born approximation is valid, calculation of higher (in u) corrections to any diagrammatic function reduces to writing down all the two-crosses correlators of the type shown in Figure 6b. The matrix element for the dashed line is already defined by eq.(48). (Extension to the non-Born approximation is rather straightforward).

Results [17] show that in the isotropic case the average (over defects) Green functions $\bar{\mathcal{G}}, \bar{\mathcal{F}}$, and $\bar{\mathcal{F}}^+$ are related to their expressions of pure superconductor by a simple relation:

$$\left. \begin{array}{c} \bar{\mathcal{G}}(x-x') \\ \bar{\mathcal{F}}(x-x') \\ \bar{\mathcal{F}}^+(x-x') \end{array} \right\} \Rightarrow \left. \begin{array}{c} \mathcal{G}(x-x') \\ \mathcal{F}(x-x') \\ \mathcal{F}^+(x-x') \end{array} \right\} \cdot \exp\left(-\frac{|\mathbf{r}-\mathbf{r}'|}{2l}\right) \tag{50}$$

Clearly, the gap, being a point-like object (see (13,13')), remains the same as in the pure sample.

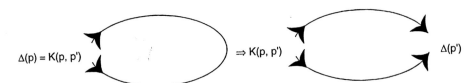

Figure 7. Diagrammatic form for self-consistency equation expressing Δ in terms of the $\mathcal{F}$-function.

Before to discuss defects in a more general fashion, it is helpful to explain why "the Anderson's Theorem" works well in the isotropic BCS-case. In Figure 7 it is shown diagrammatically how the self-consistency equation for the gap, eq.(39), is reduced to the equation for temperature of transition, T_c, in the limit $\Delta \to 0$ (L.H.S. of eq.(42)). The first order (in concentration, $1/\tau$) corrections due to defects are drawn in Figure 8. The first two self-energy parts have already been calculated in (49). The third insert is of the form:

$$\frac{n}{(2\pi)^3} \int |u(\mathbf{p};\mathbf{p}')|^2 \, \mathcal{G}(\mathbf{p}')\mathcal{G}(-\mathbf{p}')\Delta(\mathbf{p}')d^3\mathbf{p}' \Rightarrow$$
$$\Rightarrow 2\pi i \frac{n}{(2\pi)^3} \int |u(\mathbf{p};\mathbf{p}')|^2 \, \Delta(\mathbf{p}') \frac{dS_{\mathbf{p}'}}{v_F(\mathbf{p}')} \tag{51}$$

Here lies the reason why "the Anderson Theorem", strictly speaking, is valid for the isotropic model only- at integration along the Fermi surface, $S_{\mathbf{p}'}$, the angular dependence in (51), $\Delta(\mathbf{p}) = \Delta(T_c)\psi(\mathbf{p})$, would differ from that one in (49). In the isotropic case $\psi(\mathbf{p}) \equiv 1$, and the third diagram in Figure 8 exactly cancels contributions from the two terms with "self-energy" parts on the Green functions line. (We omit these simple calculations here). Any non-trivial dependence in $\psi(\mathbf{p})$ violates this exact cancelation. In practice, T_c of common superconductors is not very sensitive to defects, indeed, (at low concentrations), proving once again that isotropic model is a good approximation.

If the symmetry of superconducting order parameter becomes broken, especially, if the gap function, $\Delta(\mathbf{p})$, possesses nodes, or even lines of zeroes, as illustrated by the above example of the "d-wave"-pairing (45), the symmetry of the third diagram in Figure 8 is different from that one in Figure 6b, and no cancelation must be expected. Thus, in the simplest case of isotropic *impurity*, the impurity potential in (50, 51), $u(\mathbf{p},\mathbf{p}') \equiv u_0$, the integral over $S_{\mathbf{p}'}$ in eq.(51) is zero *identically*, merely because $\Delta(\mathbf{p})$, being a non-identical representation of the lattice group, changes sign under some symmetry operations.

Destroying effects of impurities on superconductivity are especially well-known for paramagnetic impurities- foreign ions preserving magnetic moments (spins) then embedded into the host lattice[19]. The scattering potential includes now the new term:

$$\hat{u}(\mathbf{p},\mathbf{p}') = u(\mathbf{p},\mathbf{p}')\hat{1} + u_s(\mathbf{p},\mathbf{p}')(\hat{\sigma}\cdot\mathbf{S}) \tag{52}$$

(Here **S**- spin of impurity).

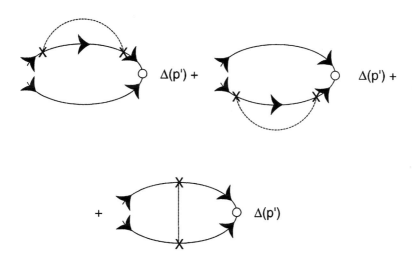

Figure 8. Corrections to the self-consistency equation in Fig.7, linear in the impurity concentration.

From the symmetry point of view, the second term in (52) violates the time-reversal symmetry for electrons of the host metallic matrix. Mathematically, one is to recall that, according eqs.(21),(22), $\hat{\Delta}_{\alpha\beta}(\mathbf{p}) = \hat{I}_{\alpha\beta}\Delta(\mathbf{p})$ originally includes spin dependence in the form of the matrix $\hat{I}$ which is to be taken into account at calculations of the third diagram in Figure 8. Therefore the second term in (52) explicitly shows up leading to non-cancelation of all contributions in Figure 8. It results in a new parameter, $1/\tau_s$ (its definition is quite similar to (49), except it includes $\mid u_s(\mathbf{p},\mathbf{p}')\mid^2 S(S+1)$). The temperature of transition $T_c(x)$ for paramagnetic alloys as a function of $\tau_s T_{c0}$ was obtained in [19]. At finite concentration of paramagnetic impurities

$$T_{c0}\tau_{scr} = \frac{2\gamma}{\pi} \sim 1.13 \tag{53}$$

($\ln\gamma = C$- the Euler's constant) superconductivity disappears, $T_c(x_{cr}) = 0$.

The second term in (52) flips spins of electron forming a pair with $S = 0$. Correspondingly, destroying influence of paramagnetic impurities on superconductivity is referred to as the "pair-breaking" effects of magnetic moments. "Pair-breaking" effects of non-magnetic impurities on anisotropic pairing may be also considered as destroying of a specific symmetry of the pair wave function.

A subtle phenomenon of "gapless superconductivity" discovered in [19] for paramagnetic alloys, results in a linear-in-T-contribution into specific heat of such an alloy in a narrow range below the critical concentration. In case of anisotropic pairing, say, with lines of nodes in the order parameter, the density of states, $\nu_s(\varepsilon) \sim \nu_N(\varepsilon_F)(\varepsilon/T_c)$, resulting, for instance, in a T^2-dependence of the electronic specific heat. An essential difference with the case of paramagnetic impurities in an "s-wave" superconductor is that "gapless" regime for the latter sets on at a *finite* impurity concentration, while lines of nodes (zeroes) are destroyed by an arbitrary small concentration of non-magnetic defects[20].

CONCLUSION

It is hardly possible to conclude our introduction into exciting field of superconductivity without mentioning the current situation in High T_c cuprates or Heavy Fermions (HF). Without any attempt to review experiment in some detail (hence, no references below), a few comments can still be done, although they may often reflect the author's feelings.

The first question is, of course, whether superconductivity itself (disregarding its mechanism) is in these systems a BCS-like phenomenon. Phenomenological analysis agrees reasonably well with what one would expect from the theory, even from notions of the Ginsbarg-Landau theory. Recall that the mean field character of the microscopic approach is based on participation of many electrons in forming of the Cooper pair: the coherence length $\xi_0 \sim \hbar v_F/T_c$ is to be large compared to interatomic distances. This is very much so in common superconductors. However, estimates for cuprates give $\xi_0 \sim 20 \text{Å}$ and in HF-$\xi_0 \sim 30 - 50 \text{Å}$, in both cases only factor six with respect to the lattice constant. Add to that the pronounced layered structure of cuprates which poses a question of dimensionality in the cuprates-case. Meanwhile, in both systems a rather sharp transition can be obtained provided a special care of the sample quality. Indeed, transition widths $< 1K$ are often seen in cuprates. With $T_c \sim 100K$ it is not in contradiction with our estimate for fluctuations range (47):

$$\left(\frac{\Delta T}{T_c}\right) \sim (\xi_0 p_F)^{-4} \qquad (47')$$

(i.e., factor six is not so bad yet!)

Next is the problem: strong coupling vs weak coupling. Our opinion is that in cuprates, again, $T_c \sim 100K$, would be still a low transition temperature when compared with the magnetic scale, $J \sim 0.15$ eV. The gap of the BCS-like value ($\Delta \sim 2 \div 3T_c$) is directly seen in photoemission and tunneling experiments. True, that similar energy scales ("spin gap", "pseudogap", etc.) are seen in the cuprates' normal state.

For HF T_c is usually very small ($\lesssim 1K$). It is unclear, however, what serves as a high energy scale: unusual properties of HF in normal state often develop below $T^* \sim 10K$. Very unusual phase diagram of UPt$_3$ may be described best in terms of some unconventional order parameter. Let us repeat without any proof that structure of the BCS-like theory even for anisotropic pairing, seems, is never capable to secure the necessary symmetry. Specific heat jumps at T_c in HF often exceed restrictions imposed by anisotropic BCS-like schemes[12].

Among other qualitative features, one should mention absence of the Hebel-Slichter peak in NMR-relaxation rate at T_c in case of anisotropic pairing, zeroes of the gap function, "half-flux" at interfaces- the two last ones follow from the symmetry considerations only. It also seems that low energy fermionic spectrum (nodes and lines of nodes) may be a property of the superconducting phase of the corresponding symmetry and be present at low temperatures independently on details of a microscopic theory.

Sensitivity to defects may be better suited for verifying of theoretical predictions. Both in cuprates and HF T_c is in fact, rather sensitive to substitutions and sample quality. Experimental problem with that direction lies in the fact that substitutions in HF materials often affect their magnetic properties and the heavy fermion behavior. In cuprates, in addition, this is complicated by interplay with doping and doping dependence of T_c. The most transparent effect to study "pair-breaking" effects in cuprates, so far- dependence of the normal density of states on magnetic field in mixed state[21].

As a matter of fact, there exist many indications concerning superconductivity in cuprates that the BCS-scheme remains in force whatever is the mechanism. It is not so for HF materials, first of all, because experimental activity there is incomparable with current efforts in cuprates.

References

1. J. Bardeen, L. N. Cooper, and J. R.Schrieffer, *Phys. Rev.* 108:1175 (1957).
2. L. N. Cooper, *Phys. Rev.* 104:1189 (1957).
3. L. P. Gor'kov, *JETP* 7:505 (1958).
4. A. A. Abrikosov, L. P. Gor'kov, and I. E. Dzyaloshinskii. *Methods of Quantum Field Theory in Statistical Physics*, Dover, New York (1975).
5. E. M. Lifshitz and L. P. Pitaevskii. *Statistical Physics; Part 2*, Pergamon Press, New York (1980).
6. E. Maxwell, *Phys. Rev.* 78:447 (1950); C.A. Reynolds, B. Serin, W. H. Wright, and L.B . Nesbitt, *Phys. Rev.* 78:487 (1950).
7. L. P. Gor'kov, *JETP* 9:1364 (1959).
8. V. L. Beresinskii, *JETP Lett.* 20:287 (1974).
9. E. Abrahams, A. Balatsky, D. J. Scalapino, and J. R. Schrieffer, *Phys. Rev.* B52: 1271 (1995).
10. L. C. Hebel and C. P. Slichter, *Phys. Rev.* 113:1504 (1959).
11. G. M. Eliashberg, *JETP* 11:696 (1960).
12. V. L. Pokrovskii, *Zh. Exp. Teor. Fiz.* 40:641 (1961); V. L. Pokrovskii and N. S. Ryvkin, *ibid*, 43:92 (1962).
13. G. E. Volovik and L. P. Gor'kov, *JETP Lett.* 39:674 (1984); *JETP* 61:843 (1985).
14. L. P. Gor'kov, *Sov. Sci. Rev. A Phys.* 9:1 (1987); M. Sigrist and K. Ueda, *Rev. Mod. Phys.* 63:239 (1991).
15. L. P. Gor'kov and T. K. Melik-Barkhudarov, *Zh. Exp. Teor. Fiz.* 45:1493 (1963).
16. P. W. Anderson, *J. Phys Chem. Solids* 11:26 (1959).
17. A. A. Abrikosov and L. P. Gor'kov, *JETP* 8:1090; *JETP* 9:220 (1959).
18. S. F. Edwards, *Phil. Mag.* 3:1020 (1958).
19. A. A. Abrikosov and L. P. Gor'kov, *JETP* 12:1243 (1961).
20. L. P. Gor'kov and P. A. Kalugin, *JETP Lett.* 41:153 (1985).
21. G. E. Volovik, *JETP Lett.* 58:469 (1993).

THE ISOTOPE EFFECT IN SUPERCONDUCTORS

A. Bill,[1] V.Z. Kresin,[1] and S.A. Wolf[2]

[1] Lawrence Berkeley Laboratory, University of California,
Berkeley, CA 94720, USA
[2] Naval Research Laboratory, Washington D.C. 20375-5343

Abstract

We review some aspects of the isotope effect (IE) in superconductors. Our focus is on the influence of factors not related to the pairing mechanism. After summarizing the main results obtained for conventional superconductors, we review the effect of magnetic impurities, the proximity effect and non-adiabaticity on the value of the isotope coefficient (IC). We discuss the isotope effect of T_c and of the penetration depth δ. The theory is applied to conventional and high-T_c superconductors. Experimental results obtained for $YBa_2Cu_3O_{7-\delta}$ related materials (Zn and Pr-substituted as well as oxygen-depleted systems) and for $La_{2-x}Sr_xCuO_4$ are discussed.

1 INTRODUCTION

Historically, the isotope effect (IE) played a major role in unravelling the questions related to the origin of the effective attractive interaction between charge-carriers which leads to the superconducting state. The theoretical considerations of Fröhlich[1] and the experimental discovery of the IE of T_c in mercury[2] pointed towards the contribution of lattice dynamics to the instability of the normal state. Since then, the IE has often been considered as a measure of the contribution of phonons to the pairing mechanism. Furthermore, it is generally assumed that only those thermodynamical quantities that depend explicitely on the phonon frequencies (as T_c or the order parameter Δ in the BCS model) display an IE.

One can show[3, 4, 5, 6, 7], however, that such understanding of the isotope effect is incomplete and leads, therefore, to confusion; this is particularly true for the high-temperature oxides. Indeed, several factors not related to the pairing mechanism can alter the value of the isotope coefficient (IC). Moreover, these factors are not necessarily related to lattice dynamics. Here we focus our attention on three such factors: magnetic impurities, the proximity effect and non-adiabatic charge-transfer as it occurs in high-T_c superconductors.

In addition, we also show [6] that fundamental quantities such as the penetration depth δ of a magnetic field also display an IE because of the three factors mentioned above. Note that this effect occurs despite the fact that δ does not explicitely depend on quantities related to lattice dynamics.

It results from our considerations that the value of the IC (even its absence) does not allow any *a priori* conclusion about the pairing mechanism. Nevertheless, it remains an interesting effect that enables one to determine the presence of magnetic impurities, proximity effect or non-adiabaticity in the system. The calculation of the IC and its comparison with experimental results was also used in previous works to determine the value of the Coulomb repulsion $\mu^\star$ or the relative weight of different electron-phonon coupling strengths in superconductors (see below). It can thus be used as a tool for the characterization of superconductors.

This review is mainly based on our papers[3, 4, 5, 6, 7]. We apply the theory to analyse the oxygen isotope effect of T_c in Zn and Pr-doped (YBCZnO and YPrBCO) as well as oxygen-depleted $YBa_2Cu_3O_{7-\delta}$ (YBCO). We also review calculations of the penetration depth isotope effect. The latter superconducting property is a good example of a quantity that does not directly depend on phonon frequencies and yet can display a substantial isotopic shift [6]. In this context, we also discuss recent experimental results obtained for $La_{2-x}Sr_xCuO_4$ (LSCO).

The structure of the paper is as follows. In section 2 we present some of the early results related to the isotope effect and show that even for conventional superconductors the relation between the isotope effect and the pairing mechanism is not simple. We also discuss the applicability of these early results to the description of high-temperature superconductors.

The remaining sections are devoted to the description of new, unconventional isotope effects. In section 3 we study the influence of magnetic impurities on the value of the isotope effect. We show that adding magnetic impurities to a superconductor can enhance the isotope coefficient α of T_c and induce a temperature-dependent isotope effect of δ. We show that both isotope effects are universal functions of T_c. That is, $\alpha(T_c)$ and $\beta(T_c)$ are independent of any adjustable parameter. We discuss Zn-doped $YBa_2Cu_3O_{7-\delta}$ in this context.

Section 4 is concerned with the influence of a normal layer on the isotope effect of a superconductor. We show that due to the proximity effect, the isotope coefficient of T_c is linear in the ratio of the normal to the superconducting film thicknesses ρ. Furthermore, the proximity effect induces a temperature and ρ-dependent isotope coefficient of the penetration depth.

Section 5 reviews the concept of the non-adiabatic isotope effect introduced in Ref. [3] and further discussed in Refs. [4, 5, 6, 7]. We show that in systems as the high-temperature superconductors where charge-transfer processes between reservoir and CuO_2-planes occur via ions that display a non-adiabatic behaviour, the charge-carrier density in the planes depend on the ionic mass. This leads to the unconventional non-adiabatic contribution to the IE of T_c. It is also interesting that non-adiabaticity induces an isotopic shift of the penetration depth δ.

In the last section (Sec. 6) we apply our theory to the oxygen isotope effect (OIE) in high-temperature superconductors of the YBCO-family. Focus is set on the oxygen isotope effect because most experimental studies have been performed on the oxygen ion which is the lightest in the CuO_2 plane. The situation is less clear in the case of copper isotopic substitution where an effect has also been observed[8, 9]. This latter

case will be discussed in more detail elswhere.
We conclude the review in section 7.

2 THE ISOTOPE EFFECT: DISCOVERY, CONVENTIONAL VIEW

The isotope effect of superconducting critical temperature T_c is best described in terms of the isotope coefficient (IC) α defined by the relation $T_c \sim M^{-\alpha}$, where M is the ionic mass. Under the assumption that the shift ΔT_c induced by isotopic substitution ($M \rightarrow M^\star$) is small compared to T_c, one can write

$$\alpha = -\frac{M}{\Delta M}\frac{\Delta T_c}{T_c}, \qquad (1)$$

where $\Delta M = M - M^\star$ is the difference between the two isotopic mass. If the superconductor is composed of different elements, one defines a *partial* isotope coefficient α_r as in Eq. (1), but where M is replaced by M_r, the mass of element r that is substituted for its isotope. In addition, one defines the total isotope coefficient by

$$\alpha_{tot} = \sum_r \alpha_r = -\sum_r \frac{M_r}{\Delta M_r}\frac{\Delta T_c}{T_c}. \qquad (2)$$

Besides T_c there are other quantities that display an isotope effect. In the next sections we focus on the isotope shift of the penetration depth δ. In analogy to the IC of T_c (denoted α) we define the isotope coefficient β of the penetration depth by the relation

$$\beta = -\frac{M}{\Delta M}\frac{\Delta \delta}{\delta}. \qquad (3)$$

One should note that there is a conventional BCS-type isotope effect of the penetration depth related to its temperature dependence $\delta^{-2} \sim (1 - T^4/T_c^4)$. Because T_c displays an isotope effect, the penetration depth is also shifted upon isotopic substitution. The effect becomes strong as one approaches T_c. The present article is not concerned with this trivial effect.

Let us first discuss what values of the isotope coefficients α and β can be expected for different types of superconductors. Table 1 shows characteristic values of the isotope coefficient for different types of superconductors. Very different values of the IC have been observed (in the range from -2 to $+1$). One notes that some systems have a negligible coefficient, some display even an "inverse isotope coefficient" ($\alpha < 0$) and some take values greater than 0.5, the value predicted by Fröhlich and the BCS model (for a monoatomic system). The purpose of the next paragraphs is to describe shortly different theoretical models allowing one to understand the coefficients observed (see Table 1 and Refs. [10, 11]). We begin with the description of conventional superconductors and discuss the relevance of these models for high-T_c materials. In the following sections we then introduce new considerations about the IE allowing one to give a consistent picture of the IE in high-temperature superconductors.

Table 1. Experimental values of the isotope coefficient of T_c (see also Refs. [11, 20]). The letters in the last column correspond to Ref. [2].

Superconductor	α	Reference (see [2])
Hg	0.5 ± 0.03	a
Tl	0.5 ± 0.1	a, and [11]
Cd	0.5 ± 0.1	b
Mo	0.33 ± 0.05	c
Os	0.21 ± 0.05	d,e
Ru	0.0	e
Zr	0.0	e
PdH(D)	-0.25	[33, 34]
U	-2	[12]
La$_{1.85}$Sr$_{0.15}$CuO$_4$	0.07	[54]
La$_{1.89}$Sr$_{0.11}$CuO$_4$ (^{16}O $-$ ^{18}O subst.)	0.75	[54]
K$_3$C$_{60}$ (^{12}C $-$ ^{13}C subst.)	0.37 or 1.4	[40]

2.1 Monoatomic Systems

Let us begin with the case of a monoatomic BCS-type superconductor. Substituting the atoms by an isotope affects the phonon dispersion (for a monoatomic lattice $\Omega \sim M^{-1/2}$, where Ω is a characteristic phonon frequency). Thus, any quantity that depends on phonon frequencies is affected by isotopic substitution. Assuming that the electron-electron pairing interaction is mediated by phonons and neglecting the Coulomb repulsion between electrons, the BCS theory predicts that $T_c \propto \Omega$ (see below, Eq. (4) for $\mu^* = 0$). Thus, for a monoatomic system the isotope coefficient of T_c given by Eq. (1) is $\alpha = 0.5$.

As seen in Table 1, the IC of most monoatomic non-transition metals is approximately equal to 0.5. Many other systems, on the other hand, deviate from this value. In particular, transition metals and alloys display values that are smaller than, or equal to 0.5. PdH displays an inverse isotope effect $\alpha_{PdH} \simeq -0.25$ [33] (there is also one report on a large inverse IC of Uranium $\alpha_U \simeq -2$[12]). Furthermore, high-T_c materials can have values both smaller and larger than 0.5, depending on the doping.

2.2 The Coulomb Interaction

One of the first reasons advanced to explain the discrepancy between theory ($\alpha = 0.5$) and experiment was that the BCS calculation described above did not take properly into account the Coulomb repulsion between charge-carriers. It was shown in Refs. [13, 14, 15, 16] that inclusion of Coulomb interactions leads to the introduction of the pseudo-potential μ^* in the BCS equation for T_c:

$$T_c = 1.13\Omega \exp\left(-\frac{1}{\lambda - \mu^*}\right), \qquad (4)$$

where μ^* is the pseudo-potential given by

$$\mu^* = \frac{\mu}{1 + \mu \ln\left(\frac{E_F}{\Omega}\right)} \qquad (5)$$

μ is the Coulomb potential and E_F is the Fermi energy. One notes that Ω is present in μ^* and thus in the exponent of Eq. (4). The value of α can thus be substantially decreased from 0.5. Indeed, from Eq. (1) and (4) one obtains

$$\alpha = \frac{1}{2}\left\{1 - \left(\frac{\mu^*}{\lambda - \mu^*}\right)^2\right\} \tag{6}$$

The isotope coefficient is shown as a function of λ for different values of μ^* in Fig. 1. Naturally, only the case $\lambda > \mu^*$ is relevant, since superconductivity is otherwise suppressed.

The effect of Coulomb interactions can be included in a similar way in the strong-coupling Eliashberg theory. Using the formula derived by McMillan [17]:

$$T_c = \frac{\Omega}{1.2} \exp\left(-\frac{1.04(1+\lambda)}{\lambda - \tilde{\mu}}\right), \tag{7}$$

with $\tilde{\mu} \equiv (1 + 0.62\lambda)\mu^*$ and μ^* is given by Eq. (5), one obtains the following result for the isotope coefficient:

$$\alpha = \frac{1}{2}\left\{1 - \frac{1.04(1+\lambda)\tilde{\mu}\mu^*}{[\lambda - \tilde{\mu}]^2}\right\} \tag{8}$$

A similar result has also been derived in Ref. [10]. Fig. 1 shows the dependence described by Eq. (8) and allows one to compare the weak and strong-coupling cases, that is, Eqs. (6) and (8).

An expression for T_c that is valid over the whole range of coupling strengths (from weak to very strong couplings) has been derived by one of the authors in Ref. [18]:

$$T_c = \frac{0.25\Omega}{\sqrt{e^{2/\lambda_{eff}} - 1}} \tag{9}$$

with

$$\lambda_{eff} = \frac{\lambda - \mu^*}{1 + 2\mu^* + \lambda\mu^* t(\lambda)}, \tag{10}$$

and $t(\lambda) \simeq 1.5 exp(-0.28\lambda)$ [18]. As usual, the parameter μ^* is given by Eq. (5). The isotope coefficient resulting from Eq. (1), (9) and (10) reads

$$\alpha = \frac{1}{2}\left\{1 - \frac{\mu^{*2}}{\lambda_{eff}\left(1 - e^{-2/\lambda_{eff}}\right)}\left[\frac{1}{\lambda - \mu^*} + \frac{2 + \lambda t(\lambda)}{3 + \lambda t(\lambda)}\right]\right\} \tag{11}$$

This isotope coefficient is also displayed in Fig. 1 for two values of the parameter μ^*. Note that for all models, because of the presence of μ^* the isotope coefficient of T_c is lowered with respect to the BCS value for $\alpha = 0.5$. One recovers the BCS result asymptotically at large λ (or T_c/Ω). Only Eq. (11), however, is valid in the strong coupling limit ($\lambda > 1$).

An interesting general feature of the results presented above is that the strongest deviation from the BCS monoatomic value 0.5 occurs when μ^* (or $\tilde{\mu}$) is of the order of λ (see Eqs. (6), (8) or (11)). Thus, excluding anharmonic or band-structure effects, a *small* isotope coefficient is correlated to a *low* T_c. This is indeed observed in most conventional superconductors. In certain cases one even obtains a *negative* (also called *inverse*) isotope coefficient ($\alpha < 0$). For realistic values of the parameters, T_c should not

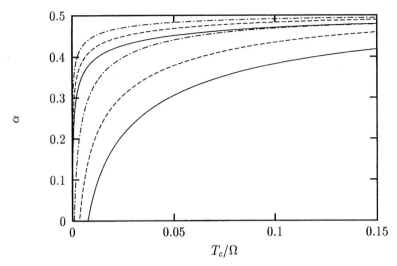

Figure 1. Isotope coefficient α as a function of T_c/Ω (i.e. λ) for different values of μ^*. Solid lines from Eq. (6), dashed lines from Eq. (8) and dash-dotted lines from Eq. (11). For each type of line, the upper curve corresponds to $\mu^* = 0.1$ and the lower to $\mu^* = 0.2$.

exceed $T_c \sim 1K$ for this inverse effect to be present. Several systems display this inverse isotope effect, among them PdH (where H is replaced by its isotope D) [33, 34] and Uranium (where ^{235}U is replaced by ^{238}U) [12]. We discuss this case below (sec. 2.3).

The situation encountered in high-temperature superconductors is very unusual. Indeed, one observes that the isotope coefficient has a *minimum* value at optimal doping (*highest* T_c). The explanation of a small α for oxygen substitution coinciding with a high T_c is likely to be related to the contribution of the oxygen modes to the pairing.

The main conclusion of this section is that the Coulomb interaction and its logarithmic weakening leads to the deviation of the isotope coefficient from the value $\alpha = 0.5$ and to the non-universality of α.

2.3 Band Structure Effects; Transition Metals

Up to now we have only considered the effect on T_c induced by the isotopic shift of phonon frequencies. Furthermore, we have considered the case where the phonon frequency appears explicitely in the expression for T_c (as prefactor and in μ^*; see Eqs. (4), (7) and (9)). Band structure effects have first been considered to explain the discrepancies observed between the results following from the strong-coupling McMillan equation (Eq. (7)) and the IC measured in transition metals. Several other generalizations of the basic model have been considered, especially in the context of high-temperature superconductors.

Let us first describe the situation encountered in transition metals. As has been shown in Refs. [15, 16], the two-square model for which (5) is derived leads to values of the IC that are only $10 - 30\%$ lower than 0.5. However, the deviation is much stronger for some transition metals, leading even to a vanishing coefficient for Ruthenium. As shown in Ref. [19] it is necessary to take into account the band structure of transition metals. The presence of a "metallic" s-band and a narrow d-band, and the associated peaked structure of the electronic density of states (DOS) has an impact on the value of μ^*. Furthermore, E_F is much smaller in transition metals than in non-transition metals and leads thus to a smaller effective screening of the Coulomb interaction. These facts have been included in Ref. [19] by modifying the two square-well model used to derive Eq. (5). The values obtained from the modified expression of α (see Eq. (9) in Ref. [19]) are in good semi-quantitative agreement with the experimental results.

One should add that the two-band structure is important even in dirty transition-metal superconductors, although the superconducting state is characterized by only one gap in this case (see Ref. [11] for further discussions on transition metals).

The example of transition metals shows that a quantitative description of the IC requires, among other things, a precise knowledge of the band structure. This has also been suggested in Ref. [12, 20] to explain the large inverse isotope effect observed in Uranium. In this context, it would be interesting to determine the uranium isotope effect in heavy-fermion systems (one should note, however, that there is only one report of the isotope effect in uranium[12]).

Another effect of the band structure arises if one assumes that the electronic density of states varies strongly on a scale given by Ω, the BCS energy cutoff (e.g., in the presence of a van Hove singularity). To obtain the BCS expression (4) we assumed that the electronic density of states (DOS) is constant in the energy intervall $[-\Omega, \Omega]$ around the Fermi energy. λ is then given by $N(E_F)V$, where N is the DOS at the Fermi level and V is the attractive part of the electron-electron effective interaction. In a more general case, however, one has to consider the energy dependence of the DOS. Though a purely electronic parameter, the energy dependence of the DOS appears to influence the isotope effect[21]. One can understand this effect qualitatively within a crude extension of the BCS model. Instead of taking the DOS at the Fermi level, one replaces $N(E_F)$ by $< N(\varepsilon) >_\Omega$, the average value of the DOS $N(\varepsilon)$ over the interval $[-\Omega, \Omega]$ around E_F. Obviously, this average depends on the cutoff energy Ω of the pairing interaction. Isotopic substitution modifies Ω which affects $< N(\varepsilon) >_\Omega$, and thus $\lambda = < N >_\Omega V$ and T_c.

A more general analysis of this effect within Eliashberg's theory was given in Ref. [21]. They find that whereas the IC can reach values above 0.5, its minimal value obtained for reasonable choices of the parameters never reaches the small values (~ 0.02) found in optimally doped high-T_c compounds. However, more recent studies of the influence of van Hove singularities on the isotope coefficient show that such small values can be obtained, because in this scenario the cutoff energy for the effective interaction between charge-carriers is given by $min(E_F - E_{vH}; \Omega)$ (where E_{vH} is the energy of the van Hove singularity and Ω is a characteristic phonon energy) [22, 23, 24]. If the cutoff energy is given by $E_F - E_{vH}$ and is thus electronic in origin, T_c displays no shift upon isotopic substitution. Taking into account the Coulomb repulsion it has been shown[24] that one can even obtain a negative IC.

Another way to extend the results presented earlier is to consider an anisotropic Eliashberg coupling function $\alpha^2 F(\mathbf{q}, \omega)$. Generally it is assumed that the system is isotropic and the coupling function can be averaged over the Fermi surface, leading to a $\mathbf{q}$-independent Eliashberg function. If the system is strongly anisotropic (as is the case of high-T_c superconductors), the average may not lead to an accurate description of the situation. The isotope effect has been studied for different $\mathbf{q}$-dependent form factors entering $\alpha^2 F$ in Ref. [25]. They show that a small isotope coefficient as observed in $YBa_2Cu_3O_{7-\delta}$ can be obtained for anisotropic systems. Nevertheless, to obtain high critical temperatures at the same time, they are forced to introduced another, electronic pairing mechanism. Such a situation is further discussed below (see Sec. 2.6).

Note that the theories presented in this section give only a qualitative picture of the situation encountered in high-temperature superconductors and do presently not account for the isotope effect observed in various systems (see also Refs. [21, 26]).

2.4 Polyatomic Systems

Until now we focused mainly on the study of monoatomic systems in which the attractive interaction leading to the formation of pairs is mediated by phonons. All previous effects are naturally also encountered in polyatomic systems. However, the presence of two or more elements in the composition of a superconductor has several direct consequences for the isotope effect that we summarize in the following. The first consequence of a polyatomic system is that the characteristic frequency Ω which determines the value of T_c depends on the mass of the different ions involved $\Omega = \Omega(M_1, M_2, \cdots)$. Obviously, the dependence $\Omega \sim M_r^{-\alpha_r}$ for element $r = 1, 2, \ldots$ must not be equal to $\alpha_r = 0.5$. It is thus not surprising if the partial isotope coefficient (obtained by substituting one type of ions for its isotope; see Eq. (3)) differs from the textbook value 0.5.

Let us consider the example of a cubic lattice with alternating masses M_1 and M_2[27, 28]. The acoustic and optical branches can be calculated analytically[27]. From this one obtains the following partial isotope coefficients:

$$\alpha_1 = \frac{1}{2\Omega^2 M_1}\left\{\bar{K} \pm \frac{\bar{K}^2 M_{12}^- + 2\bar{K}_L/M_2}{\left(\Omega \mp \bar{K} M_{12}^+\right)}\right\}, \tag{12}$$

$$\alpha_2 = \frac{1}{2\Omega^2 M_2}\left\{\bar{K} \pm \frac{-\bar{K}^2 M_{12}^- + 2\bar{K}_L/M_1}{\left(\Omega \mp \bar{K} M_{12}^+\right)}\right\}, \tag{13}$$

where $M_{12}^- = M_1^{-1} - M_2^{-1}$, $M_{12}^+ = M_1^{-1} + M_2^{-1}$. The upper sign stands for the Debye frequency and the lower sign has to be considered when the characteristic frequency is given by the optical phonon. $\bar{K} = \sum_{x,y,z} K_i$ is the sum of the force constants and $\bar{K}_L = \sum_{x,y,z} K_i \cos(qL)$ ($q = 0$ for the optical branch and $q = \pi/L$ for acoustic mode; L is the lattice constant). One notes that the total isotope coefficient is given by $\alpha_{tot} = \alpha_1 + \alpha_2 = 0.5$ whether one takes the acoustical or the optical phonon as the characteristic energy for the determination of T_c.

Given the previous result, one can ask if there exists a maximal value of the IC. It is often stated in the literature that 0.5 is the maximal value that the isotope coefficient can reach within a harmonic phonon, and electron-phonon induced pairing model. Although to the knowledge of the authors none of the systems studied so far seems to contradict this assertion, one should note that there is no proof of this statement. It was shown in Ref. [29] that for a polyatomic system $\alpha_{tot} \equiv \sum_r \alpha_r = 0.5$ (defined in Eq. (2)) *if* one assumes that $\mu^* = 0$ and all masses of the unit cell are subject to *the same* isotopic shift (i.e., $\Delta M_r/M_r$ takes the same constant value for each element r of the system, see Eq. (2)). These assumptions hold approximately for certain systems as, e.g., the Chevrel-phase Mo_6Se_8 material[29], but are certainly not valid for example in high-temperature superconductors. It is thus not clear if the *total* isotope coefficient, Eq. (2), is indeed always 0.5. The general proof of such a statement requires the knowledge of the polarization vectors and their derivatives with respect to the isotopic masses, both of which have to be calculated for each specific system studied.

Another important remark concerns the value of the (very small) IC for optimally doped high-T_c superconductors. Experimentally one measures generally the partial oxygen isotope effect (in some cases also the Cu and Ba IE). Since the unit cell of a high-temperature superconductor contains many different atoms, one expects values of the partial IC that are significantly smaller than 0.5. A crude estimate can be given by observing that under the assumption of a same contribution of each atom to the isotope coefficient (which is certainly not the case as mentioned above) the value of the isotope effect for oxygen in YBCO would be $\alpha \approx 0.5/N \approx 0.04$, where N is the number of atoms per unit cells ($N = 13$ in the case of YBCO). Since there are 6 − 7 oxygen atoms in the unit cell, one should multiply α_{ox} by this number.

If, in addition to the multiatomic structure of the system, one takes into account the effect of Coulomb interactions and/or the fact that the charge-carriers may strongly couple to certain phonons only one can obtain values of the order observed in high-temperature oxides. The second possibility has for example been studied in the context of the Chevrel-phase compound Mo_6Se_8[29] or organic superconductors[30]. It has also been studied for high-T_c superconductors in Ref. [31, 32]. They conclude that the oxygen isotope effect can be well described for reasonable parameters in the case of $La_{2-x}Sr_xCuO_4$, but that unphysical values of the coupling have to be considered to describe the partial oxygen isotope effect of $YBa_2Cu_3O_{7-\delta}$. A careful study of the effect of Coulomb interactions on the IE of high-T_c superconductors has not been done yet.

On thus concludes from the previous considerations that it is not impossible to explain the small values of the IC observed in conjuction with high T_c's in high-temperature materials, if one considers the combined effect of a polyatomic system, the Coulomb interaction and the band structure.

2.5 Anharmonicity

Anharmonic effects play an important role in materials such as PdH(D)[33, 34], Mo_6Se_8[29] or, according to some theories[26, 35, 36, 37], in high-temperature superconductors. Among other consequences the presence of anharmonicities affects the value of the isotope coefficient. We present here two different aspects of the anharmonic isotope effect: anharmonicity of the characteristic phonon mode (for PdH) and volume effects (for Molybdenum).

Quantities that are independent of ionic masses in the harmonic approximation, can become mass dependent in the presence of anharmonicity. We mention three such properties that are of interest for the isotope effect in superconductors: the lattice force constants K_i[34], the unit-cell volume [38, 39] and the electron-phonon coupling function λ[26, 35]. In the first case, it was shown that the account of Coulomb interactions alone (see above) cannot explain the inverse ($\alpha < 0$) hydrogen isotope effect observed in PdH (see Table 1). On the other hand, a change of 20% of the lattice force constants when replacing H for Deuterium (D) in PdH was infered from neutron-scattering data[34]. This results from large zero-point motion of H as compared to D. Taking this fact into account in the Eliashberg formalism allows one to obtain quantitative agreement with the experiment[34].

Another effect due to the presence of anharmonicity is the isotopic volume effect[38, 39]. This effect was discussed in the context of Molybdene and is related to the difference in zero-point motion of the two isotopes. In such a system one can write the IC as $\alpha = \alpha_{BCS} + \alpha_{vol}$ with

$$\alpha_{vol} = B \frac{M}{\Delta M} \frac{\Delta V}{V} \frac{\Delta T_c}{\Delta P}, \qquad (14)$$

where $B = -V \partial P / \partial V$ is the bulk modulus, P is the pressure and $\Delta V = V^\star - V$ is the volume difference induced by isotopic substitution. In the case of Molybdenum the volume isotope effect amounts to $\alpha_{vol} \approx 0.09$ and accounts for $\sim 27\%$ of the total isotope coefficient, which is not a negligible effect.

The presence of a volume isotope effect has also been suggested in the case of PdH, C_{60} materials as well as in Pb[38].

The other major effect of anharmonicity is the appearence of an ionic-mass dependent electron-phonon coupling function λ[26, 35, 42]. In this case not only the prefactor Ω of Eq. (4), but also the exponent depends on ionic masses. As stressed earlier in the context of band structure effects, this fact can lead to a strong deviation from $\alpha_{BCS} = 0.5$, even for monoatomic systems. An explicit expression for the motion of the oxygen in a simple double-well potential was obtained in Ref. [35] for high-T_c superconductors. The model was extended further in Refs. [35]b and [26] and it was shown that the isotope coefficient can be much smaller than 0.5 with moderate values of the coupling constant or exceed 0.5 for (very) strong couplings (the function $\alpha(\lambda)$ goes through a minimum at $\lambda \sim 1$; see Ref. [35]). Regarding high-T_c superconductors, the anharmonic model could explain qualitatively the behaviour of the isotope coefficient if one considers the fact that the coupling function λ also depends on doping (see Ref. [43] for this dependency). Assuming that these systems are strong-coupling superconductors, and that the coupling decreases upon doping, one can describe qualitatively the behaviour of $\alpha(T_c)$ for underdoped systems. However, it is difficult to obtain quantitative agreement especially in the optimally doped and strongly underdoped regime. In the first case the coupling has to be intermediate to obtain small α's but then T_c is also small. In the second, strongly underdoped case, the IC can exceed 0.5, but the coupling has to be very strong. One way to solve this problem is to assume that the superconducting pairing is mediated by an additional, non-phononic channel (see Ref. [42] and below).

The influence of anharmonicity on the electron-phonon coupling λ and on the hopping parameter (for a tight-binding type of lattice) has also been considered in the context of $^{12}C \leftrightarrow ^{13}C$ isotope substitution in A_3C_{60} materials (A =Na, Ru; see, e.g., Ref. [40] and references therein). Experimentally the isotope coefficient was shown to vary between 0.37 to 1.4, depending on the isotopic substitution process (in the first case each C_{60} molecule contains an equal amount of ^{12}C and ^{13}C isotopes whereas in the second case the system contains C_{60}-molecules that are composed of either pure ^{12}C or pure ^{13}C atoms) [40].

2.6 Non-Phonon and Mixed Mechanisms

The superconducting transition can also be caused by a non-phonon mechanism. Historically, the introduction of the electronic mechanisms [41] started the race for higher T_c's. The pairing can be provided by the exchange of excitations such as excitons, plasmons, magnons etc... (see, e.g., Ref. [28]). In general, one can have a combined mechanism. An electronic channel can provide for an additional contribution to the pairing, as is the case, e.g., in a phonon-plasmon mechanism [44]. In the context of the isotope effect, the mixed mechanism may provide for an explanation for the unusual occurence of high T_c's and small isotope coefficients α (see above).

Let us initially consider the case where only electronic excitations mediate the pairing interaction and the Eliashberg function $\alpha^2 F$ can be approximated by a single peak at energy Ω_e, the characteristic electronic energy. The theory considered in weak coupling yields then a relation of the form $T_c \simeq \Omega_e e^{-1/\lambda}$. For an electronic mechanism Ω_e is independent of the ionic mass. Therefore, the isotope coefficient of T_c vanishes for such cases.

If, on the other hand, the superconducting state is due to the combination of phonon and a high-energy excitations (as, e.g., plasmons [44]) then one has to include both excitations in $\alpha^2 F$. In this case, the simplest model considers an Eliashberg function composed of two peaks, one at low energies (Ω_0) for the phonons and one at high energies (Ω_1) for the electronic mechanism. For such a model, one obtains the following expression for T_c[45]:

$$T_c = 1.14\Omega_0^{f_0}\Omega_1^{f_1}\exp\left(-\frac{h(\rho_i,\Omega_i)}{\rho_0+\rho_1}\right) \tag{15}$$

where h is a slow varying function of ρ_i and Ω_i, $f_i = \rho_i/(\rho_0+\rho_1)$ and $\rho_i = \lambda_i/(1+\lambda_0+\lambda_1)$ ($i = 1, 2$). In weak coupling, $h \approx 1$ and the exponent is independent of the ionic masses. The resulting isotope coefficient for this mixed mechanism takes then the form:

$$\alpha = \frac{f_0}{2} = \frac{1}{2}\left\{1 - \frac{\lambda_1}{\lambda_0+\lambda_1}\right\} \tag{16}$$

If only the phonon mechanism is active $\lambda_1 = 0$ and one recovers the BCS value 0.5. Otherwise, the presence of the non-phononic mechanism reduces the value of the isotope coefficient, while enhancing T_c. For $\lambda_0 \ll \lambda_1$ one has $\alpha \approx 0$. A joint mechanism would thus allow one to explain the small values of the IC for optimally doped high-T_c superconductors since the coupling to non-phononic degrees of freedom would provide for high T_c's while it would reduce, together with the Coulomb repulsive term, the value of the IC.

Another way to include non-phononic contributions to the pairing mechanism, is to consider a negative effective Coulomb term μ^* in the Eliashberg theory[28, 46, 31]. The pseudo-potential μ^* can thus be seen as an effective attractive pairing potential contribution and thus supports superconductivity. Such a negative μ^* is for example obtained in the plasmon model [44, 28, 46]. One notes that the small values of the partial isotope coefficient in optimally doped high-T_c materials can easily be explained within such a model.

Several other non-phononic as well as combined mechanisms have been proposed [31, 42, 47, 48] that also lead to a reduction of the isotope effect. We refer to the literature for the details.

2.7 Isotope Effect of Properties other than T_c

All previous considerations were concerned with the isotope effect of the superconducting critical temperature T_c. One can ask if there are other properties displaying an isotope effect in conventional superconductors. Naturally, every quantity depending directly on phonon frequencies will display such an effect. Let us consider here a property that will be studied further in the next sections, namely the penetration depth of a magnetic field δ. In the weak-coupling London limit the penetration depth is given by the well-known relation

$$\delta^2 = \frac{mc^2}{4\pi n_s e^2} = \frac{mc^2}{4\pi n\varphi(T/T_c)e^2} \tag{17}$$

where m is the effective mass. n_s is the superconducting density of charge carriers, related to the normal density n through $n_s = n\varphi(T/T_c)$. The function $\varphi(T/T_c)$ is a universal function of T/T_c. For example, $\varphi \simeq 1 - (T/T_c)^4$ near T_c, whereas $\varphi \simeq 1$ near $T = 0$ (in the absence of magnetic impurities; their influence is discussed in section 3).

Eq. (17) does not depend explicitely on phonon frequencies. Nevertheless, it can display an isotopic dependence through T_c. In fact, this dependency is common to all superconducting properties within the BCS theory. Indeed, all major quantities such as heat capacity, penetration depth, critical field, thermal conductivity etc... can be expressed as universal functions of T_c. As a result, all these quantities display a trivial isotopic shift, caused by the isotopic shift in T_c. The value of the shift is growing as one approaches T_c. Furthermore, the shift vanishes for $T \to 0$. As will be shown in section 3 and following the situation is very different in high-temperature superconductors and manganites. For example, the influence of non-adiabaticity on charge-transfer processes results in an unconventional ionic mass dependence of the charge-carrier concentration $n = n(M)$, which, according to Eq. (17), leads to a new isotope effect (see sec. 5).

So far, we have presented various theories explaining the deviation of the measured isotope coefficient in conventional superconductors from the value $\alpha = 0.5$ derived within the BCS model. The remainder of this review is devoted to the study of three factors that are not related to the pairing interaction but affect the isotope coefficient: magnetic impurities, proximity contacts, and non-adiabatic charge-transfer processes. We show that they can strongly modify the value of the isotope coefficient and can even induce an isotopic shift of superconducting properties such as the penetration depth δ. In the last part of this review, we then apply the theory to the case of the oxygen isotopic substitution in high-temperature superconductors.

3 MAGNETIC IMPURITIES AND THE ISOTOPE EFFECT

The presence of magnetic impurities strongly affects various properties of a superconductor. It has been shown[49, 50] that because of magnetic impurity spin-flip scattering processes, Cooper pairs are broken and thus removed from the superconducting condensate. Several properties are affected by the pair-breaking effect. Magnetic scattering leads to a decrease of the critical temperature T_c, the energy gap, the jump in the specific heat at T_c, etc... At some critical impurity concentration $n_M = n_{M,cr}$ superconductivity is totally suppressed. Moreover, there exists an impurity concentration $n_M = n_{M,g} < n_{M,cr}$ (for conventional superconductors $n_{M,g} = 0.9 n_{M,cr}$, see Ref. [49, 50]) beyond which the superconducting state is gapless. Pair-breaking also leads to an increase of the penetration depth δ. We show in the following sections that the presence of magnetic impurities leads to a change of the isotope coefficient of T_c and δ.

3.1 The Critical Temperature

The change of T_c induced by magnetic impurities is described in the weak-coupling regime by the Abrikosov-Gor'kov equation[49]:

$$\ln\left(\frac{T_{c0}}{T_c}\right) = \psi\left(\frac{1}{2} + \gamma_s\right) - \psi\left(\frac{1}{2}\right) \quad , \tag{18}$$

where $\gamma_s = \Gamma_s/2\pi T_c$ and T_c (T_{c0}) is the superconducting critical temperature in the presence (absence) of magnetic impurities. $\Gamma_s = \tilde{\Gamma}_s n_M$ is the spin-flip scattering amplitude and is proportional to the magnetic impurity concentration n_M ($\tilde{\Gamma}_s$ is a constant).

It is easy to derive a relation between the isotope coefficient α_0 in the absence of magnetic impurities [$\alpha_0 = -(M/\Delta M)(\Delta T_{c0}/T_{c0})$, Eq. (1)] and its value in the presence of magnetic impurities [$\alpha = -(M/\Delta M)(\Delta T_c/T_c)$]. From Eqs. (1) and (18) one obtains[51, 4, 5, 52]:

$$\alpha_m = \frac{\alpha_0}{1 - \psi'(\gamma_s + 1/2)\gamma_s} \quad , \tag{19}$$

where ψ' is the derivative of the psi function and is a positive monotonous decreasing function of γ_s. The fact that α_0 and α are not identical relies on the essential feature that the relation between T_{c0} and T_c (Eq. (18)) is non-linear. It is interesting to note that magnetic scattering decreases T_c (reduction of the condensate) but *increases* the value of the IC ($\alpha_m > \alpha_0$; see Eq. (19)).

An important consequence of Eqs. (18) and (19) is that one can write the IC as a universal function $\alpha(T_c)$. This relation does not contain any adjustable parameter, but depends solely on the measurable quantities α_0 and T_c. The universal curve is shown in Fig. 2. Note also that α_m can, in principle, exceed the value $\alpha_{BCS} = 0.5$. Thus, the

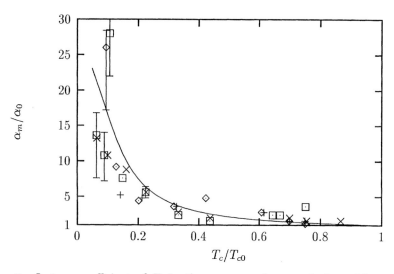

Figure 2. Isotope coefficient of T_c in the presence of magnetic impurities. Solid line: Universal dependence $\alpha(T_c)$ (normalized to α_0; see Ref. [4]); Points: Isotope effect for YBa$_2$(Cu$_{1-x}$Zn$_x$)$_3$O$_7$ obtained with various experimental techniques and normalized to $\alpha_0 = 0.025$ (from Ref. [56, 4]).

observation of large values of α are not necessarily related to the pairing mechanism.

3.2 Zn-doped YBa$_2$Cu$_3$O$_{7-\delta}$

Zn substitution for Cu in the CuO$_2$ planes of YBCO has attracted a lot of interest, since it leads to a drastic decrease in T_c [53] and a strong increase of the penetration depth [54] (superconductivity is destroyed with $\sim 10\%$ of Zn). In addition, this decrease of T_c is accompanied by an increase of the isotope coefficient (see Ref. [53, 56] and Fig. 2). We think that the peculiar behaviour of Zn doping is related to the pair-breaking effect. It has been established by several methods (see Refs. [53, 54]) that Zn substitution leads to the formation of local magnetic moments ($\sim 0.63\mu_B$/Zn, where μ_B is the Bohr magneton) in the vicinity of Zn.

Fig. 2 displays the experimental results obtained for the oxygen isotope coefficient of Zn-doped YBCO[56, 4] (normalized to $\alpha_0 \simeq 0.025$[57]). One can see from this figure that the agreement between the theoretical dependence $\alpha(T_c)$ and the experiment is very good. Note that the uncertainty in the data is growing as $T_c \to 0$ (see Ref. [56]).

3.3 The Penetration Depth

The isotope effect of the penetration depth is more complicated than the isotope effect of T_c, since it appears to be temperature dependent. Here we present the main results near T_c and at $T = 0$ and refer the reader to Ref. [5, 6] for details. Near T_c, the penetration depth is given in second order of the superconducting order parameter $\Delta \equiv \Delta(T, \Gamma_s)$ by[50]:

$$\delta^{-2} = \sigma \frac{\Delta^2}{T_c} \zeta(2, \gamma_s + \frac{1}{2}) \quad , \tag{20}$$

where $\sigma = 4\sigma_N/c$ (σ_N is the normal state conductivity), $\zeta(z, q) = \sum_{n \geq 0} 1/(n+q)^z$ and γ_s was defined in Eq. (18). One can calculate the isotope effect from Eqs. (3), (20) and the analytical expression for Δ near T_c (see Refs. [50, 5]):

$$\Delta^2 = 2\Gamma_s^2(1-\tau)\frac{1 - \bar{\zeta}_2 + (1-\tau)\left[\frac{1}{2} - \bar{\zeta}_2 + \bar{\zeta}_3\right]}{\bar{\zeta}_3 - \bar{\zeta}_4} \equiv 2\Gamma_s^2 \frac{N_1}{D_1} \tag{21}$$

with $\bar{\zeta}_z = \gamma_s^{z-1} \zeta(z, \gamma_s + 1/2)$, $z = 1, 2, \ldots$ and $\tau = T/T_c$.

As mentioned in Sec. 2, δ experiences a trivial BCS isotope effect near T_c. Since we are only interested in the unconventional IE resulting from the presence of magnetic impurities, we substract the BCS isotope effect by calculating the isotope coefficient $\tilde{\beta}_m$ of $\delta(T, \Gamma_s)/\delta(T, 0)$. Near T_c this coefficient can be written as $\tilde{\beta}_m = \beta_m - \beta_0$, where β_m (β_0) is the isotope coefficient of δ in the presence (absence) of magnetic impurities. One can show[6] that (near T_c) $\tilde{\beta}_m$ can be rewritten in the form:

$$\tilde{\beta}_m = (R_1 - R_0)\alpha_m \quad , \tag{22}$$

where

$$R_1 = -\frac{1}{2}f_1 = -\frac{1}{2}\left(\frac{N_2}{N_1} - \frac{D_2}{D_1} + 2\frac{\bar{\zeta}_3}{\bar{\zeta}_2} - 1\right) \quad , \tag{23}$$

$$R_0 = -\frac{1}{2}\frac{\alpha_0}{\alpha_m}f_0 = -\frac{1}{2}[1 - \psi'(\gamma_s + 1/2)\gamma_s]f_0 \quad , \tag{24}$$

and $f_0 = (3 - \tau^2)/(1-\tau)(3-\tau)$. The functions N_1, D_1 are defined in Eq. (21) and

$$N_2 = 3(1-\tau)^2\left[\bar{\zeta}_2 - 2\bar{\zeta}_3 + \bar{\zeta}_4\right] + \tau(2-\tau) - \bar{\zeta}_2$$
$$D_2 = 2\left(3\bar{\zeta}_4 - \bar{\zeta}_3 - 2\bar{\zeta}_5\right) \quad .$$

This relation is valid near T_c (where Δ is small) and for impurity concentrations such that $\Delta T_c / T_c \ll 1$.

One notes first that the isotope coefficient of δ is proportional to the IC of T_c. As for the isotope coefficient of T_c (Eq. (19)) all quantities can either be obtained from experiment (e.g., α_0, T_{c0} and T_c) or calculated self-consistently using Eqs. (18) and (19). There is thus no free parameter in the determination of $\tilde{\beta}_m$. Fig. 3 displays the IC $\tilde{\beta}_m$ (normalized to α_0) as a function of T_c (i.e. on the concentration n_M) for fixed values of the temperature. Setting $\alpha_0 = 0.025$, one obtains the relation $\alpha(T_c)$ expected for YBCZnO. It would be interesting to measure the isotope effect of δ near T_c in this system, to verify our predictions.

An interesting property of Eq. (22) that is shown in Fig. 3 is the fact that the isotope coefficient of δ is temperature dependent; an unusual feature in the context

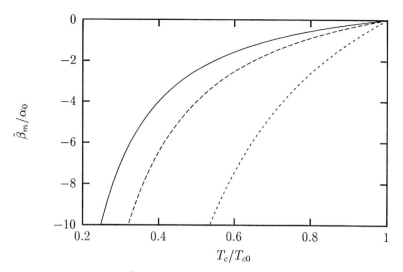

Figure 3. Isotope coefficient $\tilde{\beta}_m$ near T_c (normalized to α_0) as a function of T_c (i.e. of magnetic impurity concentration) for $T/T_c = 0.75$ (solid line), 0.85 (dashed) and 0.95 (dotted). Multiplying by $\alpha_0 = 0.025$ gives the IC expected for $YBa_2(Cu_{1-x}Zn_x)_3O_{7-\delta}$.

of the isotope effect. We will see in Sec. 4 that a temperature dependent IC is also observed for a proximity system.

As can be seen by comparing Figs. 2 and 3, the qualitative behaviour of $\tilde{\beta}_m(T_c)$ is similar to $\alpha_m(T_c)$ [Eq. (19) with $\alpha_0 = \alpha_{ph}$] but with opposite sign. Furthermore, in the absence of magnetic impurities one has $\tilde{\beta}_m = 0$, whereas $\alpha_m(T_c) = \alpha_{ph}$.

Note finally, that since $R_1 - R_0 < 0$, the IC of the penetration depth is always negatif when magnetic impurities are added to the system. This conclusion might not hold in certain cases if the non-adiabatic channel is also included (then one has $\beta_{tot} = \beta_m - \beta_0 + \beta_{na}$ and β_{na} is positive, see Sec. 5).

All previous considerations have been done near T_c. At $T = 0$, the BCS contribution arising from the T_c dependency of the superconducting charge-carrier density n_s (see Eq. (17)) vanishes, because $\varphi(T = 0) = 1$. The only contribution to the IE is thus due to magnetic impurities. In the framework of the Abrikosov-Gor'kov theory one can write the isotope coefficient at $T = 0$ in the form:

$$\beta_m = R_0 \alpha_0 \quad , \qquad (25)$$

where R_0 is given in appendix. Note that R_0 is a negative function of Γ_2 (the direct scattering amplitude) and $\Gamma_s = \Gamma_1 - \Gamma_2$ (the spin-flip, or exchange scattering amplitude) as defined by Abrikosov and Gor'kov[49]. The IC of δ is thus negative both near T_c and at $T = 0$. This has to be seen in contrast to the IC of T_c which is always positive (see Fig. 2). Fig. 4 shows the universal relation $\beta(T_c)$ for two values of Γ_2. One remembers that T_c is determined by the magnetic impurity concentration.

There are two major differences between the results obtained near T_c (Fig. 3) and at $T = 0$ (Fig. 4). First, the direct scattering amplitude Γ_2 appears only at $T = 0$. Secondly, the IC near T_c is proportional to α_m given by Eq. (19), whereas the IC at $T = 0$ is function of α_0, the IC of T_{c0} in the absence of magnetic impurities. This difference is due to the fact that $\varphi = 1$ at $T = 0$ (see Eq. (17)) and the penetration depth does consequently not depend on T_c (the critical temperature in the presence of magnetic impurities).

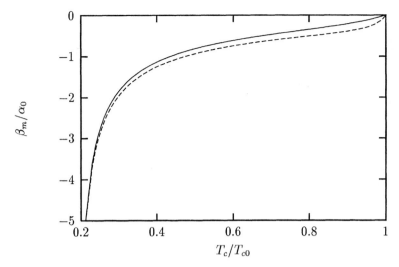

Figure 4. Isotope coefficient of the penetration depth at $T = 0$ in the presence of magnetic impurities (normalized to α_0). There is no free parameter in the relation $\beta_m(T_c)$. Solid and dashed lines are for $\Gamma_2/\Gamma_s = 10$ and 50 respectively (see Ref. [6]).

As shown in Fig. 2, the IC of T_c fits the data of Zn-doped YBCO. On the other hand, there are no experimental data for conventional superconductors. Furthermore, the magnetic impurity contribution to the IE of δ shown in Fig. 3 has never been measured. It would thus be interesting to perform measurements of these effects, especially since they can be described by universal relations.

4 ISOTOPE EFFECT IN A PROXIMITY SYSTEM

In this section we consider another case, a proximity system, in which a factor not related to lattice dynamics induces a change of the isotope coefficient of T_c and δ. Consider an $S - N$ sandwich where S (N) is a superconducting (normal) film. The value of the critical temperature T_c and the penetration depth δ differ substantially from the values T_{c0} and δ_0 of the isolated superconductor S[58, 59]. We show in the following that the proximity effect also influences the isotope coefficient of T_c and δ.

4.1 The Critical Temperature

Let us consider a proximity system composed of a weak-coupling superconductor S of thickness L_S and a metal or a semiconductor N of thickness L_N (e.g. Nb-Ag). In the framework of the McMillan tunneling model[17], which can be used when $\delta < L_N \ll \xi_N$ ($\xi_N = h v_{F;N}/2\pi T$ is the coherence length of the N-film as defined in Ref. [60]), the critical temperature T_c of the $S - N$ system is related to the critical temperature T_{c0} of S by the relation[58]:

$$T_c = T_{c0} \left(\frac{\pi T_{c0}}{2\gamma u}\right)^\rho, \quad \rho = \frac{\nu_N L_N}{\nu_S L_S} \quad , \qquad (26)$$

where $\gamma \simeq 0.577$ is Euler's constant. The value of u is determined by the interplay of the McMillan tunneling parameter $\Gamma = \Gamma_{SN} + \Gamma_{NS}$ and the average phonon frequency

Ω. For an almost ideal $S-N$ contact ($\Gamma \gg \Omega$) one has $u \simeq \Omega$. In the opposite limit, when $\Gamma \ll \Omega$ one obtains $u \simeq \Gamma$ (see Ref. [58]).

Let us first consider the case $\Gamma \gg \Omega$. In the BCS model, one has $T_{c0} \propto \Omega$. Thus, according to Eq. (26) T_c and T_{c0} have the same dependency on the ionic mass in this limit (because T_{c0}/u is then independent of Ω). This results in the simple relation $\alpha_{prox} = \alpha_0$ for the IC of T_c (α_0 is the isotope coefficient of T_c for the isolated S-film).

The opposite situation where $\Gamma \ll \Omega$ is more interesting since $u = \Gamma$ is independent of ionic masses. Using Eqs. (1) and (26) one obtains for the isotope coefficient of T_c in the limit $\Gamma \ll \Omega$:

$$\alpha_{prox} = \alpha_0 \left(1 + \frac{\nu_N L_N}{\nu_S L_S}\right) \quad . \tag{27}$$

One notes first that whereas the presence of a normal film on the superconductor decreases T_c, the film induces an *increase* of the isotope coefficient of T_c. The same is true for the presence of magnetic impurities and for the non-adiabatic IE when $\partial T_c/\partial n > 0$ (see next section). The second interesting feature of Eq. (27) is that one can modify the value of α_{prox} by changing the thicknesses of the films. For example, if $\nu_N/\nu_S = 0.8$ and $L_N/L_S = 0.5$ then $\alpha_{prox} = 0.28$. By increasing the thickness of the normal film such that $L_N = L_S$, one obtains $\alpha_{prox} = 0.36$.

We stress the fact that, as in the previous section on magnetic impurities, the change of the IC is due to a factor not related to lattice dynamics. Therefore, there is no reason for α_{prox} to be limited to values below 0.5.

To the best of our knowledge the change in the IC caused by the proximity effect has never been measured, even in conventional superconductors. It would be interesting to carry out such experiments in order to observe this phenomenon.

4.2 The Penetration Depth

The proximity effect also affects the shielding of a magnetic field. The most dramatic effect of the normal layer on the penetration depth is seen in the low-temperature regime ($T/T_c \lesssim 0.3$). Although the penetration depth of a pure conventional superconductor is only weakly temperature dependent in this regime ($\delta^{-2} \sim \varphi = 1 - T^4/T_c^4$) the presence of the normal layer induces a temperature dependence through the proximity effect. For the same $S-N$ proximity system as considered above the penetration depth is given by[59]

$$\delta^{-3} = a_N \Phi \quad , \tag{28}$$

where a_N is a constant depending only on the material properties of the normal film (it is ionic-mass independent) and

$$\Phi = \pi T \sum_{n \geq 0} \frac{1}{x_n^2 p_n^2 + 1} , \quad p_n = 1 + \varepsilon t \sqrt{x_n^2 + 1} \quad . \tag{29}$$

$x_n = \omega_n/\epsilon_S(T)$ with $\omega_n = (2n+1)\pi T$ (the Mastubara frequencies) and $\epsilon_S(T)$ is the superconducting energy gap of S. In the weak-coupling limit considered here, $\epsilon_S(0) = \varepsilon \pi T_{c0}$ with $\varepsilon \simeq 0.56$. The dimensionless parameter $t = \ell/S_0$ with $\ell = L_N/L_0$ and $S_0 = \Gamma/\pi T_{c0}$ ($\Gamma \sim 1/L_0$ is the McMillan parameter[17]). L_N and L_0 are the thickness of the normal film and some arbitrary thickness, respectively (in the following we take $L_0 = L_S$, the thickness of the superconducting film). From Eqs. (3) and (28) one obtains the IC of the penetration depth for the proximity system:

$$\beta_{prox} = -\frac{2\alpha_0}{3\Phi} \sum_{n>0} \frac{x_n^2 p_n^2}{x_n^2 p_n^2 + 1} \left(1 - \frac{\varepsilon t}{p_n\sqrt{x_n^2 + 1}}\right) \quad , \qquad (30)$$

where α_0 is again the IC of T_{c0} for the superconducting film S alone. This result has three interesting features. First, as for the case of magnetic impurities, the IC of T_c and δ have opposite signs. This observation is valid as long as one does not mix the different channels presented in this work (magnetic impurities, proximity effect and non-adiabaticity). The addition of a non-adiabatic contribution (next section) may lead to a different conclusion. Secondly, one can see from Eq. (30) that β_{prox} depends on the proximity parameter t. One can thus modify the IC either by changing the ratio $\ell = L_N/L_S$ or the McMillan tunneling parameter Γ (e.g. by changing the quality of the interface; see Ref. [6]). Fig. 5 shows this dependence for different temperatures. One

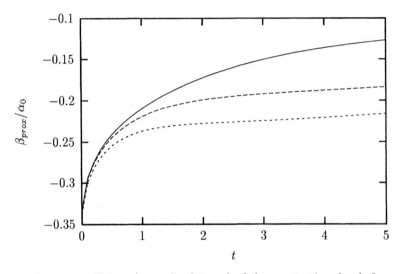

Figure 5. Isotope coefficient (normalized to α_0) of the penetration depth for a proximity system as a function of t, for $T/T_c = 0.1$ (solid), 0.2 (dashed), 0.3 (dotted). See Ref. [6].

notes that, contrary to the value of α_{prox}, the IC of the penetration depth decreases with increasing ratio ℓ. Finally, one notes that the IC β_{prox} is temperature-dependent. This unconventional feature was also observed for the IC of δ in the presence of magnetic impurities near T_c. Fig. 6 shows the temperature dependence for different values of the parameters. The trend is similar to the case of magnetic impurities: $|\beta_{prox}|$ increases with increasing T. Note, however, that the two effects are calculated in very different temperature ranges (near T_c and near $T = 0$). A complete description of this effect can be found in Ref. [5].

5 NON-ADIABATIC ISOTOPE EFFECT

The non-adiabatic isotope effect introduced in Ref. [3] was used in Refs. [3, 4, 5, 6, 7] to describe the unusual behavior of the isotope coefficient in several high-temperature superconductors. The theory also allowed us to describe the large isotopic shift of the ferromagnetic phase transition in manganites [7]. The concept of the non-adiabatic IE relies on the fact that in the above-mentioned systems charge-transfer processes involve

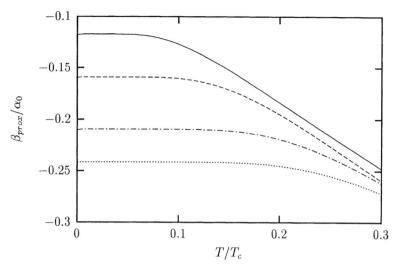

Figure 6. Isotope coeff. β_{prox} (normalized to α_0) for a proximity system as a function of T/T_c for $S_0 = 0.2$: $l = 1$ (solid) $l = 0.5$ (dashed) and $S_0 = 1$: $l = 1$ (dash-dotted) $l = 0.5$ (dotted). See Ref. [5].

a non-adiabatic channel. In other words, the electronic ground state of the group of atoms over which the charge-transfer takes place is degenerate, leading to a dynamic Jahn-Teller (JT) effect ($E_{JT} \leq \hbar\omega$, where E_{JT} is the JT energy; see, e.g., Ref. [61]).

The concept of the non-adiabatic isotope effect can be understood by considering two examples: the apex oxygen in high-T_c oxides and the motion of charge-carriers in manganites. Let us first consider high-T_c materials. This system can be seen as a stack of charge reservoirs (CuO chains) and conducting subsystems (CuO$_2$ planes). Charges are transfered from the reservoir to the conducting layers through the apex oxygen that bridges the two subsystems. Several experiments have shown[62, 63, 64] that the apex oxygen displays a non-adiabatic behaviour. The ion oscillates between two close positions that correspond to two configurational minima of the potential energy surface. These two minima are due to the Jahn-Teller effect since we are dealing with the crossing of electronic terms (see Fig. 7).

The Jahn-Teller effect leads to a "double-well" type potential (which should not be confused with the double-well appearing when the crystal is anharmonic; here we consider energy terms crossing and for each term we use the harmonic approximation). This structure has a strong impact on charge-transfer processes in these materials. Indeed, the motion of charges from the chains to the planes occurs through the apex oxygen. Thus, the charge-transfer process involves the motion of the non-adiabatic ion. The immediate consequence of this observation is that the density of charge-carriers n in the CuO$_2$-planes (the conducting subsystem) depends on the mass of the non-adiabatic ions involved $n = n(M)$ (see below).

The situation encountered in manganites is different from the example just described. In these materials, there is no transfer between a reservoir and a conducting subsytem. Instead, the motion of charge-carriers occurs on the Mn-O-Mn complex that displays a non-adiabatic behaviour[65]. As a consequence, the electron hopping involves the motion of the ions and depends on their mass. It was shown in Ref. [7] that this fact can account for the unexpectedly low ferromagnetic critical temperature $T_{c,f}$[66] (which is determined by the hopping of the electrons between Mn ions) as well as for the huge isotopic shift of $T_{c,f}$[67].

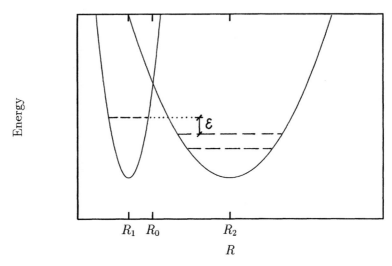

Figure 7. Potential energy surface (PES) as a function of the configurational coordinate R. The crossing of electronic terms leads to a dynamical Jahn-Teller distortion (two minima of the PES).

Note that although the charge-transfer processes are different in high-temperature superconductors and in manganites, the formalism described below and in Ref. [68] applies equally well to the two systems. In the following we refer to the example of the apex oxygen. The case of manganites can be mapped by replacing "reservoir" and "conducting subsystem" by "Mn ions".

It is important to realize that the dependence $n(M)$ is specific to systems where the motion of charges occurs through non-adiabatic ions. In metals, the charge-carriers are not affected by the motion of the ions because of the validity of the adiabatic approximation. We also stress the fact that the apex oxygen is given as a simple example but, as will be shown below, the theory is not limited to this case. Finally, one should note that a phenomenological theory of the IC based on the assumption that $n = n(M)$ has been proposed in Ref. [69].

To demonstrate that in the presence of non-adiabatic charge-transfer one has $n = n(M)$, one best uses the so-called diabatic representation[70] (see also Ref. [68]). Here we only recall the main steps of the calculation. Let us assume that, because of the degeneracy of electronic states (or Jahn-Teller crossing), the potential energy surface $E(R)$ (R is the relevant configurational coordinate) of the group of non-adiabatic ions (such as the apex O, in-plane Cu and O for high-T_c materials or Mn and O ions in manganites) has two close minima (see Fig. 7). In the diabatic representation the total wave-function $\Psi(\mathbf{r},\mathbf{R},t)$ ($\mathbf{r}$ is the electronic coordinate) is written as a linear combination of the wave-functions in the two crossing potential surfaces. The total wave-function cannot be decomposed as a product of electronic and lattice wave-functions. On the other hand, one can write the total wave-function as a sum of symmetric and antisymmetric terms. The energy splitting between these symmetric and antisymmetric terms, which corresponds to the inverse lifetime of oscillations between the configurations (minima of the potential energy surface), has the form[3, 5, 7, 68]:

$$H_{12} = <\Psi_1|H_e|\Psi_2> \simeq L_0 F_{12} \quad , \tag{31}$$

where H_e is the electronic part of the total Hamiltonian and

$$L_0 = L(\mathbf{R}_0) = \int d\mathbf{r}\, \psi_1^*(\mathbf{r},\mathbf{R}) H_e \psi_2(\mathbf{r},\mathbf{R}) \quad , \tag{32}$$

$$F_{12} = \int d\mathbf{R}\, \Phi_1^*(\mathbf{R}) \Phi_2(\mathbf{R}) \tag{33}$$

The last equality in Eq. (31) is obtained under the assumption that the electronic wave-function ψ_i ($i = 1, 2$) is a slowly varying functions of $\mathbf{R}$. L_0 can then be evaluated at $\mathbf{R}_0$, the crossing of electronic terms, and taken out of the integral over $\mathbf{R}$. It does not depend on ionic masses. On the other hand, the important Franck-Condon factor (33) depends on the lattice wave-functions $\Phi_i(\mathbf{R})$, and thus on ionic masses. Given the perturbation, Eq. (31), one can calculate the probability of finding the charge-carrier in the conducting layer [3, 5, 7, 68]. Note that non adiabaticity affects the bandwidth of the reservoir and conducting bands rather than the chemical potential (which is the same because of thermodynamic equilibrium). Thus, in the case of YBCO, the plane and chain bands have the same chemical potential, but different widths. This affects the charge-carrier concentration in the two subsystems.

Qualitatively, the charge transfer can be visualized as a multi-step process. First, the charge carrier can move from the reservoir to the group of ions (e.g. from the chains to the apical oxygen in YBCO). Then, the non-adiabatic ions tunnel to the other electronic term ($\Psi_1 \to \Psi_2$). As a final step, the charge carrier can hop to the conducting layer (from the apical oxygen to the CuO_2 planes). The crucial point of the theory is that as a result of this multi-step process, the probability P of finding the charge-carrier in the conducting subsystem depends on Eq. (31) and thus on the Franck-Condon factor $F_{12} = F_{12}(M)$, Eq. (33), that depends on ionic masses. In other words the charge-carrier concentration n, which is proportional to the probability P, depends on the ionic mass M. It is this unusual dependence $n = n(M)$, found in high-temperature superconductors and manganites, that is responsible for the unconventional isotope effect of T_c and δ.

5.1 The Critical Temperature

Given that $n = n(M)$, we can write the isotope coefficient of T_c as $\alpha = \alpha_{ph} + \alpha_{na}$, where $\alpha_{ph} = (M/T_c)(\partial T_c/\partial \Omega)(\partial \Omega/\partial M)$ is the usual (BCS) phonon contribution (Ω is a characteristic phonon energy) and the non-adiabatic contribution is given by:

$$\alpha_{na} = \gamma \frac{n}{T_c} \frac{\partial T_c}{\partial n} \quad , \tag{34}$$

where the parameter $\gamma = -M/n(\partial n/\partial M)$ has a weak logarithmic dependence on M (see Ref. [3]). This parameter should not be confused with Euler's constant. Eq. (34) shows that the IC of T_c depends on the doping of the conducting layer and on the relation $T_c(n)$. This result was used in Refs. [3, 4] to analyse the IE of high-temperature superconductors (see below).

5.2 The Penetration Depth

The concept of the non-adiabatic isotope effect relies on the fact that the charge-carrier concentration depends on the ionic mass because of the Jahn-Teller crossing of electronic terms. Obviously, any quantity that depends on $n(M)$ will also display an unconventional isotope effect. One such quantity is the penetration depth of a magnetic field δ given in Eq. (17). Note that the relation $\delta^{-2} \sim n_s$ is also valid in the strong-coupling case (see, e.g., Refs. [5]). From Eq. (3) one has

$$\beta \equiv -\frac{M}{\delta}\frac{\partial \delta}{\partial n_s}\frac{\partial n_s}{\partial M} = \frac{M}{2n_s}\frac{\partial n_s}{\partial M} \tag{35}$$

As mentioned in Sec. 2.7, because of the relation $n_s = n\varphi(T)$, one has to distinguish two contributions to β. There is a usual (BCS) phonon contribution, β_{ph}, arising from the fact that $\varphi(T/T_c)$ depends on ionic mass through the dependency of T_c on the characteristic phonon frequency. This BCS contribution was discussed in Sec. 2. In the present paper we focus on the non-trivial manifestation of isotopic substitution arising from the isotope dependence of the charge-carrier concentration n. Such an effect can even be observed in the low-temperature region where $\varphi \simeq 1$. From Eq. (35) and the relation $n_s = n\varphi(T)$ it follows that

$$\beta = \beta_{ph} + \beta_{na} = \frac{M}{2\varphi(T)}\frac{\partial \varphi(T)}{\partial M} + \frac{M}{2n}\frac{\partial n}{\partial M} \quad , \tag{36}$$

where $n(M)$ is the normal-state charge-carrier concentration. Comparing Eq. (34) and the second term of the right hand side of Eq. (36) one infers that $\beta_{na} = -\gamma/2$ and thus establish a relation between the non-adiabatic isotope coefficients of T_c and δ:

$$\alpha_{na} = -2\beta_{na}\frac{n}{T_c}\frac{\partial T_c}{\partial n} \tag{37}$$

This result holds for London superconductors. The equation contains only measurable quantities and can thus be verified experimentally. It is interesting to note that β_{na} and α_{na} have opposite signs when $\partial T_c/\partial n > 0$ (which corresponds to the underdoped region of high-T_c materials).

Note, finally, that in the presence of magnetic impurities the relation $n_s = n\varphi(T)$ remains valid, but $\varphi(T)$ depends now on the direct scattering amplitude Γ_2 defined in Sec. 3.3 (this results, e.g., in the inequality $n_s(T=0) < n$ in the gapless regime). As a consequence, magnetic impurities affect the first term in Eq. (36) (that now depends on Γ_2), but leaves β_{na} and thus Eq. (37) unchanged.

6 OXYGEN ISOTOPE EFFECT IN HIGH-T_c MATERIALS

In this section we briefly discuss experiments done on high-temperature superconductors in the light of the theory exposed in sections 3 and 5. A detailed study of the oxygen isotope effect in high-T_c materials can be found in Refs. [3, 4, 5, 6].

The oxygen isotope effect of T_c has been observed in a number of experiments[71, 57, 56, 72, 73]. Here we discuss the results obtained on Pr-doped and oxygen-depleted YBa$_2$Cu$_3$O$_{7-\delta}$ (YPrBCO and YBCO respectively). We also discuss shortly the isotope effect of the penetration depth observed in La$_{2-x}$Sr$_x$CuO$_4$ (LSCO). The analysis of the isotope effect in Zn-doped YBCO has already been presented in Sec. 3.

6.1 The Critical Temperature

Let us first consider the isotope coefficient of T_c for Pr-doped YBCO (YPrBCO). Several experiments have established that the Pr replaces Y which is located between the two

CuO$_2$ planes of a unit cell. The doping affects YBCO mainly in two ways. First, it was shown that because of the mixed-valence state of Pr, holes are depleted from the CuO$_2$ planes (see, e.g., Ref. [74]). Secondly, as for Zn substitution, praesodimium changes the magnetic impurity concentration of the system[56, 74]. Regarding the IE, these experimental facts imply that through the first effect, the non-adiabatic channel is activated, whereas the second effect leads us to consider the magnetic impurity channel for the calculation of α. The IE is therefore described by $\alpha_t = \alpha_{ph} + \alpha_{m+na}$, where $\alpha_{ph} = 0.025$[57] is the phonon contribution, and α_{m+na} is defined by Eq. (19) with α_0 given by Eq. (34). The resulting expression depends on two parameters γ and $\tilde{\Gamma}_s$, characterizing the non-adiabatic and magnetic impurity channels, respectively. The latter quantity has been extracted from a fit to the relation $T_c(x)$ (x describes the Pr doping)[74] and is $\tilde{\Gamma}_s = 123K$. The parameter $\gamma = 0.16$ is determined from the mean-square fit to the experimental data. The result is shown in Fig. 8. The theory

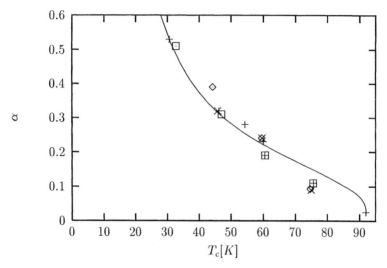

Figure 8. Dependence of the isotope coefficient α on T_c for Y$_{1-x}$Pr$_x$Ba$_2$Cu$_3$O$_{7-\delta}$. Theory: solid line with $\gamma = 0.16$, $\tilde{\Gamma}_s = 123K$, $\alpha_0 = 0.025$; Experiment: dc magnetization, resistivity and ac susceptibility from Refs. [71, 56].

is in good agreement with the experimental data. Note that the negative curvature at high T_c's reflects the influence of the non-adiabatic channel (magnetic impurities give a contribution with opposite curvature), whereas the positive curvature seen at low T_c is due to the presence of magnetic impurities[5].

Oxygen-depleted YBCO is very similar to the previous case in that oxygen depletion both introduces magnetic impurities into the system and removes holes from the CuO$_2$ planes. The same equations as above can thus be used, however, with appropriate values of the parameters. The result is shown in Fig. 9, together with the experimental data from Ref. [57]. Note the sharp drop observed near $T_c = 60K$. This drop and the peak above it are related to the presence of a plateau in the dependence $T_c(n)$ near $T_c = 60K$ (see, e.g., Ref. [57]). Since Eq. (34) contains the derivative $\partial T_c/\partial n$, α_{na} will be nearly zero at 60K. As one increases T_c, the slope of $T_c(n)$ jumps to a large value, and decreases again to zero as $T_c \to 90K$. This behavior of $T_c(n)$ explains the maximum appearing in $\alpha_{na}(T_c)$ above 60K. The peak structure is thus specific to oxygen-depleted YBCO because neither YPrBCO nor YBCZnO display such a plateau in the relation $T_c(n)$. It would be interesting to verify this result, by measuring the oxygen isotope effect between 60 and 90K.

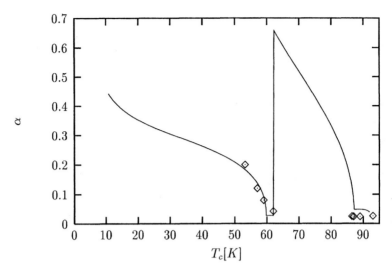

Figure 9. Dependence of the isotope coefficient α on T_c for $YBa_2Cu_3O_{6+x}$. Theory: solid line with $\gamma = 0.28$, $\tilde{\Gamma}_s = 15K$, $\alpha_0 = 0.025$; Experiment: points from Ref. [57].

The main difference between Pr-substituted (Fig. 8) and oxygen-depleted YBCO (Fig. 9) lies in the fact that the doping affects different ions of the system. Magnetic impurities are introduced at different sites (on Y for the first material and in the chains or the apical oxygen position for the second system). Furthermore, non-adiabatic charge-transfer processes involve mainly Pr as well as Cu and O of the planes in YPrBCO, whereas they involve mainly the apical O in oxygen-depleted YBCO (to a lower extent, chain and in plane O and Cu are also involved).

Contrary to YPrBCO, it is not possible to extract the value of $\tilde{\Gamma}_s$ (the contribution of magnetic impurities). The best fit to the few data gives $\gamma = 0.28$ and $\tilde{\Gamma}_s = 15K$. Two important points have to be noted concerning these values. First, because of the limited data available, the values may vary, although the order of magnitude will remain (see also Ref.[6]). The other remark concerns the value of $\tilde{\Gamma}_s$ for YPrBCO and YBCO. From these values one concludes that oxygen depletion introduces a smaller amount of magnetic moments than Pr doping. It would be interesting to perform more isotope effect experiments on oxygen-depleted YBCO and to determine the effective magnetic moment per depleted oxygen through other means.

6.2 The Penetration Depth

The only experimental observation of the isotopic shift of the penetration depth has been done on $La_{2-x}Sr_xCuO_4$ (LSCO)[73]. There are no data available on YBCO-related materials. Since only Pr-doped YBCO has one free parameter (γ) we present the results of our theoretical calculations only in this case. The oxygen-depleted case was studied as a function of the two parameters γ and $\tilde{\Gamma}_s$ in Ref. [5].

Let us begin with the case of LSCO. This case is simpler to study than YBCO related materials, since no significant amount of magnetic impurities has been detected in this material. Our analysis can thus be carried out with Eqs. (34) or (37). The isotope coefficient has been measured for Sr concentrations near $x \approx 0.11$ and $x \approx 0.15$. The first concentration corresponds to the region where $T_c(x)$ experiences a small dip. The origin of this dip is not well-established but is probably related to electronic inhomogeneities and structural instability. Since many factors affect T_c at this Sr concentration, it is difficult to interpret correctly the isotope effect of T_c and δ.

The second concentration at which an isotope shift of δ has been measured is at optimal doping (T_c is maximal). The experimental shift $\Delta\delta/\delta \simeq 2\%$ and Eq. (3) allows one to determine the value of $\beta_{na} \simeq 0.16$ and thus $\gamma \simeq 0.32$ through Eq. (37). This result is in good agreement with the values obtained for Pr-doped and O-depleted YBCO (see Ref. [6] for a discussion on LSCO).

Let us now turn to the isotope coefficient of the penetration depth for YBCO-related materials. From the evaluation of the parameter γ (see Eq. (34) and above) and from Eq. (37) one obtains the non-adiabatic contribution to the IC of the penetration depth in Pr-doped and O-depleted YBCO. One obtains $\beta_{na} = -0.08$ in the first case and $\beta_{na} = -0.14$ in the second.

Using Eqs. (19), (22) and (34), as well as the values of the parameters γ and $\tilde{\Gamma}_s$ given above, one can calculate the IC resulting from the contributions of magnetic impurities and non-adiabaticity for YPrBCO near T_c. The result is shown in Fig. 10 for three different temperatures. One notes that contrary to the case of magnetic

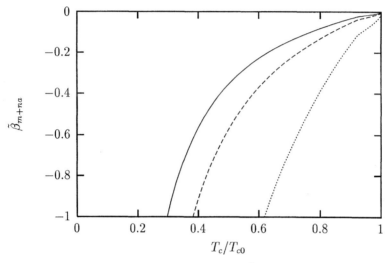

Figure 10. Dependence of the isotope coefficient $\tilde{\beta}_{m+na}$ on T_c for $T/T_c = 0.75$ (solid line), 0.85 (dashed), 0.95 (dotted). $\alpha_0 = 0.025$, $\gamma = 0.16$, and $\tilde{\Gamma}_s = 123K$ (parameters for $Y_{1-x}Pr_xBa_2Cu_3O_{7-\delta}$).

impurities alone (one sets $\gamma = 0$) studied in Sec. 3 (Figs. 2 and 3), the isotope effect of the penetration depth does not have the same qualitative behaviour as the IC of T_c (compare Figs. 9 and 10). The change of curvature does not take place at the same value of T_c (in the case of $\tilde{\beta}_{m+na}$ the change occurs at high T_c's and is barely visible on the figure; see Ref. [6] for a discussion of this point). One should measure the IC of δ for YPrBCO and YBCO, since it would allow one to give a better estimate of the parameters and to test the theory.

7 CONCLUSIONS

We have reviewed different aspects of the isotope effect. In a first part, we have summarized the main theories developed to explain the isotope effect in conventional superconductors and we have discussed their relevance for high-T_c oxides. In particular, we have considered the effect of the Coulomb interaction, band structure (van Hove singularities), multiatomic compounds, anharmonicity and non-phononic mechanisms on the value of the isotope coefficient. These effects can account for most experimental data obtained on conventional superconductors. For example, they allow one to describe the deviation from the standard BCS value $\alpha = 0.5$ in transition metals or the inverse isotope effect ($\alpha < 0$) found in PdH. The results show clearly that a vanishing or even a negative isotope coefficient can be obtained within conventional superconductivity where the pairing between charge-carriers is mediated by the electron-phonon interaction. Generally, however, the small isotope coefficient is obtained only in conjunction with low T_c's. On the other hand, one can also obtain values of the isotope coefficient that are larger than $\alpha = 0.5$ when anharmonic or band-structure effects are present in the system.

In a second part, we have analyzed the effect of magnetic impurities, proximity contacts and non-adiabatic charge-transfer processes on the isotope coefficient (see also Refs. [3, 4, 5, 6, 7]). An important common feature of these factors is that they are not related to the pairing mechanism but, nevertheless, strongly affect the isotope effect. On notes that $\alpha > 0.5$ is allowed in all three channels. Furthermore, these factors can induce a non-trivial isotopic shift of quantities such as the penetration depth δ. In the case of magnetic impurities and the proximity effect, the isotope effect of the penetration depth is temperature dependent. This phenomenon has not been investigated experimentally yet. In the presence of non-adiabaticity, we have established a relation between the isotope shift of T_c and δ for London superconductors.

The theory presented in the second part of this review allowed us to describe the unconventional behavior of the isotope coefficient in $YBa_2Cu_3O_7$ related systems (Zn and Pr-substituted as well as oxygen-depleted materials). The case of Zn-substituted YBCO was described by involving solely the magnetic impurity channel and did not require any fitting parameter. The theoretical curve also applies to conventional superconductors.

Our calculations suggest several experiments both on conventional and high-temperature superconductors. In particular, it would be interesting to measure the change of the isotope coefficient induced by a proximity system or magnetic impurities in conventional superconductors to test our theory. Furthermore, more experiments have to be done on high-temperature superconductors so as to determine more precisely the parameters appearing in the theory.

8 Acknowledgment

A.B. is grateful to the Naval Research Laboratory and the Swiss National Science Foundation for the support. The work of V.Z.K. is supported by the U.S. Office of Naval Research under contract No. N00014-96-F0006.

9 APPENDIX

We derive the explicit form of the function R_0 given in Eq. (25) (see Ref. [6]). The penetration depth calculated by Skalski et al. at zero temperature is given by $\delta^{-2} = -(4\pi n e^2/mc^2)\tilde{K}(\omega = 0, \mathbf{q} = 0)$ with[50]

$$\tilde{K}(0,0) = -\frac{1+\bar{\Gamma}\bar{\eta}^{-3}}{\bar{\eta}}\left[\frac{\pi}{2} - \frac{f(\bar{\eta})}{R(\bar{\eta})}\right] + \bar{\Gamma}\bar{\eta}^{-3}\left[\frac{2}{3}\bar{\eta} - \frac{\pi}{4}\bar{\eta} + 1\right] \qquad (38)$$

for $\bar{\Gamma}, \bar{\eta} < 1$ (with $f(\bar{\eta}) = \arccos\bar{\eta}$) or $\bar{\Gamma} < 1$, $\bar{\eta} > 1$ (with $f(\bar{\eta}) = \mathrm{arcosh}\bar{\eta}$) and

$$\tilde{K}(0,0) = -\frac{1+\bar{\Gamma}\bar{\eta}^{-3}}{\bar{\eta}}\left\{\frac{\pi}{2} - 2\frac{\bar{\Gamma}-1}{R(\bar{\Gamma})} - R^{-1}(\bar{\eta})\left(\mathrm{arcosh}\bar{\eta} - 2\mathrm{artanh}\mathcal{R}\right)\right\} \qquad (39)$$
$$+\bar{\eta}^{-3}\left\{\left(\frac{2}{3}\bar{\eta}^2+1\right)\left(\bar{\Gamma}-R(\bar{\Gamma})\right) - \frac{1}{2}\bar{\eta}R(\bar{\Gamma})\left(\frac{2}{3}\bar{\eta}-1\right) - \bar{\eta}\bar{\Gamma}\left(\frac{\pi}{4} - \frac{\bar{\Gamma}-1}{R(\bar{\Gamma})}\right)\right\}$$

for $\bar{\Gamma}, \bar{\eta} > 1$. We have introduced the notation $\bar{\Gamma} = \Gamma_s/\Delta$, $\bar{\eta} = \eta\bar{\Gamma} = \Gamma_2/\Delta$, $\eta = \Gamma_2/\Gamma_s$, $R(x) = \sqrt{|1-x^2|}$ with $x = \bar{\Gamma}, \bar{\eta}$ and $\mathcal{R} = [(\bar{\Gamma}-1)(\bar{\eta}-1)/(\bar{\Gamma}+1)(\bar{\eta}+1)]^{1/2}$. $\Delta \equiv \Delta(T=0, \Gamma_s)$ is the order parameter in the presence of magnetic impurities. Eqs. (38),(39) are valid when $\Gamma_s \ll \Gamma_2$. These two scattering amplitudes, Γ_2 and Γ_s ($\Gamma_s = \Gamma_1 - \Gamma_2$), defined by Abrikosov and Gor'kov[49], describe the direct and exchange scattering, respectively. One can calculate the magnetic impurity contribution to the IC at $T=0$ from Eq. (3) in a straightforward way using Eqs. (38) and (39). The result can be written as

$$\beta_m(T=0) = -\frac{\alpha_\Delta}{2}\frac{K_1 + K_2}{\tilde{K}(0,0)} \qquad (40)$$

where α_Δ is defined below, $\tilde{K}(0,0)$ is given by Eqs. (38), (39) and

$$K_1 = -\frac{1+3\bar{\Gamma}\bar{\eta}^{-3}}{\bar{\eta}}\left(\frac{\pi}{2} - \frac{f(\bar{\eta})}{R(\bar{\eta})}\right) \pm \frac{1+\bar{\Gamma}\bar{\eta}^{-3}}{R(\bar{\eta})^2}\left(\frac{\bar{\eta}}{R(\bar{\eta})}f(\bar{\eta}) - 1\right) \qquad (41)$$
$$K_2 = \bar{\Gamma}\bar{\eta}^{-3}\left(2 - \frac{\pi}{4}\bar{\eta}\right)$$

for $\bar{\Gamma}, \bar{\eta} < 1$ (upper sign, $f(\bar{\eta}) = \arccos\bar{\eta}$) or $\bar{\Gamma} < 1$, $\bar{\eta} > 1$ (lower sign, $f(\bar{\eta}) = \mathrm{arcosh}\bar{\eta}$) and

$$K_1 = -\frac{1+3\bar{\Gamma}\bar{\eta}^{-3}}{\bar{\eta}}\left\{\frac{\pi}{2} - 2\frac{\bar{\Gamma}-1}{R(\bar{\Gamma})} - \frac{1}{R(\bar{\eta})}(\mathrm{arcosh}\bar{\eta} - 2\mathrm{artanh}\mathcal{R})\right\} - \frac{1+\bar{\Gamma}\bar{\eta}^{-3}}{\bar{\eta}}\left\{2\frac{\bar{\Gamma}(\bar{\Gamma}-1)}{R(\bar{\Gamma})^3}\right.$$
$$\left. -\frac{1}{R(\bar{\eta})}\left[\frac{\bar{\eta}}{R(\bar{\eta})^2}(\mathrm{arcosh}\bar{\eta} - 2\mathrm{artanh}\mathcal{R}) + \frac{\mathcal{R}}{1-\mathcal{R}^2}\left(\frac{\bar{\Gamma}^2}{R(\bar{\Gamma})^2} + \frac{\bar{\eta}^2}{R(\bar{\eta})^2}\right) - \frac{\bar{\eta}}{R(\bar{\eta})}\right]\right\} \qquad (42)$$
$$K_2 = \frac{3}{\bar{\eta}^3}\left(\frac{2}{3}\bar{\eta}^2 + 1\right)[\bar{\Gamma} - R(\bar{\Gamma})] - \frac{1}{2}\bar{\eta}R(\bar{\Gamma})\left(\frac{2}{3}\bar{\eta}^2 - 1\right) - \bar{\eta}\bar{\Gamma}\left(\frac{\pi}{4} - \frac{\bar{\Gamma}-1}{R(\bar{\Gamma})}\right)$$
$$+ \frac{1}{\bar{\eta}^3}\left\{\left[\frac{\bar{\Gamma}}{R(\bar{\Gamma})}\left(\frac{2}{3}\bar{\eta}^2 + 1\right) - \frac{4}{3}\bar{\eta}^2\right][\bar{\Gamma} - R(\bar{\Gamma})]\right.$$
$$\left. + \frac{1}{2}\bar{\eta}\frac{\bar{\Gamma}^2}{R(\bar{\Gamma})}\left(\frac{2}{3}\bar{\eta}^2 - 1\right) + 2\bar{\eta}\bar{\Gamma}\left[\frac{\pi}{4} - \frac{\bar{\Gamma}-1}{R(\bar{\Gamma})}\left(1 + \frac{\bar{\Gamma}}{2R(\bar{\Gamma})^2}\right)\right]\right\}$$

for $\bar{\Gamma}, \bar{\eta} > 1$ and $\Delta T_c/T_c(\bar{\Gamma}-1) \ll 1$. The last condition expresses the fact that the calculation is not valid in the immediate vicinity of $\bar{\Gamma} = 1$. Eq. (40) contains α_Δ which is the IC of the order parameter Δ. In strong-coupling systems, α_Δ has to be calculated numerically using Eliashberg's equations. Here we calculate the IC in the framework of the BCS model where α_Δ can be calculated analytically. Indeed, from the relations

$$\ln\left(\frac{\Delta}{\Delta_0}\right) = \begin{cases} -\frac{\pi}{4}\bar{\Gamma} & : \bar{\Gamma} \leq 1 \\ -\ln[\bar{\Gamma} + R(\bar{\Gamma})] + \frac{R(\bar{\Gamma})}{2\bar{\Gamma}} - \frac{\bar{\Gamma}}{2}\arctan R(\bar{\Gamma})^{-1} & : \bar{\Gamma} > 1 \end{cases}, \qquad (43)$$

derived by Abrikosov and Gor'kov [$\Delta_0 = \Delta(T = 0, \Gamma_s = 0)$ is the order parameter in the absence of magnetic impurities] one obtains

$$\alpha_\Delta = \alpha_{\Delta_0} \begin{cases} \left(1 - \frac{\pi}{4}\bar{\Gamma}\right)^{-1} & : \bar{\Gamma} \leq 1 \\ \left[1 - \frac{\bar{\Gamma}}{2} \arctan R(\bar{\Gamma})^{-1} - \frac{R(\bar{\Gamma})}{2\bar{\Gamma}}\right]^{-1} & : \bar{\Gamma} > 1 \end{cases} \quad . \quad (44)$$

In the BCS approximation one further has $\alpha_{\Delta_0} = \alpha_0$, where the last quantity was defined before as the IC of T_{c0}, that is, in the absence of magnetic impurities.

Finally, one obtains Eq. (25) with R_0 given by:

$$R_0 = -\frac{\alpha_\Delta}{2\alpha_0} \frac{K_1 + K_2}{\tilde{K}(0,0)} \quad (45)$$

and α_Δ is given by by Eq. (44).

References

[1] H. Fröhlich, *Phys. Rev.* **79**, 845 (1950).

[2] (a) E. Maxwell, *Phys. Rev.* **78**, 477 (1950); C.A. Reynolds, B. Serin, and L.B. Nesbitt, *Phys. Rev.* **84**, 691 (1951); (b) J.L. Olsen, *Cryogenics* **2**, 356 (1963); (c) B.T. Matthias, T.H. Geballe, E. Corenzwit, and G.W. Hull Jr., *Phys. Rev.* **128**, 588 (1962); (d) R.A. Hein, and J.W. Gibson *Phys. Rev.* **131**, 1105 (1963); (e) E. Bucher, J. Müller, J.L. Olsen, and C. Palmy *Phys. Lett.* **15**, 303 (1965)

[3] V.Z. Kresin, and S.A. Wolf, *Phys. Rev. B* **49**, 3652 (1994); and in *Anharmonic Properties of High-T_c Cuprates*, p. 18, D. Mihailovic, G. Ruani, E. Kaldis, K.A. Müller, Eds., World Scientific (1995).

[4] V.Z. Kresin, A. Bill, S.A. Wolf, and Yu.N. Ovchinnikov, *Phys. Rev. B* **56**, 107 (1997); *J. Supercond.* **10**, 267 (1997).

[5] A. Bill, V.Z. Kresin, and S.A. Wolf, *Z. Phys. Chem.* **201**, 271 (1997); *Z. Phys. B*, in press.

[6] A. Bill, V.Z. Kresin, and S.A. Wolf, *preprint*.

[7] V.Z. Kresin, and S.A. Wolf, *Phil. Mag. B* **76**, 241 (1997).

[8] J.P. Franck, *Physica Scripta* **T66**, 220 (1996); J.P. Franck, S. Harker, and J.H. Brewer, *Phys. Rev. Lett.* **71**, 283 (1993); J.P. Franck, and D.D. Lawrie, *Physica C* **235-240**, 1503 (1994); *J. Supercond.* **8**, 591 (1995); *J. Low Temp. Phys.* **105**, 801 (1996).

[9] G.M. Zhao, V. Kirtikar, K.K. Singh, A.P.B. Sinha, and D.E. Morris, *Phys. Rev. B* **54**, 14956 (1996).

[10] G. Gladstone, M.A. Jensen, and J.R. Schrieffer, Superconductivity in transition metals, in: *Superconductivity*, R.D. Parks, ed., Marcel Dekker, New York (1967).

[11] E.A. Lynton. *Superconductivity*, Methuen, London (1969).

[12] R.D. Fowler, J.D.G. Lindsay, R.W. White, H.H. Hill, and B.T. Matthias, *Phys. Rev. Lett.* **19**, 892 (1967).

[13] N. Bogolyubov, N. Tolmachev, and D. Shirkov. *A New Method in the Theory of Superconductivity*, Cons. Bureau, New-York (1959).

[14] I.M. Khalatnikov, and A.A. Abrikosov, *Adv. in Physics* **8**, 45 (1959).

[15] J.C. Swihart, *Phys. Rev.* **116**, 45 (1959); *IBM J. Res. Develop.* **6**, 14 (1962).

[16] P. Morel and P.W. Anderson, *Phys. Rev.* **125**, 1263 (1962).

[17] W.L. McMillan, *Phys. Rev.* **167**, 331 (1968); **174**, 537 (1968).

[18] V.Z. Kresin, *Phys. Lett. A* **122**, 434 (1987).

[19] J.W. Garland, *Phys. Rev. Lett.* **11**, 114 (1963).

[20] R. Meservey, and B.B. Schwartz, in Ref. [10].

[21] E. Schachinger, M.G. Greeson, and J.P. Carbotte, *Phys. Rev. B* **42**, 406 (1990); J.P. Carbotte, and E.J. Nicol *Physica C* **185-189**, 162 (1991).

[22] J. Labbé, and J. Bok, *Europhys. Lett.* **3** 1225 (1987).

[23] A.A. Abrikosov, *Physica C* **233** (1994) 102.

[24] T. Hocquet, J.-P. Jardin, P. Germain, and J. Labbé, *Phys. Rev. B* **52** (1995) 10330.

[25] T. Dahm, D. Manske, D. Fay, and T. Tewordt, *Phys. Rev. B* **54**, 12006 (1996).

[26] V.H. Crespi, and M. L. Cohen, *Phys. Rev. B* **48**, 398 (1993).

[27] A.A. Maradudin, E.W. Montroll, G.H. Weiss, and I.P. Ipatova. *Theory of Lattice Dynamics in the Harmonic Approximation*, Acad. Press, New York (1971).

[28] V.Z. Kresin, H. Morawitz, and S.A. Wolf. *Mechanisms of Conventional and High-T_c Materials*, Oxford Univ. Press, New York (1993).

[29] D. Rainer, and F.J. Culetto, *Phys. Rev. B* **19**, 2540 (1979); F.J. Culetto, and F. Pobell, *Phys. Rev. Lett.* **40**, 1104 (1978).

[30] P. Auban-Senzier, C. Bourbonnais, D. Jerome, C. Lenoir, and P. Batail, *Synthetic Metals* **55-57**, 2542 (1993); J.C.R. Faulhaber, D.Y.K. Ko, and P.R. Briddon, *Synthetic Metals* **60**, 227 (1993).

[31] B. Ashauer, W. Lee, D. Rainer, and J. Rammer, *Physica B* **148**, 243 (1987).

[32] T.W. Barbee III, M.L. Cohen, L.C. Bourne, and A. Zettl, *J Phys. C* **21**, 5977 (1988).

[33] B. Stritzker, and W. Buckel, *Z. Phys.* **257**, 1 (1972); T. Stoskiewicz, *Phys. Status Solidi A* **11**, K123 (1972).

[34] B.N. Ganguly, *Z. Phys.* **265**, 433 (1973); *Z. Phys. B* **22**, 127 (1975); B.M. Klein, E. N. Economou, and D.A. Papaconstantopoulos *Phys. Rev. Lett.* **39**, 574 (1977); D.A. Papaconstantopoulos, B.M. Klein, E. N. Economou, and L.L. Boyer *Phys. Rev.* **17**, 141 (1978); R.J. Miller, and C.B. Satterthwaite *Phys. Rev. Lett.* **34**, 144 (1975); B.M. Klein, and R.E. Cohen *Phys. Rev. B* **45**, 12405 (1992); M. Yussouff, B.K. Rao, and P. Jena *Solid State Comm.* **94**, 549 (1995).

[35] S.L. Drechsler, and N.M. Plakida *Phys. Stat. Sol.* **144**, K113 (1987); T. Galbaatar, S.L. Drechsler, N.M. Plakida, and G.M. Vujiçi'c, *Physica C* **176**, 496 (1991).

[36] K.A. Müller, *Z. Phys. B* **80**, 193 (1990).

[37] M. Cyrot et al., *Phys. Rev. Lett.* **72**, 1388 (1994); D.S. Fisher, A.J. Millis, B. Shraiman, and R.N. Bhatt, *Phys. Rev. Lett.* **61**, 482 (1988).

[38] L. Jansen, *private communication*. We thank Prof. Jansen for drawing our attention to this effect.

[39] T. Nakajima, T. Fukamachi, O. Terasaki, and S. Hosoya, *J. Low Temp. Phys.* **27**, 245 (1977).

[40] N.W. Ashcroft, and M. Cyrot, *Europhys. Lett.* **23**, 605 (1993); Yu.N. Garstein, A.A. Zakhidov, and E.M. Conwell, *Phys. Rev. B* **49**, 13299 (1994); A.P. Ramirez et al., *Phys. Rev. Lett.* **68**, 1058 (1992); T.W. Ebbsen el al., *Nature* **355**, 620 (1992); P. Auban-Senzier et al., *Synthetic Metals* **55-57**, 3027 (1993).

[41] W.A. Little, *Phys. Rev.* **134**, A1416 (1964); V. Ginzburg, *Sov. Phys.-JETP* **20**, 1549 (1965); B. Geilikman, *Sov. Phys.-JETP* **48**, 1194 (1965).

[42] H.-B. Schlütter, and C.-H. Pao, *Phys. Rev. Lett.* **75**, 4504 (1995); *J. Supercond.* **8**, 633 (1995).

[43] V.Z. Kresin, and H. Morawitz, *Solid State Comm.* **74**, 1203 (1990).

[44] V.Z. Kresin, and H. Morawitz, *Phys. Rev. B* **37**, 7854 (1988)

[45] B.T. Geilikman, V.Z. Kresin, and N.F. Masharov, *J. Low Temp. Phys.* **18**, 241 (1975).

[46] S.A. Wolf, and V.Z. Kresin, in Ref. [3]b, p. 232.

[47] S. Banerjee, A.N. Das, and D.K. Ray, *Phys. Lett. A* **214**, 89 (1996); *J. Phys. C* **8**, 11131 (1996); S. Sil, and A.N. Das, *J. Phys. C* **9**, 3889 (1997).

[48] F. Marsiglio, R. Akis, and J.P. Carbotte, *Solid State Comm.* **64**, 905 (1987)

[49] A. Abrikosov, and L. Gork'ov, *Sov. Phys. JETP* **12**, 1243 (1961).

[50] S. Skalski, O. Betbeder-Matibet, and P.R. Weiss, *Phys. Rev.* **136**, 1500 (1963).

[51] J. Carbotte et al. *Phys. Rev. Lett.* **66** (1991) 1789.

[52] S.P. Singh et al., *J. Supercond.* **9**, 269 (1996); K. Hanzawa, *J. Phys. Soc. Japan* **63**, 2494 (1994); S.P. Singh et al., *J. Supercond.* **9**, 269 (1996).

[53] S. Zagoulev et al. *Phys. Rev. B* **52**, 10474 (1995); *Physica C* **259**, 271 (1996).

[54] C. Panagopoulos, J.R. Cooper, N. Athanassopoulou, and J. Chrosch, *Phys. Rev. B* **54** (1996) 12721.

[55] R.B. Schwarz, P.J. Yvon, and D. Coffey, in *Studies of High Temperature Superconductors*, vol. 9, Narlikar, ed., Nova Science Publ., New York (1992); M.K. Crawford, M.N. Kunchur, W.E. Farneth, M.McCarron III, and S.J. Poon, *Phys. Rev. B* **41**, 282 (1990); B. Batlogg et al., *Phys. Rev. Lett.* **59**, 912 (1987);T.A. Faltens et al., *Phys. Rev. Lett.* **59**, 915 (1987).

[56] G. Soerensen, and S. Gygax , *Phys. Rev. B* **51**, 11848 (1995).

[57] D. Zech, K. Conder, H. Keller, E. Kaldis, and K.A. Müller, *Physica B* **219&220**, 136 (1996).

[58] V. Kresin, *Phys. Rev. B* **25**, 157 (1982).

[59] V. Kresin, *Phys. Rev. B* **32**, 145 (1985).

[60] J. Clarke, *Proc. R. Soc. London*, Ser. A **308**, 447 (1969).

[61] L. Salem. *The Molecular Orbital Theory of Conjugated Systems*, Benjamin, New York (1966).

[62] R.P. Sharma, T. Venkatesan, Z.H. Zhang, J.R. Liu, *Phys. Rev. Lett.* **77**, 4624 (1997).

[63] J. Mustre de Leon et al., *Phys. Rev. Lett.* **64**, 2575 (1990); L. Gasparov et al., *J. Supercond.* **8**, 27 (1995); G. Ruani et al., *Solid State Comm.* **96**, 653 (1995); A. Jesowski et al., *Phys. Rev. B* **52**, 7030 (1995).

[64] D. Haskel, E.A. Stern, D.G. Hinks, A.W. Mitchell, and J.D. Jorgensen, *Phys. Rev. B* **56** (1997) 521.

[65] R.P. Sharma, G.C. Xiang, C. Ramesh, R.L. Greene, and T. Venkatesan, *Phys. Rev. B* **54**, 10014 (1996).

[66] G. Jonker, and J. van Santen, *Physica* **16**, 337 (1950).

[67] G.M. Zhao, K. Conder, H. Keller, and K.A. Müller, *Nature* **381**, 676 (1996).

[68] V.Z. Kresin, *this volume*.

[69] T. Schneider, and H. Keller, *Phys. Rev. Lett.* **69**, 3374 (1993); *Int. Journ. Mod. Phys. B* **8**, 487 (1993).

[70] T.F. O'Malley, *Phys. Rev.* **162** (1967) 98; *Adv. Atomic Molec. Phys.* **7**, 223 (1971).

[71] J.P. Franck, J. Jung, M.A-K. Mohamed, S. Gygax, and G.I. Sproule, *Phys. Rev. B* **44**, 5318 (1991); in *High-T_c Superconductivity, Physical Properties, Microscopic Theory and Mechanisms*, J. Ashkenazi et al. eds. (Plenum Press, New-York, 1991), p. 411.

[72] H.J. Bornemann and D.E. Morris, *Phys. Rev. B* **44**, 5322 (1991).

[73] G.-M. Zhao and D.E. Morris, *Phys. Rev. B* **51**, 16487 (1995); G.-M. Zhao, K.K. Singh, A.P.B. Sinha, and D.E. Morris, *Phys. Rev. B* **52**, 6840 (1995); G.-M. Zhao, M.B. Hunt, H. Keller, and K.A. Müller, *Nature* **385**, 236 (1997).

[74] M.B. Maple, C.C. Almasan, C.L. Seaman, S.H. Han, *J. Supercond.* **7**, 97 (1994).

HYPERFINE INTERACTIONS IN METALS

Walter D. Knight

Department of Physics
University of California
Berkeley, CA 94720, USA

INTRODUCTION

We deal with the magnetic interactions between conduction electrons and nuclei of the metal atoms. The electrons are polarized by an external magnetic field, and expose the nuclear magnetic moments to a hyperfine field which in the simplest case involves s-electrons. This is usually called the "contact" interaction, which is applied to the nucleus while they are both in the same place, in contrast to "action at a distance" as between two separate ions.

The electron polarization depends on the temperature-independent paramagnetism, which Pauli (Pauli, 1926) discovered in the first application of the recently developed Fermi-Dirac statistics to a physical problem. Not long afterward, Fermi and Segrè worked out the analysis of the contact interaction (Fermi and Segrè, 1933) one summer, when Segrè became disgruntled after failing to win some prize (Segrè, 1993).

In the following we will present briefly a discussion of hyperfine interactions and nuclear magnetic resonance (NMR) in metals, as well as indirect exchange interactions between nuclear moments at a distance, as mediated by second-order electron-nuclear couplings.

NMR AND THE STRUCTURE OF METALS

The metal NMR shift refers to the relative shift K in NMR frequency for atoms of a metal compared with the same atoms in a non-metallic environment. The observed shift reflects the local magnetic field at the metal atom nucleus caused by the magnetization of the conduction electrons. For example, the average local field in sodium metal augments the applied resonance field by approximately one part per thousand. In non-metallic sodium chloride the local field is small in comparison.

Conduction electrons in a metal are delocalized, i.e., no electron is uniquely associated with any particular atom more than another, but all exist in eigenstates within an energy band, whose highest occupied state lies at the Fermi energy E_F. At lower energies the electron wave functions tend to be mainly s-states, while at higher energies there are increasing admixtures of higher ($p, d, ...$) states. For s-states the probability density P_F at the nucleus ($R=0$) is high, and since each electron carries a spin and magnetic moment μ_B, the contact interaction momentarily produces significant local magnetic fields at the nuclei, as the electron explores the lattice. The hyperfine field in atomic sodium is approximately 39 T. This is an atomic property, independent of external fields. In zero applied field, the average contact hyperfine field in a metal is zero, because the nuclei are exposed to equal numbers of spin-up and spin-down electrons. In an externally applied magnetic field the conduction electron system acquires a magnetization which is proportional to the magnetic field and susceptibility. The appropriate susceptibility, as implied earlier, is the Pauli temperature-independent susceptibility, χ_p. The fractional change in local field or observed NMR frequency is (Townes et al., 1950)

$$\frac{\Delta v}{v} = \frac{\Delta B}{B} = \frac{8\pi}{3}\chi_p V_0 P_F, \qquad (1)$$

where χ_p is the Pauli susceptibility per unit volume, V_0 is the volume per atom in the metal, P_F is the average value of the s-electron probability at the nucleus, B is the applied magnetic field, and v is the observed resonance frequency. The average probability P_F is taken for electrons at the Fermi energy E_F. The local field of p-electrons is usually 10% or less of the contact field, and produces small anisotropic shifts in non-cubic metals.

To summarize some typical parameters: the hyperfine field in a free sodium atom is 39 T; the corresponding contact hyperfine field in metallic sodium is smaller, approximately 30 T, mainly because of p admixtures in the wave functions at the Fermi surface; the small Pauli susceptibility reflects the small number of unpaired spins; the average local field is 1.1×10^{-3} T for an applied field of 1 T. Both Eq. (1) and experiment give closely the same result.

Values of K have been measured for most of the metallic elements, and recorded results range from K=0.026% to 2.7% for Li and Hg, respectively (see, e.g., Fig. 1). The measurement of K is a sensitive means for probing, noninvasively, electronic and geometric structures of solid and liquid metallic systems. For noncubic crystals K depends on crystal orientation with respect to the field.

ORBITAL AND CORE-POLARIZATION SHIFTS

In addition to the contact interaction associated with the Pauli paramagnetism χ_p, the NMR shifts relating to the orbital (χ_v) and core-polarization (χ_d) magnetizations are important, particularly for the heavy transition metals. The Van Vleck temperature-independent orbital paramagnetism (Kubo, 1956) is given by

$$\chi_v \sim 2\mu_B^2 \frac{N}{E} \qquad (2)$$

where N is the number of d electrons per unit volume, and E is the average separation between mixed (filled and empty) states in the same energy band.

The total magnetic interaction can be written simply as the sum of three terms

$$\chi = \chi_p + \chi_v + \chi_d. \qquad (3)$$

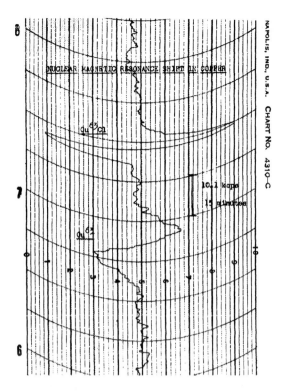

Figure 1. The NMR shift is +23 kHz at 10 Mhz resonance frequency, in a mixed powder sample of Cu and CuCl. The experiment was performed at Brookhaven National Laboratory, 1949.

The last term in Eq. (3) represents the temperature-dependent d-states' contribution to the electron-nucleus interaction. It tends to be relatively large, paramagnetic, and temperature-dependent for transition metals with narrow d bands and low Fermi temperatures (Cohen, 1959). It represents the polarization of the core s electrons coupling with the d electrons. This interaction greatly enhances, and in some cases reverses, the contact hyperfine field.

For example (Clogston *et al.*, 1964), the predominant interaction in platinum metal produces a large negative NMR shift, K_{Pt}=-3.5%. It is also believed (Clogston *et al.*, 1962) that the major contribution to the NMR shift in vanadium results from the Van Vleck temperature-independent term, and accounts for the fact that the NMR shift for vanadium does not change at the superconducting transition temperature.

A recent Green's function analysis (Ebert *et al.*, 1986) points out that Eq. (3) implies independence among the several terms, while in fact the terms are not strictly independent, and it is in principle necessary to include exchange and correlation effects in the calculations. It is also clear from their results and comparison of several calculations with Eq. (3) that the latter provides useful but imprecise estimates for magnetic susceptibilities and shifts. Nevertheless, the theoretical results (Ebert *et al.*, 1986) support the interpretations for the observed shifts in vanadium and platinum (Clogston *et al.*, 1964).

Further analysis of NMR shifts in intermetallic compounds (Clogston and Jaccarino, 1961) illuminates the interplay of the terms in Eq. (3).

INDIRECT EXCHANGE INTERACTIONS BETWEEN NUCLEAR MOMENTS

Nuclear magnetic resonance techniques were developed rapidly in the years following the successful observation of NMR by Felix Bloch and his collaborators at Stanford University and by Edward Purcell and his collaborators at Harvard University (Bloch et al., 1946; Purcell et al., 1946). The first publications of both groups appeared in the same volume of the Physical Review. There followed a period of development by these and other groups during which the basic techniques were explored along with the investigation of the significance of the experiments. Then (Townes et al., 1950) reports appeared which indicated that the methods of NMR were directly applicable to the study of the properties of materials in the solid and liquid states. Shortly thereafter, experiments revealed evidence for indirect coupling between nuclear magnetic moments in molecules (Ramsey and Purcell 1952; Hahn and Maxwell, 1952) and in metals (Bloembergen and Rowland, 1955). These are second-order effects describing the isotropic hyperfine coupling between two nuclear spins ($I_1 \cdot I_2$). In a simplified scenario for metals, nuclear spin #1 interacts with a conduction electron spin, which subsequently interacts with a distant nuclear spin #2, whose hyperfine field then correlates with that of #1. The effect was first noticed in the NMR linewidth of medium and heavy metals Ag (Jeffries, 1952) and Tl (Bloembergen and Rowland, 1955). The results were puzzling and occupied much time among theoreticians. I remember watching Kittel and Teller somewhere beyond left field during a softball game at a physics picnic. They threw up their hands and shook their heads, with no immediate results. Finally, papers (Ruderman and Kittel, 1954) and (Yosida, 1957) produced memorable solutions.

NMR AND SUPERCONDUCTIVITY

With the publication of the BCS theory of superconductivity, (Bardeen et al., 1957), the well-know behavior of type I and type II superconductors regarding critical temperature T_c, Meissner effect of flux exclusion, energy gap, excited states, and singlet s pairing were understood on a microscopic scale. A good summary is given by McLaughlin (1976).

In particular, the NMR shift in metals was observed to approach zero at T=0 for the simple metal Al (Hammond and Kelly, 1967; Fine et al., 1969). For heavier, more complex metals such as Hg (Reif, 1956) and Sn (Androes and Knight, 1961) the shift was reduced considerably, but remained finite at T=0, as a result of reduction of the mean free path for electrons in small particles. The result is a mixing of up and down spin states because of spin-orbit interactions (Yosida, 1958). A similar effect was observed in small particles of normal metal Cu (Yee and Knight, 1975).

We thought we understood superconductors, until high temperature superconductivity (HTS) was discovered (Bednorz. and Müller, 1986). An example of the much-studied cuprates is $YBa_2Cu_3O_7$, with a critical temperature of 93 K, a complex anisotropic structure, and convenient nuclear moments for NMR and NQR (nuclear quadruple resonance) studies.

NMR studies, and NQR studies as well, provide information concerning the local environments of the atomic nuclei in these important HTS materials. Extensive studies (Barrett and Slichter, 1966; Slichter et al., 1996) present a picture of the current state of affairs.

The cuprates have parallel planes which contain copper and oxygen ions. Removal of all of the oxygen from the O(1) sites of $YBa_2Cu_3O_7$ mentioned above results in $YBa_2Cu_3O_6$, which is an antiferromagnetic insulator with a Néel temperature of 400 K. The copper ions Cu^{2+}, with a hole in the $3d$ state and a spin of 1/2, plus strong intra-plane coupling, give the appearance of the Heisenberg two-dimensional antiferromagnet. It is postulated that the development of the holes destroys the long range order and permits development of the superconductivity. Other related systems are believed to have similar antiferromagnetic couplings. Most of the data derived from these experiments involved NMR and NQR measurements of resonance line shifts and spin relaxation times. Mention is also made of transferred hyperfine couplings among nearby nuclei.

In conclusion, it appears that the systems under study are too complicated to permit unique solutions to be developed yet. Some of the evidence favors a picture of microscopic d wave pairing. Anisotropic s wave pairing also appears to be a possibility.

ACKNOWLEDGMENTS

I would like to thank Kyler Kuehn for assistance with manuscript preparation. This work was supported by the Berkeley Faculty Committee on Research.

REFERENCES

Androes, G., and Knight, W., 1961, Phys. Rev. **121**:779
Bednorz, J., and Müller, K., 1986, Z. Phys. B **64**:89
Bardeen, J., Cooper, L., and Schrieffer, J., 1957, Phys. Rev. **108**:1175
Barrett, S., and Slichter, C.P., *Encyclopedia of Nuclear Magnetic Resonance*, Eds. D.M. Grant and R.K.Harris, Wiley, London 1996, pp. 2379-2389
Bloembergen, N. and Rowland, T., 1955, Phys. Rev. **97**:1679
Carter *et al.*, *Metallic Shifts in N.M.R.*, Part I, Pergamon Press, Oxford 1977
Clogston, A. and Jaccarino, V., 1961, Phys. Rev. **121**:1357
Clogston, A., *et al.*, 1962, Phys. Rev. Lett. **9**:262
Clogston, A., et al., 1964, Phys. Rev. **134**:A650
Cohen, M.H., *et al.*, 1959, Proc. Phys. Soc. (London) **A73**:811
Ebert, H. *et al.*, 1986, J. Phys. F **16**:1133
Fermi, E. and Segrè, E., 1933, Zeit. Phys. **82**:729
Fine, H.L. *et al.* 1969, Phys. Lett. **29A**:366
Fröhlich, H. 1937, Physica **6**:406
Hahn, E.L. and Maxwell, D.E., 1952, Phys. Rev. **88**:1070
Hammond, R. and Kelly, G., 1967, Phys. Rev. Lett. **18**:156
Jeffries, C.D., 1952, unpublished
Kubo, R. and Obata, O., 1956, J. Phys. Soc. Japan **11**:547
Kubo, R., 1962, J. Phys. Soc. Japan **17**:1975
MacLaughlin, D., 1976, Solid State Physics **31**:1-69
Pauli, W., 1926, Zeit. Phys. **41**:81
Purcell, E., Torrey, H., and Pound, R., 1946, Phys. Rev. **69**:37
Ramsey, N. and Purcell, E., 1952, Phys. Rev. **85**:143
Reif, F., 1956, Phys. Rev. **102**:1417
Rowland, T.J., *Nuclear Magnetic Resonance in Metals*, Progress in Materials Sci., Vol. 9, Ed. B. Chalmers, Pergamon, Oxford, 1961, pp. 1-91
Ruderman, M. and Kittel, C., 1954, Phys. Rev. **96**:99
Segrè, E., *A Mind Always in Motion, the Autobiography of Emilio Segrè*, University of California Press, Berkeley, 1993, p. 81
Slichter, C.P. *et al.* 1996, Phil. Mag. **B74**:545

Townes, C., Herring, C., and Knight, W., 1950, Phys. Rev. 77:851
Yee, P. and Knight, W.D., 1975, Phys. Rev. B **11**:3261
Yosida, K., 1957, Phys. Rev. **106**:893
Yosida, K., 1958, Phys. Rev. **110**:769

PEROVSKITE OXIDES:
A RICH AND FASCINATING CRYSTAL CLASS FAMILY

A. Bussmann-Holder

Max-Planck-Institut für Festkörperforschung
Heisenbergstr. 1
70569 Stuttgart, Germany

ABSTRACT

Perovskite oxides ABO_3 exhibit an enormously rich phase diagram which covers properties like charge density wave instability, ferro- and antiferroelectricity, ferro- and antiferromagnetism, superconductivity, structural instability, and metal insulator transitions. The different properties of ABO_3-systems have been modeled by different approaches, which were either based on purely ionic models, or purely electronic models. In this contriubtion it is aimed to show that the strong electron-lattice interaction, present in all these compounds, may be the driving force in producing the above outlined richness of observed phases.

INTRODUCTION

Perovskite oxides have the general formula ABO_3 with A being alkali, earth alkali or rare earth metal, and B a transition metal. At high temperatures the structure is cubic with the A ions occupying the cube corners, the oxygen ions face-centered and the transition metal is in the center of the cube surrounded by the oxygen octahedra. Structural distortions are always observed in ABO_3 systems and are related to the canting rotation of the octahedra, the off-center displacement of the transition metal, the elongation of the octahedra, the displacement of the A-lattice or combined effects. Depending on the transition metal properties ferro- and antiferromagnetic ground states result, ferro- and antiferroelectric polar phases appear, charge density wave instabilities and charge ordered states are observed, and superconductivity is found. A summary of part of these properties is listed in Table 1, where ground state properties of the corresponding material are listed.

The richness of phases observed in ABO_3 systems clearly requires that any model for perovskites has to include strong lattice anharmonicity and, in addition, a crucial interplay between the lattice and the electrons. Thus, purely ionic concepts, as have been considered for ferro- and antiferroelectric systems,[1] lose their meaning and p-d hybridization and covalency have to be taken into account.[2]

In the following, introductory remarks will be given with respect to the special role of the oxygen ion played in these systems, followed by a summary of the present understanding of ferro- and antiferroelectrics. CMR materials are then shown to be understood on the basis of strong electron-lattice coupling, and finally, a brief account of superconductivity in perovskites is given together with conclusions.

Table 1. Summary of oxide perovskites with various groundstates.

$NaNbO_3$	
$PbTiO_3$	
$KNbO_3$	ferroelectric (FE)
$BaTiO_3$	
$PbZrO_3$	antiferroelectric (AFE)
$KTaO_3$	incipient FE
$SrTiO_3$	structural phase transition (SPT), incipient FE, superconducting (SC)
$BaBiO_3$	charge density wave instability (CDW)
$BaPb_{1-x}Bi_xO_3$	SPT, SC, metal-insulator-transition (M-I-T)
$LaMnO_3$	antiferromagnetic (AFM)
$CaMnO_3$	ferromagnetic (FM)
$La_{1-x}Ca_xMnO_3$	AFM, FM, M-I-T, SPT, collossal magnetoresistant (CMR)

POLARIZABILITY AND STRUCTURAL INSTABILITY

The rich phenomenology of oxide perovskites is rather unique and substantially differs from halide perovskites which have the same cubic structure, but show neither ferro-, antiferromagentism, ferro-, antiferroelectricity, superconductivity, nor any other of the above mentioned instabilities. This fact is rather surprising as F^- has the same closed shell configuration as O^{2-}, i.e., $1s^2 2s^2 2p^6$. Adopting a naive point of view, one would expect that both types of perovskites show similar structural and electronic instabilities. But opposite to F^-, the doubly negative charged oxygen ion is unstable as a free ion.[3] In a crystal a certain stabilization is achieved through the Coulomb interaction with the neighboring ions, but this stabilization is incomplete, i.e., the O^{2-} $2p^6$ state is virtual only.[4] The consequences of this peculiar property are that O^{2-} prefers to localize holes in order to minimize its energy. While in ferroelectrics this tendency remains dynamical only, in CMR and superconducting systems the hole localizes at the oxygen ion lattice site and strong p-d hybridization arises. Clearly this phenomenology requires to treat oxide perovskites not as a purely ionic compound, but electron-lattice effects are dominating their dynamic and static properties even if the considered systems are nominally ionic insulators, e.g., ferro- and antiferroelectrics.

FERROELECTRICITY/ANTIFERROELECTRICITY

Ferro- and antiferroelectric systems are characterized, in the displacive limit, by the softening of a transverse optic $q = 0 (q = \frac{2\pi}{2})$ lattice mode which freezes out at the phase transition temperature $T_F(T_{AF})$ and determines the low temperature structure.[1] Together with the softening the static dielectric constant diverges at $T_F(T_{AF})$. The low temperature phase is polar where the permanent polarization can be aligned and reversed by the application of an external electric field. The Landau-Ginzburg-order parameter is given by the spontaneous polarization, which is zero above T_F and follows a $(T_F - T)^{1/2}$-dependence below T_F. As ferro- antiferroelectricity are in general being observed in nominally ionic systems, the general understanding of this phenomenon was based on lattice anharmonicity and double-well potentials.[5,6] Even though these approaches yielded a qualitative agreement with experimental data, they always failed quantitatively. A quantitatively correct approach to Raman spectra, soft modes, inelastic neutron scattering was first given by Migoni[2], who modeled the lattice dynamics phenomenologically on the basis of a nonlinear model (NSM) with onsite anharmonicity at the oxygen ion lattice site, in order to account for the peculiar properties of O^{2-}. The NSM has since then been studied in detail by either using a self-consistent phonon approximation or by studying exact solutions of the nonlinear model.[7–10] Especially the exact solutions turned out to be highly interesting as new types of solutions, as compared to ϕ_4 or sine-Gordon models, emerged with exciton, kink, breather, soliton and pulse character.[10,11] From these solutions, statics and dynamics of domain walls, their formation and stability could be obtained. Also it turned out that the local double-well could, globally, change its character to a ϕ_6 potential if the time scale changes.[12] This is, of course, of importance to new experimental techniques like EXAFS, PDF, NMR, which work on a different time and length scale then conventional scattering experiments like e.g., Raman scattering, inelastic neutron scattering, and X-ray. It became especially clear that in the vicinity of T_F the particle dynamics are governed by two time scales and order-disorder behavior may coexist with displacive features.[8] The above phenomenology has been confirmed by first-principle methods,[13] where it was found that the p-d hybridization of oxygen p and transition metal d-electrons trigger the lattice instability. Even in systems like $NaNbO_3$, which show strong order-disorder behavior, this hybridization remains crucial in understanding the dynamics.[14] In order to put the phenomenological model on microscopic grounds the NSM can be mapped by a Tomonoga transformation onto an electron-phonon interaction Hamiltonian which contains, besides onsite and intersite electron-phonon couplings, also higher-order density-density multiphonon interactions:[15]

$$H = \sum_{\substack{i=1,2 \\ q,q',q'',q'''}} \left\{ \frac{P_{q,i}^2}{2} + \frac{\omega_{q,i}^2}{2} Q_{q,i}^2 + \frac{g_4}{4} Q_{q,1} Q_{q',1} Q_{q'',1} Q_{q''',1} - \frac{1}{2} V_{qq'} Q_{q,1} Q_{q',1} \right\}$$

$$+ \frac{1}{\sqrt{N}} \sum_{\substack{i+1,2 \\ q}} g_{2,i}(q) n_{q,i} \sqrt{\frac{2M_i \omega_{q,i}}{\hbar}} Q_{-q,i} + \frac{1}{\sqrt{N}} \sum g_4(q) n_{q,1} \sqrt{\frac{2M_i \omega_{q,1}}{\hbar}} Q_{-q,1}$$

$$\times \left\{ n_{q',1} Q_{-q'} \sqrt{\frac{2M_i \omega_{q',1}}{\hbar}} + Q_{-q'',1} Q_{-q''',1} \frac{2M_i \omega_{q'',1} \omega_{q''',1}}{\hbar} \right\} \quad (1)$$

P_i, Q_i are q-dependent momentum and conjugate displacement coordinates of ion i with mass M_i, $V_{qq'}$ represents the nearest neighbor interaction between oxygen ions ($i = 1$) and the term proporational to g_4 is the onsite lattice anharmonicity. The electron-phonon interaction contains the linear onsite ($g_{2,1}$) and intersite ($g_{2,2}$) couplings and has, in addition, higher-order density-density multiphonon contributions proportional to g_4. With the definitions:

$$\lambda_i = \frac{2g_i^2}{N\hbar\pi\epsilon_F\omega_i}$$

$$I = \int_0^1 \frac{1}{E_k} \tanh \frac{E_k}{2kT} dk$$

$$E_k^2 = \pi\epsilon_F \left\{ \lambda_{2,2}^{1/2} \omega_{q,2} Q_{q,2} + \lambda_{2,1}^{1/2} \omega_{q,1} Q_{q,1} \left\{ 1+ \sqrt{\frac{\lambda_4}{\lambda_{2,1}}} (\omega_{q',1} Q_{q',1} + \omega_{q'',1} Q_{q'',1} Q_{q''',1}) \right\} \right\} \quad (2)$$

The equations of motion are obtained as:

$$M_1 \ddot{Q}_{q,1} = \omega_{q,1}^2 Q_{q,1} + g_4 Q_{q',1} Q_{q'',1} Q_{q''',1} - V_{q,q'} Q_{q',1}$$
$$- \lambda_{2,1} \epsilon_F \omega_{q,1}^2 I \left(1 + \frac{\lambda_4}{\lambda_{2,1}} (\epsilon_F \omega_{q,1}^2 I + \omega_{q'',1} Q_{q'',1} \omega_{q''',1} Q_{q''',1}) \right)$$
$$M_2 \ddot{Q}_{q,2} = -\lambda_{2,2} \epsilon_F \omega_{q,2}^2 I Q_{q,2}$$

$$Q_{q,1} \left(\omega_{q,1}^2 + g_4 Q_{q',1} Q_{q'',1} Q_{q''',1} \right) = -\lambda_{2,2} \epsilon_F \omega_{q,2}^2 I Q_{q,2}$$
$$- \lambda_{2,1} \epsilon_F \omega_{q,1}^2 I Q_{q,1} \left\{ 1 + \frac{\lambda_4}{\lambda_{2,1}} (\epsilon_F \omega_{q',1} I + \omega_{q'',1} Q_{q'',1} \omega_{q''',1} Q_{q''',1}) \right\} \quad (3)$$

with ϵ_F being the Fermi energy. At T = 0 and with $q = q'+q'' = q'''$, $\hbar(\omega_{q'}+\omega_{q''}+\omega_{q'''}) = \hbar\omega_q$ the coupled equations of motion can be solved self-consistently for arbitrary q, and the corresponding displacement coordinates Q_1, Q_2 together with the frequencies are obtained. The dispersion of the soft optic mode and the coupled acoustic mode resemble closely the previous results from self-consistent phonon theory,[16] except that strong anomalies in the acoustic mode at small q-vector appear. The effect of temperature is simulated by varying the local double-well potential depth, and it is found that with decreasing temperature the anomalies become much more pronounced (Fig. 1). Experimentally, similar anomalies have been observed in e.g. $SrTiO_3$,[17] and have been interpreted as the onset of a new quantum coherent state. The present results shed doubt on this interpretation and an alternative suggestion would be to consider anharmonic mode-mode coupling to be the origin of this finding. Interestingly, the solutions of the equation of motion admit to investigate the q-dependence of an effective electron-phonon coupling. It is observed that the effective coupling strongly varies with q, thus leading to nanoscale structures in real space which induce dynamical stripe formation and domain patterns.[15] Even for a system like $SrTiO_3$, which is an incipient ferroelectric only, fluctuating polarized regions exist at all temperatures which have been investigated, indicating strong precursor effects and order-disorder-type dynamics.

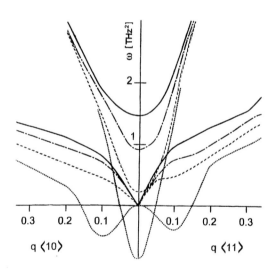

Fig. 1. Summary of oxide perovskites with various groundstates.

CMR Materials

The renewed interest in the perovskite manganites[18] originates in the finding of the colossal magnetoresistance,[19] which has a high potential of applicability. The phase diagram of e.g. $La_{1-x}Ca_xMnO_3$ ranges for small x from ferromagnetic insulator to ferromagnetic metal. At $x \sim 0.5$ the system becomes charge ordered and undergoes a transition to an antiferromagnetic insulator.[20] The conventional understanding is based on the double-exchange model (DEM) where mobile d-electrons hop in the ferromagnetic background, thus avoiding the strong Hund's rule exchange energy.[21] Recent new experiments, however, revealed that the DEM is not sufficient in understanding the properties of CMR materials and electron-lattice interaction effects have to be included.[22,23] Specifically, the finding of a large isotope effect on the ferromagnetic transition temperature T_c pointed to the importance of Jahn-Teller coupling in manganites.[24] Most interestingly the authors found that the observed isotope effect correlates with the average ionic radius $\langle r_A \rangle$ of the cationic sublattice. Similar effects of the cationic radius have been reported in Ref. 25 where the transition temperature T_c and the resistance have been observed to scale with $\langle r_A \rangle$. In contrast to both of these experiments a systematic investigation of the isotope effect on T_c revealed that $\langle r_A \rangle$ is not the relevant scaling parameter but the distortion of the Mn-O-Mn bond angle.[26]

In the following it will be shown that the isotope effect on T_c results from two competing interactions which are related to the local Jahn-Teller distortion and phonon mediated intersite-interband interactions.[27] While the first effect clearly forwards large isotope coefficients, the latter reduces it and may even reverse its sign and induce a metal-insulator transition. As the second interaction acts intersite and interband, Mn-O-Mn bond angle distortions are driven by it, which are dynamic for small enough interactions, but become static if its strength is large.

The model Hamiltonian combines the DEM with the Jahn-Teller coupling[28] and includes additional intersite electron-phonon interactions. It is also closely related to Eq. 1 except that in the CMR systems the band energies and the Coulomb interaction have to be included explicitly:

$$H = \sum_{k,i} \epsilon_{ki} c^+_{ki} c_{ki} + \sum_{k,k'} t_{pd} \left(c^+_{kp} c_{k'd} + c^+_{kd} c_{k'p} \right) +$$

$$\sum_{k,k'} U_d n_{kd\uparrow} n_{k'd\downarrow} + \sum_{kk'} V_{pd}(n_{kp} n_{k'd} + n_{kd} n_{k'p}) +$$

$$\sum_{q,i} \hbar\omega_{q,i} (b^+_{qi} b_{qi} + 1/2) - \sum_k c^+_{ki} c_{ki} \sum_q \hbar\omega_{qi} \left\{ \lambda^{i*}_{JT} b_{qi} + \lambda^i_{JT} b^+_{qi} \right\}$$

$$\sum_{k,k'} \left\{ c^+_{kp} c_{k'd} + c^+_{kd} c_{k'p} \right\} \sum_q \hbar\omega_{q,pd} \left\{ F^*_q \gamma^*_{q,pd} b_q + F_q \gamma_{q,pd} b^+_q \right\} \quad (4)$$

where ϵ_{ki} (i = p,d) is the k-dependent band energy with density $n_k = c^+_k c_k$. t_{pd} is the p-d hopping integral, U the onsite Coulomb repulsion, and V_{pd} the degree of p-d hybridization. The lattice energy $\hbar\omega_q$ is assumed to be pseudoharmonic with phonon creation and annihilation operators b^+_q, b_q, respectively. Anharmonicity is certainly important in the manganites, but for the evaluation of T_c and the corresponding isotope effect, this can be treated on a meanfield level leading to renormalized temperature dependent phonon energies. The electron-phonon interaction consists of the onsite Jahn-Teller coupling λ_{JT} and the intersite interaction with coupling degree F_q, which is usually either not considered nor neglected.[29,30] In the limit of $U \gg t_{pd}$, Eq. 1 reduces to the Anderson model Hamiltonian[21] and an antiferromagnetic insulating ground state results. In this regime the phonon energies and electron-phonon couplings are negligible and the Hamiltonian can be mapped onto the t-J model. With increasing doping, U is screened, t_{pd} becomes active and an insulator metal transition takes place. In this regime the electron-phonon interactions gain importance and may act substantially on electron- and spin degres of freedom, thus leading to static and dynamic polaron formation. A decoupling of lattice and electron spin coordinates can be carried through by a homogeneous Long-Firsov[30] transformation: $\sim H = e^{-S}He^{S}$ with $S = \sum c_k^+ c_k (b_q^+ \gamma_q - b_q \gamma_q^*)$ which correctly describes the squeezing of band energies and the dressing of phonons. The band energies are renormalized like $\sim \epsilon_{ki} = \epsilon_{ki} - \frac{1}{2N}\sum_q \hbar\omega_{qi}|\lambda_{JT}|^2$ and the hopping integral becomes $\sim t_{pd} = \{t_{pd} - \frac{1}{2N}\sum \hbar\omega_q |\lambda_{IB}|^2\} e^{-\phi_T}$ with $\phi_T = \frac{1}{2N}\sum |\lambda_{JT}^2 f(q)|$, $f(q)$ being a function of scattering angle, and $\lambda_{IB} = F_q \gamma_q$. While the Jahn-Teller coupling exponentially reduces the hopping integral and is important also for large t_{pd}, the interband interaction gains importance only in the regime where t_{pd} is small, i.e. at the borderline to the antiferromagnetic insulating state.

In order to calculate the ferromagnetic transition temperature T_c, the hole is assumed to propagate in a mean band, and using the virtual crystal approximation, T_c is given by:

$$kT_c = \frac{1}{15}\frac{(2S-1)(4S+1)}{S(2S+1)}(-\tilde{\epsilon}) \tag{5}$$

where $\tilde{\epsilon}$ is the Fourier transform of $\tilde{t}_{pd}$ where $\tilde{t}_{pd} = \tilde{t}_{pd}(S_o + 1/2)/2S + 1$, and S_o being the total spin/cell. The calculated T_c as a function of λ_{JT} for various λ_{IB} is shown in Fig. 2. Besides the effect of λ_{JT} to decrease T_c rapidly, increasing λ_{IB} substantially favors this effect. With the definition $\alpha = -\frac{d\ln T_c}{d\ln m}$ the isotope effect has been calculated

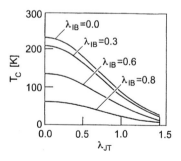

Fig. 2. Ferromagnetic transition temperature T_c as formation of Jahn-Teller coupling λ_{JT} for various λ_{IB}.

again as a function of λ_{JT} for various λ_{IB} (Fig. 3). From Fig. 3 the data of Ref. 25 can be understood as long as λ_{IB} is zero. In accordance with experiment α increases with decreasing T_c and follows the approximate dependence $\alpha \approx \lambda_{JT}^3$. With finite λ_{IB} this dependence is shifted down and α reverses sign for small λ_{JT} and small λ_{IB}, which gets more pronounced when λ_{IB} increases. As λ_{IB} is active only close to the metal-insulator transition at the antiferromagnetic boundary, increasing T_c's with increasing isotope mass are expected and eventually it could be possible to induce antiferromagnetism through isotopic substitutions.

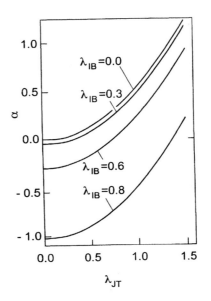

Fig. 3. Isotope coefficient α as a function of Jahn-Teller coupling λ_{JT} for various λ_{IB}.

SUPERCONDUCTING SYSTEMS

Finally, superconductivity in perovskite-type systems is addressed shortly, only. As the high-temperature superconductors (HTSC)[31] are structurally related to perovskite oxides, it would be suggestive to assume that lattice effects[32] are important in the pairing mechanism. Even though there exist considerably vast experimental data which underline the importance of electron-lattice effects, the pairing mechanism is still under discussion and there seems to be also strong experimental support that antiferromagnetic spin fluctuations drive the electron/hole pairing. Neither experimentally nor theoretically clear-cut conclusions are available at present, which could evidence the symmetry of the superconducting order parameter. A possible understanding of the microscopic pairing mechanism could eventually arise if the complete phase diagram of the copper oxides is investigated in more detail. It is well known, e.g., that the underdoped parent compounds of HTSC are antiferromagnetic insulators. But the phase boundary is rather poorly characterized and also the antiferromagnetic order is not well defined. Similarly, the finding of pseudo-gaps and the insulator/metal regimes are not investigated in deep detail. In order to understand the superconducting properties of the cuprates, certainly also the other ground states should be understood.

Based on theoretical considerations and the knowledge of oxide perovskites, a specific experiment is suggested which could eventually contribute substantially to the understanding of the microscopic origin of antiferromagnetism and superconductivity in cuprates. If the hole pairing in cuprates results from electron-phonon interaction, then these interactions are also present in the insulating antiferromagnetic state. From CMR materials it is known that a substantial Jahn-Teller coupling together with phonon-mediated intersite interband interactions induces a large isotope effect on T_c and T_N. For HTSC it is concluded that in the underdoped antiferromagnetic regime the existence of an isotope effect on T_N would lead to a brakdown of theoretical models which are purely electronic like, e.g. the t-J-model. If an isotope effect on T_N is observed, then it will be difficult to argue that antiferromagnetic fluctuations alone are driving the pairing, and certainly electron-phonon interactions have to be reconsidered.[33] The suggested experiment is under preparation and dependent upon the results, the question on the pairing mechanism in HTSC might find an answer.

CONCLUSIONS

As has been shown in the foregoing paragraphs, perovskite oxides are an extremely rich and fascinating crystal class family. The various ground states which are observed in ABO_3 materials, can all be related to a crucial interplay of lattice and electrons. Many properties can be understood if the special role played by the oxygen ion O^{2-} is taken into account in modelling the dynamical and electronic properties. Even though the concept of polarizability is related to phenomenology, it clearly shows in the case of O^{2-} that delocalization of the outer p-electrons is important which induces dynamical and static p-d hybridization and strong electron-phonon coupling. In the case of ferro- and antiferoelectrics and CMR materials, this effect is well established. In superconducting systems and especially in HTSC, the mechanism is still unclear, but eventually the aforementioned experiment could play a key role in clarifying the importance of hybridization effects and electron-phonon interactions here, too.

Acknowledgments

It is a pleasure to acknowledge many stimulating discussions with K.A. Müller, A.R. Bishop, L. Genzel and A. Simon. I am very much indebted to J.P. Franck for providing his experimental results prior to publication.

References

1. M.E. Lines, A.M. Glass, *Principles and Applications of Ferroelectrics and Related Materials*, Clarendon Press (1977).
2. R. Migoni, H. Bilz, D. Bäuerle, Origin of Raman scattering and ferroelectricity in oxidic perovskites, *Phys. Rev. Lett.* 37:1155 (1976).
3. J.R. Tessmann, A.H. Kahn, W. Shockley, Electronic polarizabilities of ions in crystals, *Phys. Rev.* 92:890 (1953).
4. A. Bussmann-Holder, H. Bilz, R. Roenspiess, K. Schwarz, Oxygen polarizability in ferroelectric phase transitions, *Ferroelectrics* 25:343 (1980).
5. E. Pytte, J. Feder, Theory of a structural phase transition in perovskite-type crystals, *Phys. Rev.* 187:1077 (1969).
6. H. Thomas, in: *Structural Phase Transitions and Soft Modes*, ed. E.J. Samuelson, Universitetsvorleiget, Oslo (1971).
7. H. Bilz, G. Benedek, A. Bussmann-Holder, Theory of Ferroelectricity: The polarizability model, *Phys. Rev. B* 35:4880 (1987).
8. M. Stachiotti, R. Migoni, Lattice polarization around off-centre Li in $Li_xK_{1-x}TaO_4$, *J. Phys. Cond. Mat.* 2:4341 (1990); M. Stachiotti, A. Dobry, R. Migoni, A. Bussmann-Holder, Crossover from a displacive to an order-disorder transition in nonlinear-polarizability model, *Phys. Rev. B* 47:2473 (1993).
9. H. Bilz, H. Büttner, A. Bussmann-Holder, Nonlinear lattice dynamics of crystals with structural phase transitions, *Phys. Rev Lett.* 48:264 (1982).
10. G. Benedek, A. Bussmann-Holder, H. Bilz, Nonlinear travelling waves in ferroelectrics, *Phys. Rev. B.* 36:630 (1987).
11. A. Bussmann-Holder, A.R. Bishop, Time-dependent evolution of double-well potentials to model structural anomalies, *Phil. Mag. B* 73:657 (1996); A. Bussmann-Holder, A.R. Bishop, G. Benedek, Quasiharmonic periodic traveling wave solutions in anharmonic potentials, *Phys. Rev. B* 53:11521 (1996).
12. A. Bussmann-Holder, Soft modes and order-disorder effects in ferroelectric phase transitions, *J. Chem. Phys. Sol.* 57:1145 (1996).
13. R.E. Cohen, H. Krakauer, Lattice dynamics and origin of ferroelectricity in $BaTiO_3$: Linearized-augmented-plance-wave total-energy calculation, *Phys. Rev. B* 42:6416 (1990).
14. For a recent review, see Proceedings *Intl. Workshop of Fundamental Properties of Ferroelectrics*, Williamsburg (1997), to be published in *Ferroelectrics*.
15. A. Bussmann-Holder, A.R. Bishop, competing length scales in anharmonic lattices: Domains, stripes, and discommensurations, *Phys. Rev. B* Sept. issue (1997).
16. A. Bussmann-Holder, H. Bilz, G. Benedek, Applications of the polarizability model to various displacive-type ferroelectric systems, *Phys. Rev. B* 39:9214 (1989).
17. E. Courtens, B. Hehlen, G. Coddens, B. Hemion, New excitations in quantum paraelectrics, *Physica B* 219/220: 577 (1996); B. Hehlen, A.-L. Pérou, E. Courtens, R. Vacher, Observations of a doublet in the quasi-elastic central peak of quantum paraelectric $SrTiO_3$, *Phys. Rev. Lett.* 75:2416 (1995).
18. J.H. Jonker, Magnetic compounds with perovskite structure, *Physica* 22:707 (1956).
19. S. Jin, T.H. Tiefel, M. McCormack, R.A. Fastnacht, R. Ramesh, L.H. Chen, Thousand-fold change in resistivity in magnetoresistive La-Ca-Mn-O films, *Science* 264:413 (1994).

20. P. Schiffer, A.P. Ramierez, W. Bao, S.-W. Cheong, Low-temperature magnetoresistance and magnetic phase diagram of $La_xCa_xMnO_3$, *Phys. Rev. Lett.* 75:3336 (1995).

21. P.W. Anderson, H. Hasegawa, Considerations on double exchange, *Phys. Rev.* 100:675 (1955).

22. H. Roeder, Jun Zang, A.R. Bishop, Lattice effects in colossal-magneto-resistance manganites, *Phys. Rev. Lett.* 76:1356 (1996).

23. H.J. Millis, P.B. Littlewood, B.I. Shraiman, Double-exchange alone does not explain resistivity of $La_{1-x}Sr_xMnO_3$, *Phys. Rev. Lett.* 74:5144 (1995).

24. G.-M. Zhao, K. Conder, H. Keller, K.A. Müller, Giant oxygen isotope shift in the magnetoresistive perovskite $La_{1-x}Ca_xMnO$, *Nature*, 381:676 (1996).

25. H.Y. Hwang, S.-W. Cheong, P.G. Radaelli, M. Marezio, B. Batlogg, Lattice effects on the magnetoresistance in doped $LaMnO_3$, *Phys. Rev. Lett.* 75:914 (1995).

26. J.P. Franck, private communications.

27. A. Bussmann-Holder, A.R. Bishop, Competing interactions and the isotope effect in CMR materials, unpublished.

28. H.-K. Hoeck, H. Nickisch, H. Thomas, Jahn-Teller polarons, *Helvet. Phys. Acta* 56:237 (1983).

29. K.I. Kugel, D.I. Khomskii, Polaron effects and exchange interaction in magnetic dielectrics with Jahn-Teller ions, *Sov. Phys. JETP* 52:501 (1980).

30. I.G. Lang, Yu. A. Firsov, Kinetic theory of semiconductors with low mobility, *Sov. Phys. JETP* 16:1301 (1963).

31. J.G. Bednorz, K.A. Müller, Possible high-T_c superconductivity in the Ba-La-Cu-O system, *Z. Phys. B* 64:189 (1986).

32. See e.g. *Lattice Effects in High-T_c Superconductors*, ed. Y. Bar-Yam, T. Egami, J. Mustre de Leon, A.R. Bishop (World Scientific Press), Singapore, New Jersey, London, Hong Kong (1992).

33. A. Bussmann-Holder, A.R. Bishop, Antiferromagnetism and superconductivity, *Phil. Mag. B*, in press.

JOSEPHSON EFFECT: LOW-Tc vs. HIGH-Tc SUPERCONDUCTORS

Antonio Barone

Dipartimento di Scienze Fisiche - Università di Napoli "Federico II"

Napoli - Italia

Istituto Nazionale di Fisica della Materia

INTRODUCTION

The aim of this lecture is to review basic aspects of the phenomenology of the Josephson effect and to discuss its role as a probe of the symmetry of the order parameter in new classes of superconductors.

It is well known that the Josephson effect, besides its intrinsic significance, offers a variety of stimulating applications in different fields. Above all it represents a quite powerful tool to investigate intriguing aspects of fundamental nature of the superconductive state.

Indeed, both the family of heavy-Fermion systems as well as that of high-temperature superconducting cuprates show a variety of phenomena which can not be cast in completely defined theoretical frames.

This lecture brings together a brief outline of the basic ideas of the Josephson effect and a bird's-eye view of aspects of the phenomenology underlying experiments towards the determination of the pairing symmetry in "unconventional" superconductors. In spite of the sacrificing brevity it will be hopefully given at least the flavor of this stimulating specific issue.

JOSEPHSON EQUATIONS

A superconductor can be described as a whole by a macroscopic wave function $\Psi = \rho^{1/2} e^{\alpha}$ where $\rho = |\Psi|^2$ is the density of particles (pairs of electrons) and α the phase. In a single isolated superconductor the number N of pairs is fixed, therefore the uncertainty relation $\Delta N \Delta \alpha \cong 2\pi$ tells us that the phase α is undefined though if we fix its value at a given point it is automatically fixed at all points ("long range order"). When two superconductors, S_1 and S_2, are placed very close to each other (say at a distance of $\cong 10$ Å) then their macroscopic wave functions can overlap leading to a system of two "weakly coupled superconductors"[1-3]. In practice the separation between S_1 and S_2 can be realized by a thin dielectric barrier. If the energy involved in the coupling is greater than thermal fluctuations energy, then a "phase correlation" is established between the two superconductors and Cooper pairs can tunnel through the barrier. In this way, a supercurrent, function of the relative phase $\varphi = \alpha_{S1} - \alpha_{S2}$, between the two superconductors, can flow leading thereby to a situation in which the two coupled superconductors behave to some extent as a single superconductor [4].

The current phase $I(\varphi)$ and voltage-phase $V(\varphi)$ relations

$$I = I_c \sin \varphi \qquad (1a)$$

$$\frac{d\varphi}{dt} = \frac{2e}{\hbar} V \qquad (1b)$$

are the well known constitutive equations of the Josephson effect. I_c is the critical current. We see that V=0 implies φ = constant and therefore the possibility of a finite zero voltage current flowing through the structure (d.c. Josephson effect). In Fig.1 it is reported a sketch of the junction and a current-voltage characteristics showing at finite voltage the quasiparticle tunneling branch and at V=0 the d.c. Josephson current.

Moreover V≠0 across the structure implies the occurrence of an oscillating current $I = I_c \sin (\varphi + 2e/\hbar\ Vt)$ with frequency $\nu = 2eV/h$ (a.c. Josephson effect). It is 483.6 GHz/µV.

There is also a dependence of a relative phase φ on the spatial coordinates produced

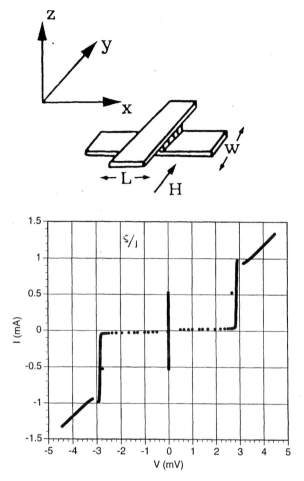

Fig.1 a) Sketch of the junction structure.

b) Current-voltage characteristics of a Josephson tunnel junction (Nb based structure with Al_xO_y tunneling barrier).

by the magnetic field H in the plane (x,y) of the junction :

$$\nabla \varphi = (\frac{2e}{\hbar c} d) \mathbf{H} \times \mathbf{n} \qquad (2)$$

with $d = \tau + \lambda_{L1} + \lambda_{L2}$, where λ_{L1} (λ_{L2}) is the London penetration depth in the superconductor $S_1(S_2)$ and τ indicates the physical thickness of the tunneling barrier. Combining 1 and 2 with the Maxwell equations, we get the general equation:

$$\frac{\partial^2 \varphi}{\partial x^2} + \frac{\partial^2 \varphi}{\partial y^2} - \frac{1}{c^2} \frac{\partial^2 \varphi}{\partial t^2} = \frac{1}{\lambda^2_J} \sin \varphi \qquad (3)$$

where

$$c = \left(\frac{1}{4\pi cd}\right)^{\frac{1}{2}} = c\left(\frac{t}{\varepsilon,d}\right)^{\frac{1}{2}} \text{ and } \lambda_J = \left(\frac{hc^2}{8\pi edJ_c}\right)^{\frac{1}{2}}$$

In one-dimension, neglecting dissipation, measuring x in units of λ_J and τ in units λ_J/c the equation can be cast in the compact form of the Sine-Gordon Equation :

$$\phi_{xx} - \phi_{tt} = \sin\phi$$

ϕ can be interpreted either as the relative phase φ or as a normalized measure of the magnetic flux Φ.

Indeed by integration of the second constitutive relation of the Josephson effect it follows :

$$\varphi = \frac{2\pi}{\Phi_0}\int Vdt = 2\pi\frac{\Phi}{\Phi_0}$$

The permanent profile solutions $\phi = \phi(x-ut)$, are of wide interest in the study of flux propagation in extended Josephson junctions.

Let us observe that equation (3) is a sort of non linear penetration equation. Indeed in the stationary limit and small ϕ it reproduces a London type equation. In our case the penetration called "Josephson penetration length" λ_J gives a measure of the distance at which Josephson currents are confined at the edge of the junction, as a consequence of the screening effect produced by the magnetic self field (i.e. generated by the currents flowing in the junctions). Depending on whether the transverse junction dimensions (width W, and length L), are large or small compared to λ_J we identify the two classes of "small" and "large" junctions respectively.

Let us confine our attention to a small junction (L,W < λ_J). It is easy to show that the maximum of the Josephson current for an externally applied magnetic field H is given by the modulus of the Fourier transform of the current density profile, that is

$$I_c(H) = \left|\int_{-\infty}^{+\infty} I(x)e^{ikx}dx\right|$$

where $k = \frac{2\pi d}{\Phi_o} H_y$. (we have assumed $H \equiv H_y$)

For instance, for a rectangular barrier junction the I_c vs. H dependence is given by a diffraction Fraunhofer-like pattern.

In terms of the magnetic flux threading the junction, $\Phi = H_y L d$, it is

$$I_c\left(\frac{\Phi}{\Phi_o}\right) = I_c(o) \left|\frac{\sin\pi\frac{\Phi}{\Phi_o}}{\pi\frac{\Phi}{\Phi_o}}\right| \qquad (4)$$

This circumstance is an obvious consequence of the wave-like nature of the order parameter. Accordingly, interference phenomena can occur as well. Two Josephson junctions connected in parallel by a superconductive path will give for the $I_c(\phi)$ dependence an interferential pattern

$$I_{ct} = I_{ct}(o)\left[\cos\frac{\Phi_e}{\Phi_o}\right] \qquad (5)$$

where Φ_e is the flux enclosed in the loop. The periodicity is given by a flux quantum Φ_o. In suitable experimental conditions a quite small fraction of the period and therefore of Φ_o can be estimated. These oversimplified arguments give the flavor of the underlying physics of the well known class of devices referred to as SQUIDs[5] which represents the most sensitive exhisting tool for measuring the magnetic fields.

The reason to recall some of such elementary concepts of the Josephson effect, confined to the basic ideas of the dependence of the Josephson current on the magnetic field, lies in the paramount importance of these measurements for the diagnostics of the junction structure such as the actual shape, non uniformities in the tunneling barrier, structural fluctuations etc. Moreover, as we shall see in the following, the Josephson effect in general and this type of measurements in particular represent a powerful probe to investigate aspects of dramatic importance for the understanding of new classes of superconductors.

SYMMETRY OF THE ORDER PARAMETER IN UNCONVENTIONAL SUPERCONDUCTORS

The problem of which symmetry of the order parameter characterizes new classes of superconductors has been, and still is, an intriguing issue. Indeed, in contrast with conventional superconductors which are characterized by an isotropic s-wave pairing, possible unconventional pairing states have been proposed for heavy-Fermions and for high-T_c superconductors.

Heavy-fermion superconductors

We shall start considering a system of weakly coupled superconductors involving materials exhibiting "heavy Fermion" behaviour[6].
Values of the effective mass excedingly large, namely hundred times larger than that of the free electron, justifies the name of this family of interesting superconducting materials. Low critical temperatures (e.g. T_c=0.5K for UBe_{13} and $CeCu_2Si$; T_c=0.85K for UPt_3) make this class of superconductors not interesting for applications, while a quite unusual behavior of both normal and superconducting state reveals intriguing aspects of the underlying physics of heavy fermions.
The decrease of the electronic heat capacity with temperature does not show the exponential behavior ($\exp(-\Delta/T)$) observed in conventional superconductors which is a signature of an energy gap near the Fermi energy E_F. Rather, a power low decrease occurs indicating the existence of nodes for the gap along lines of the Fermi surface or zeros at some points. Thus, while in the conventional superconductors the energy gap is very nearly isotropic in k space (isotropic s-wave pairing), in heavy-fermion systems occurrence of pairing with a nonzero angular momentum is expected resulting in a triplet state.
Various strategies to determine the actual symmetry of the pairing state have been considered. As we shall discuss in the following, the Josephson effect with the related diffraction and interference phenomena discussed in the previous section, allows a class of experiments which are sensitive to the phase of the superconducting order parameter

leading to a quite direct determination of the pairing state symmetry.

Following the work by Geshkenbein, Larkin and Barone[7] let us refer in particular to a structure in which one electrode is made by a conventional (S) superconductor while the other is a heavy-fermion (P) material.

The expression of the free energy of such a weakly coupled structure is given by :

$$F = Re \int \Delta_1 \Delta_2^* G_1 G_1 |T|^2 G_2 G_2 \, dk_1 \, dk_2 \qquad (6)$$

Δ_1 (Δ_2) is the order parameter in the S(P) superconductor. Indeed the order parameter can be expressed as $\Delta_{\alpha,\beta}(k\,x) = \sum_i \eta^i \Psi^i_{\alpha,\beta}(k)$ and, consequently, is :

$$F = A \, Re \, \eta_1 \eta_2^*$$

The coefficient A is an odd function of $\underline{n}$ (unit vector normal to the boundary S/P; $|T|^2$ is the tunneling matrix element ($A \cong |T|^2 \cong R^{-1}$)). Assuming Δ_2 to be pseudoscalar, for a cubic crystal symmetry (e.g. UBe$_{13}$), it can be written :

$$A_n = A \, f(n)$$

with
$$f(n) = n_x n_y n_z (n_x^2 - n_y^2)(n_y^2 - n_z^2)(n_z^2 - n_x^2) \qquad (7)$$

A rotation would imply the permutation of the various factors in (7) while a reflection would leave invariant the quantities in the parentesis but would lead a change of sign of f($\underline{n}$)

If we consider now the system of Fig.2 the total phase variation is $\varphi = \pi$ and, being $\varphi = 2\dfrac{\Phi}{\Phi_o}$, it follows $\Phi = \dfrac{\Phi_o}{2}$,. More generally, the minima of F occur at flux values $\Phi = (n+1/2)\Phi_o$.

Thus, on the basis of symmetry properties of the order parameter in heavy-fermion systems, in ref. 7 it was predicted a new vortex state in a superconductor characterized by the existence of half magnetic flux quanta. In the same reference the occurrence of such vortices in a policrystal was discussed pointing out that, due to their negative energy, they can exist even in the absence of externally applied magnetic field.

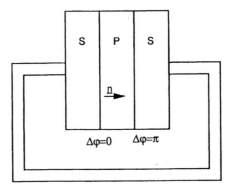

Fig.2 Sketch of a S-P-S structure (see text) closed by a superconducting loop.

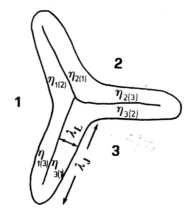

Fig.3 Josephson-like current distribution at the grain boundary. Currents and fields penetrate across the boundary confined at distance of λ_J. This situation occurs for $\lambda_L \ll \xi^2/b\xi_o$.

Thus, from this direct, though oversimplified discussion, it is clear the role that the Josephson effect can play for a quite straightforward determination of the symmetry of the order parameter.

For a description of a policrystalline sample, topological reasons suggest a quite general structure configuration as in Fig. 3 where three grains are indicated together with the values of the order parameters at their boundaries. It is obvious that while in 2D there will be a common point, in the 3D sample there will be a common line, the border line.

In this case the free energy associated with the three boundaries will be

$$F = \text{Re } A_{12}\,\eta_1\,\eta^*_2 + A_{23}\,\eta_2\,\eta^*_3 + A_{31}\,\eta_3\,\eta^*_1$$

The sign of the product $A=A_{12} A_{23} A_{31}$ is not arbitrary but fixed by the reciprocal orientation of the crystal grains. It can be shown[7] that for $A > 1$ the minimum of the free energy corresponds to a state with a vortex on the border line carrying a half magnetic flux quantum.

Following the previous cited reference, it can be seen that for boundary transparency of the order of unit ($b \cong 1$) it is $\lambda_L \gg \xi^2/b\xi_0$ and the current distribution in the vortices at distances larger than ξ_{eff} is as that in conventional vortices, while the energy of the vortex is 1/4 of that of usual ones. In the opposite limit of small b, that is $\lambda_L \ll \xi_{eff}$, the current distribution is like that occurring in Josephson junctions.

In this case it is $\partial^2\varphi/\partial x^2 = b\xi_0/\lambda_L\xi^2 \sin\varphi$ and, accordingly, currents and fields penetrate across the grain boundaries within a Josephson penetration length $\lambda_J = (\lambda_L \xi^2/b\xi_0)^{1/2}$.

Cuprate superconductors

As we have seen, measurements of the magnetic field modulation of the supercurrent in a suitable SQUID loop configuration, can be a very powerful probe to study, through instrinsic phase shift of the order parameters, possible unconventional symmetry of the pairing state of a superconductor. While, as previously discussed, the underlying idea of this experiment was first proposed in the context of heavy-Fermion systems (possible p-wave symmetry)[7], the proposal to search for d-wave symmetry in the cuprate superconductors was independentely discussed by Sigrist and Rice[8].

The symmetry of the order parameter in high-Tc superconductors has been in these years quite controversial. Possible unconventional pairing states such as anisotropic s-wave or $d_{x^2-y^2}$, as well as complex mixtures of s and d, or as $d_{x^2-y^2} + i\varepsilon d_{xy}$ (i.e. ε being the measure of id_{xy} symmetry in the order parameter) have been considered (see Fig. 4).

Anisotropy of the order parameter with fourfold rotation symmetry in the (a,b)-plane can reflect qualitatively these hypotheses.

The pioneering experiments, developed at the University of Illinois at Urbana-Champaign[9], employed YBCO crystals and thin films of a conventional (s-wave) superconductor (Pb), forming YBCO-Pb tunnel junctions at the orthogonal a and b faces of the crystal and closed by a lead loop in a d.c. SQUID configuration (see Fig.4 a,b).

Since the tunneling probability is peaked for electron wave vectors normal to the junction barrier, such an experiment is able to probe the relative phase of the superconducting order parameter in orthogonal k-space directions identifying thereby the symmetry of the pairing state.

The expected dependence of the maximum supercurrent on the applied magnetic flux is reported on the bottom of Fig.4. For s-wave symmetry a phase shift between the two

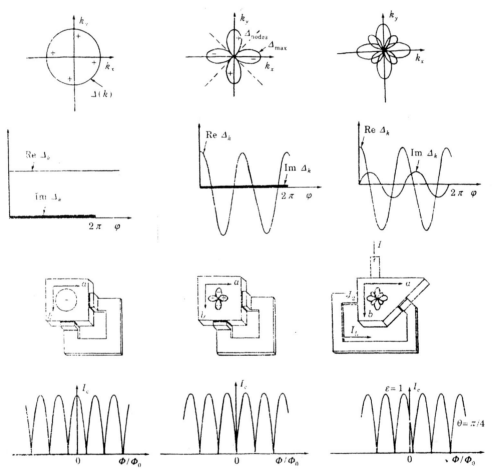

Fig.4 a)   Simple s state b) simple d_{x-y2} state c) exotic $d_{x2-y2}+i\varepsilon d_{xy}$. First row : qualitative k-polar diagrams. Second row : qualitative behavior of the phase of the order parameter. Third row : sketches of SQUID configurations. Bottom row : expected pattern of the Josephson current vs. magnetic flux.

directions a and b would be $\delta_{a,b} = 0$ since the order parameter in the YBCO is the same for both junctions.

Accordingly, the usual I_C vs. Φ pattern is predicted (Fig. 4a), while for a $d_{x^2-y^2}$ symmetry the order parameter $\Delta(k_x, k_y) = \Delta_0 (\cos(k_x a) - \cos(k_y a))$ would imply a non zero phase change $\delta_{a,b}$ leading to the shift in the pattern as in (Fig. 4b). The experimental results of ref.9 confirmed such a shift being therefore in favor of the d-wave symmetry.

To reduce possible effects of SQUID asymmetries and flux trapping, the same group performed experiments using a "corner" junction configuration[10] in which a conventional (s-wave) superconducting counterelectrode is deposited partly on the a-c and partly on the b-c faces of the YBCO crystal. Depending on whether we are dealing with s or d pairing symmetry a different situation is expected as summarized in Fig.5. The results confirm the expected I_c vs H pattern. Details of these experiments as well as a discussion, confirmations and criticisms can be found in the excellent paper by Van Harlingen[11].

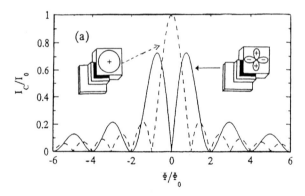

Fig.5   Josephson current vs. magnetic flux dependence for corner type junctions in the case of s (dashed line) and d (full line) pairing states.

Experimental evidence in favor of s-wave symmetry was found by the group of Dynes at U.C. San Diego[12] in planar tunnel junctions employing a high-T_c film as base electrode and Pb as a counterelectrode. The cuprate superconductor was either $YBa_2Cu_3O_{7-\delta}$ or $Y_{1-x}Pr_xBa_2Cu_3O_{7-\delta}$. A 10 Å Ag layer was also deposited before the Pb film.

The authors claimed that the observed Josephson current occurring into the c-axis, being not consistent with the orthogonality of the $d_{x^2-y^2}$ and s wave symmetries, excludes the possibility of an unconventional pairing symmetry of the superconductor cuprates. Although of good quality, as confirmed by IJ vs H measurements, these junctions would deserve further attention for what concerns the actual nature of the barrier, a possible in plane distortion of $d_{x^2-y^2}$ symmetry, the gapless behavior of YBCO in the c-direction and a possible role of the thin Ag layer. Grain boundary Josephson system configurations have also been considered to probe the interference between the order parameters of high-Tc superconductors.

In particular, Chaudari and Shan-Yu Lin[13] investigated quite complex configurations showing results supporting a s-wave symmetry scenario.

A test of the order parameter symmetry was realized in a series of experiments at IBM on flux quantization in high-T_C rings comprising two and three grain boundary Josephson junctions[14,15]. The samples were obtained by epitaxial thin films of YBCO grown on substrates properly designed as described in Fig.6. It was shown the occurrence of flux quantization in half-integer multiples of Φ_o at the tricrystal point, as consequence of the intrinsic unconventional symmetry of the order parameter in agreement with prediction of ref. 7 in the context of the heavy Fermion systems. A scanning SQUID microscope was employed to directly image the magnetic fields associated with the half flux quantization structure.

We shall not dwell further on the description of these fundamental experiments we have just mentioned in this section nor on other important ones. Rather, for the various aspects of the intriguing debate about the s vs d-wave symmetry, we address the reader to the quite extensive reviews[11,16,17] and references reported therein. A two gap behavior inferred theoretically by Kresin[18] also from various experiments and a related possible cohexistance of s and d-wave condensates, as proposed by Alex Muller[19,20] would also deserve great attention.

The implications of new theoretical ideas and experimental results on the whole underlying physics involved in high-T_c superconductivity appears to be more and more challenging[20,21] for the interpretation of both normal and superconducting state properties.

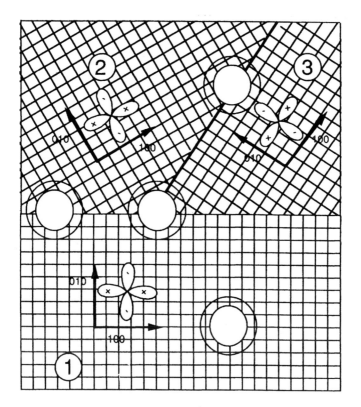

Fig.6 Sketch of the tricrystal 100 SrTiO$_3$ substrate including 4 epitaxial YBCO rings for flux quantization experiments (from Tsuei et al. [14]). Two rings contain two grain boundary weak links (1-2, 2,3), one contains three (1,2 2-3, 3-1) and one contains no grain boundary weak links.

In closing this lecture let us remark that the understanding of the precise mechanisms of superconductivity and complete microscopic theory of "unconventional" materials go beyond the symmetry properties we have discussed. Moreover I share the view of Lev Gor' kov that heavy-Fermion systems in particular should deserve a greater attention particularly on the experimental side.

References

1. B.D. Josephson, Phys. Lett 1:251 (1962).
2. P.W. Anderson, Special effects in Superconductivity in "Lectures on the many body problem" Ravello 1963, E.R. Caianiello, Ed., Academic Press, New York(1964).

3. A.Barone and G. Paternò, "Physics and Applications of the Josephson Effect" Wiley, New York(1982).

4. It is important to remind that the weak coupling between the two superconductors can be realized not necessarly by a tunneling barrier (see for instance ref. 2 chapter 7).

5. See for instance "Principles and Applications of Superconducting Quantum Interference Devices", A.Barone Ed., Singapore, World Scientific (1992).

6. F. Steglich et al., Phys. Rev. Lett. 43:1892 (1979). See also the lecture by L.P. Gor' kov in this volume.

7. V. B. Geshkenbein, A.I. Larkin and A. Barone, Phys. Rev. B36:235 (1987).

8. M. Sigrist and T.M. Rice, J.Phys. Soc. Japan 61:4283 (1992).

9. D.A. Wollman et al., Phys. Rev. Lett. 71:2134 (1993).

10. D.A. Wollman et al., Phys. Rev. Lett. 74:797 (1995) .

11. D.J. Van Harlingen, Rev. Mod. Phys. 67:515 (1995).

12. A.G. Sun, D.A. Gajewski, M.B. Maple and R.C. Dynes, Phys. Rev.Lett.72: 2267 (1994).

13. P. Chaudari and Shan-Yu Lin, Phys. Rev. Lett. 72:1084 (1994).

14. C.C. Tsuei et al., Phys. Rev. Lett. 73:593 (1994) ; Science 272:329 (1995).

15. J.R. Kirtley et al., Phys. Rev.Lett. 76:1336 (1996).

16. A. Barone, Il Nuovo Cimento, 16D:1635 (1994).

17. M.R. Beasley, IEEE Trans. Appl. Superc. 5:141 (1995).

18. V.Z. Kresin and S.A. Wolf Phys. Rev. B46:6438 (1992).

19. K. Alex Muller, Nature 377:133 (1995); J. Phys.Soc. jap. 65:3090 (1966).

20. Guo-meng Zhao, M.B. Hunt, H.Keller and K.A. Muller, Nature 385:236 (1997).

21. P.W. Anderson, Physics World, December 1995, p.37 and Letters in Physics World, January 1996, p.16.

THE SPECTRUM OF THERMODYNAMIC FLUCTUATIONS IN SHORT COHERENCE LENGTH SUPERCONDUCTORS

Andrea Gauzzi

MASPEC – CNR Institute
Via Chiavari 18/A, I – 43100 Parma, Italy.

1. INTRODUCTION

Our current understanding of phase transitions enables us to establish a *universal* correspondence between critical behaviour at the transition point and symmetry of the effective Hamiltonian $\mathcal{H}_{eff}$ of a given system, regardless of the specific nature of the microscopic interaction responsible for the transition. The universal validity of this correspondence requires that the critical behaviour is dominated by fluctuations of the order parameter having a wavelength much larger than the range of the above interaction. This is so for systems characterised by a long correlation (or coherence) length ξ of the order parameter. An example of these systems is given by superconducting metals and alloys with low transition temperature T_c, which are well described by the conventional $\mathcal{H}_{eff}$ of the BCS theory. In this paper we discuss how the above fluctuation spectrum is modified in the opposite case of short-ξ systems. This is a relevant issue, since we can derive the $\mathcal{H}_{eff}$ of a system undergoing a second order phase transition from the fluctuation spectrum. This study might elucidate the still open question of generalising the conventional BCS theory to account for the unusual properties of cuprate superconductors. Fluctuation studies are suited for determining the $\mathcal{H}_{eff}$ of complex systems, such as the above materials, because only the relevant interaction driving the transition exhibits a critical behaviour in the fluctuation region, while this interaction is usually hidden among other interactions at temperatures far above T_c.

This paper is organised as follows. We first critically review selected general aspects on fluctuation phenomena near second order phase transitions in section 2 and the universal predictions of conventional theories of critical phenomena in section 3. The content of these sections can be found in any textbook (Landau and Lifshitz, 1980; Patashinskii and Pokrovskii, 1979; Ma, 1976; Stanley, 1971). However, we prefer discussing some key concepts, with particular emphasis on the limit of validity of the theoretical results, to facilitate our discussion. In section 4 we review the predictions of conventional fluctuation theories referring to the physical properties studied in the following sections. In section 5 we re–examine the application of standard universal predictions to selected experiments on conventional BCS superconductors with long ξ. In section 6 we present a similar analysis for cuprates and highlight the controversial interpretations reported for these superconductors with short ξ. In section 7 we re–analyse a selection of the experimental results which are a subject to controversy within the framework of a generalised mean–field approach. In particular, we examine the qualitative features of the fluctuation spectrum of cuprates as determined by fluctuation measurements within such generalised approach. Finally, in section 8 we compare these features with those of the conventional spectrum of the BCS $\mathcal{H}_{eff}$ in the long–ξ limit and draw some conclusions. Several aspects discussed in this paper can be extended to other families of superconductors with short ξ, such as Bi–based oxides, Chevrel phases and alkali–doped C_{60}, and also to other second order phase transitions characterised by a short correlation length of the order parameter.

2. RELEVANCE OF THERMODYNAMIC FLUCTUATIONS NEAR SECOND ORDER PHASE TRANSITIONS

According to general results of statistical mechanics, a thermally activated deviation (fluctuation) $\delta\psi$ of the order parameter ψ of a given second order phase transition with respect to the thermodynamic equilibrium value $\langle\psi\rangle$ at temperature T has probability

$$w \propto \exp(-\delta\Phi/k_B T) \tag{1}$$

where $\delta\Phi$ is the extra–energy required to activate such deviation with respect to the value $\langle\Phi\rangle$ of Φ at thermodynamic equilibrium. Φ is the difference of the appropriate thermodynamic potential associated with the phase transition

$$\Phi \equiv \Phi(\psi) - \Phi(0) \tag{2}$$

and we assume for simplicity $\Phi(0) \equiv 0$. The quantity Φ is called *incomplete thermodynamic potential* and plays the role of the effective Hamiltonian $\mathcal{H}_{eff}$ of the system.

This paper is based on the concept of *fluctuation spectrum*, which is defined as the expectation value $\langle|\delta\psi_{k,\omega}|^2\rangle$ of the density of the fluctuation $\delta\psi$ with given wavevector $\mathbf{k}$ and with given frequency ω. We recall that $\langle|\delta\psi_{k,\omega}|\rangle = 0$ by definition of fluctuation. For simplicity, we consider the limit of slow fluctuations, corresponding to $\omega = 0$. The fluctuation spectrum is relevant to our discussion, since the type and the microscopic parameters of $\mathcal{H}_{eff}$ and the critical behaviour of all physical properties, i.e. their *critical exponents*, are derived from this spectrum. The fluctuation spectrum is expressed by the probability w as a function of $\delta\psi$ in eq. (1). To determine this probability, the exact functional dependence of Φ on ψ in eq. (2), i.e. the $\mathcal{H}_{eff}$, is required. We first consider the case of a uniform order parameter and note that the qualitative dependence of Φ on ψ above and below T_c must be like the one schematically represented in Fig. 1, according to the well–known arguments by Landau (1937). The equilibrium value $\langle\Phi\rangle$ corresponds to a zero value of $\langle\psi\rangle$ above T_c and non–zero below T_c. Under reasonable physical assumptions, the functional $\Phi = f(\psi)$ is continuous and therefore admits an expansion in powers of ψ. Such an expansion is justified for sufficiently small values of ψ, which corresponds to temperatures near T_c. For simplicity, we shall consider only the high–symmetry state above the transition ($\langle\psi\rangle = 0$, $T > T_c$). The fluctuating potential can then be written as

$$\delta\Phi = \frac{1}{2!}\left(\frac{\partial^2\Phi}{\partial\psi\partial\psi^*}\right)_{\psi=\langle\psi\rangle}|\delta\psi|^2 + \frac{1}{4!}\left(\frac{\partial^4\Phi}{\partial\psi^2\partial\psi^{*2}}\right)_{\psi=\langle\psi\rangle}|\delta\psi|^4 + ... \quad (3)$$

$$= V\left(a|\delta\psi|^2 + \beta|\delta\psi|^4 + ...\right)$$

where V is the sample volume. In eq. (3), the linear term is zero, since $\langle\Phi\rangle$ is a minimum; the cubic term is also zero if the critical points are not isolated in the phase diagram. From Fig. 1, one notes that the sign of the first coefficient a must change at the transition. It follows that, at temperatures sufficiently near T_c, the temperature dependence of this coefficient can be approximated by the linear term $a = at$, where $t \equiv (T - T_c)/T_c$ is a reduced temperature. Usually, we can neglect the quadratic term $\propto t^2$ and the higher order terms in the expansion of a in powers of t if

$$t \ll 1 \quad (4)$$

The validity of the universal predictions for the critical exponents of the mean–field theory cited in section 4 and reported in Table 1 requires the validity of this linear approximation.

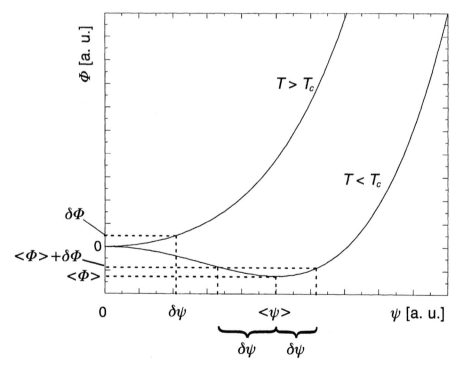

Figure 1. Schematic diagram of the dependence of the incomplete thermodynamic potential Φ of eq. (3) on the order parameter ψ in both regions above and below the transition (after Landau and Lifshitz (1980)). The fluctuation regions of $\delta\Phi$ and $\delta\psi$ are indicated by the broken lines.

2a. Weak fluctuations. Fluctuation spectrum

The Landau formulation described above is particularly powerful thanks to its general validity: in principle, every physical system can be described by an appropriate choice of the coefficients in the expansion of eq. (3), provided that a sufficiently large number of terms is taken into account. This observation is valid for both weak and strong fluctuations. The former case can be treated by mean–field theories, since these are appropriate for weakly interacting systems. In this case, it is sufficient to retain the first quadratic term in eq. (3)

$$\delta\Phi \approx aVt|\delta\psi|^2 \tag{5}$$

Weak fluctuations are also called Gaussian, since they are independent of each other in first approximation and therefore their probability follows a Gaussian distribution function. This can also be seen mathematically: by inserting eq. (5) into eq. (1), the probability w becomes a Gaussian distribution function of the statistical variable $\delta\psi$. The Gaussian approximation is found to be appropriate for weak–coupling superconductors and for certain ferroelectrics but, in general, not for magnetic transitions.

An important generalisation of the simple treatment described above consists in taking into account spatial variations of the order parameter. Such variations arise, for example, in the presence of a current flow in a superconductor. The functional in eq. (3) must then be replaced by an integral over the sample volume of a spatially varying functional density. Within the limit of slow spatial variations, it is sufficient to add a quadratic term in the first gradient of ψ to the above density. According to the original idea of Ornstein and Zernicke (Landau and Lifshitz, 1980), eq. (3) must be substituted by the following

$$\delta\Phi = \int_V \left(at|\delta\psi|^2 + \beta|\delta\psi|^4 + g^i_j \partial_i\delta\psi \partial^j\delta\psi^*\right)dV \tag{6}$$

where ∂_i are, in a superconductor, covariant gauge–invariant partial spatial derivatives

$$\partial_i \equiv \frac{\partial}{\partial x^i} - \frac{2ie}{\hbar}A_i \tag{7}$$

where $2e$ is the charge of a Cooper pair and $\mathbf{A}$ is the vector potential. It has been shown by Gor'kov (1959) that the functional of eq. (6) coincides with the BCS effective Hamiltonian in the vicinity of the transition by setting $\psi \propto \Delta$, where Δ is the superconducting energy gap. This equivalence establishes a correspondence between fluctuation spectrum and microscopic Hamiltonian of a given BCS superconductor. It would be important to extend this result to cuprates, since the microscopic Hamiltonian of these superconductors is still unknown. This method for determining the effective Hamiltonian is applicable to any second order phase transition, since eq. (6) formally describes the interaction of a charged scalar field with a vector potential. The same functional is then applicable not only to superconductors, but also to nematic–smectic–A transitions in liquid crystals (de Gennes, 1972) and to Higgs' mechanism of charged boson fields (Coleman and Weinberg, 1973).

To determine the probability of weak fluctuations in the general case of spatially–varying order parameter described by eq. (6), we again neglect the quartic term and expand the fluctuating field $\delta\psi$ into Fourier components with wavevector **k**

$$\delta\psi(\mathbf{r}) = \sum_{|\mathbf{k}| \leq |\tilde{\mathbf{k}}|} e^{i\mathbf{k}\mathbf{r}} \delta\psi_\mathbf{k} \tag{8}$$

where we have emphasized that the sum is limited to wavevectors smaller than a cutoff value $\tilde{k}$. The existence of such cutoff is due to the fact that the order parameter ψ is defined in a sufficiently large volume $\tilde{V}$ containing a number of Cooper pairs much larger than unity. If this condition were not fulfilled, it would not be possible to define the quantity $|\psi|^2$ as the expectation value of the superfluid density. The value of $\tilde{k}$ is then determined by the condition that the spatial variations of ψ contain only Fourier components with a wavelength larger than the linear dimension of $\tilde{V}$. In sections 5 and 6 we discuss the fact that the above limitation is not relevant to conventional superconductors with long ξ, while it is fundamental for short–ξ superconductors. The substitution of eq. (8) in eq. (6) yields

$$\delta\Phi = V \sum_{|\mathbf{k}| \leq |\tilde{\mathbf{k}}|} \left(at + g^i_j k_i k^j \right) |\delta\psi_\mathbf{k}|^2 \tag{9}$$

This result enables us to explicitly evaluate the **k**–dependence of the probability w of the fluctuating field $\delta\psi(r)$. By replacing the above expression in eq. (1), we find that w becomes a product of Gaussian distribution functions $w_\mathbf{k}$ of each Fourier component $\delta\psi_\mathbf{k}$

$$w \propto \prod_{\mathbf{k} \leq |\tilde{\mathbf{k}}|} \exp\left[-\frac{V}{k_B T}\left(at + g^i_j k_i k^j \right) |\delta\psi_\mathbf{k}|^2 \right] \tag{10}$$

The expectation value of the component $\delta\psi_\mathbf{k}$, i.e. the *fluctuation spectrum*, is found to be a *Lorentzian* function of **k** in a reference system where the tensor g is diagonal

$$\langle |\delta\psi_\mathbf{k}|^2 \rangle = \frac{k_B T}{2Vat} \frac{1}{\left(1 + \xi_x^2 k_x^2 + \xi_y^2 k_y^2 + \xi_z^2 k_z^2\right)} \tag{11}$$

where we have set

$$\xi_x \equiv \sqrt{\frac{g^x_x}{at}} = \xi_{x,0} t^{-1/2} \tag{12}$$

and similar expressions for the y– and z–directions. The above spectrum is called Ornstein–Zernicke spectrum and is represented in Fig. 2. The physical meaning of ξ is the (Ginzburg–Landau) correlation length of ψ, i.e. it is the characteristic length over which the correlation function of the order parameter decays significantly. The square–root divergence of ξ at the transition, which we note from eq. (12), arises from the linear approximation of the temperature dependence of the quadratic coefficient in the Ginzburg–Landau functional of eq. (6).

We finally note that the **k**–dependence of the fluctuation spectrum can be expressed in the explicit Ornstein–Zernicke form of eq. (11) thanks to the statistical independence of each fluctuation with wavevector **k** within the weak–fluctuation approximation. Within this approximation, the fluctuating field is analogous to a perfect gas of quasiparticles with wavevector **k** and with density $\langle|\delta\psi_{\mathbf{k}}|^2\rangle$. The Ornstein–Zernicke spectrum is therefore the Green function of a free fluctuating field with no self–energy correction.

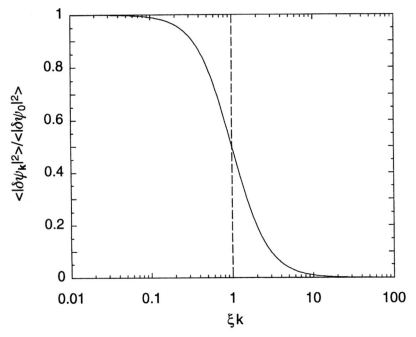

Figure 2. Dependence of the Ornstein–Zernicke fluctuation spectrum (see eq. (11)) as a function of wavevector. This spectrum is valid within the limit of weak (Gaussian) fluctuations and within the long wavelength limit $\xi\tilde{k} \gg 1$, where $\tilde{k}$ is the short–wavelength cutoff described in the text. The effect of this cutoff in the case of short coherence length ($\xi\tilde{k} \approx 1$) is represented schematically by the broken line.

2b. Strong fluctuations

In this regime, interactions between fluctuations are important. Mathematically, this circumstance implies that the quartic term and higher–order terms in the Ginzburg–Landau expansion (3) are no longer negligible. The quartic term has the meaning of two–particle interaction, the sixth-order term that of three–particle interaction, etc. Strong fluctuations imply that the uniform fluctuation $\langle|\delta\psi_{\mathbf{k}=0}|^2\rangle$ in eq. (11) becomes comparable to the average value $\langle|\psi|^2\rangle = -\alpha t/2\beta$ of the order parameter in the coherence volume $\sim\xi^3$. This condition is known as Ginzburg condition and is expressed by the inequality (Levanyuk, 1959; Ginzburg, 1960)

$$t \lesssim t_G \equiv \left[\frac{\beta k_B T_c}{\alpha^2 \xi_{x,0}\xi_{y,0}\xi_{z,0}}\right]^2 = \frac{1}{32\pi^2}\left(\frac{k_B V}{\Delta C \xi_{x,0}\xi_{y,0}\xi_{z,0}}\right)^2 \qquad (13)$$

where t_G is called Ginzburg temperature and ΔC is the specific heat jump at T_c. In the case of short ξ, one can have $t_G \gtrsim 1$, hence the Gaussian approximation is not valid at any temperature and the fluctuations are strong (or critical) at any temperature. The conditions leading to this case have been studied by Kapitulnik, Beasley, Di Castro and Castellani (1988) for cuprates. For the most studied compound $YBa_2Cu_3O_{6+x}$, with $T_c = 92$ K, we estimate $t_G \approx 3 \times 10^{-2} - 3 \times 10^{-3}$ (Gauzzi and Pavuna, 1995) by using experimental data of specific heat (Junod, 1990) and of ξ (Matsuda, Hirai and Komiyama, 1988; Lee, Klemm and Johnston, 1989; Lee and Ginsberg, 1992). We therefore expect that weak fluctuations dominate and that critical fluctuations become important below 0.3–3 K above T_c, but this question remains controversial.

The generalisation of the previous considerations to the case of strong fluctuations is straightforward. However, the partition function

$$Z = \sum_{\delta\psi} e^{-\frac{\delta\Phi(\delta\psi)}{k_B T}} \tag{14}$$

where the effective Hamiltonian $\delta\Phi$ has the Ginzburg–Landau form of eq. (6) can not be calculated explicitly in presence of the quartic term. Numerical solutions have been obtained by renormalisation group methods. These methods were first applied by Wilson (1971) and are based on perturbation theories or on approximate schemes. Besides the Ginzburg–Landau Hamiltonian, alternative Hamiltonians can be used. Classic examples are the Ising, XY or Heisenberg Hamiltonians, corresponding to the case of respectively one, two and three scalar components of the order parameter. Contrary to the case of the Ginzburg–Landau model, these are discrete models defined on a lattice and are therefore suited for numerical simulations. The XY Hamiltonian describes phase transitions with one single complex order parameter where *phase* fluctuations dominate. This is the opposite of the Ginzburg–Landau Hamiltonian, which accounts for predominant weak fluctuations of the *amplitude* of the order parameter.

3. UNIVERSALITY PRINCIPLE AND SCALING HYPOTHESIS FOR CRITICAL PHENOMENA

On the basis of the foregoing considerations, we might ask whether the choice of a particular effective Hamiltonian is important or not to account for the behaviour of a given system in the fluctuation region. In other words, we shall ask under which conditions fluctuation phenomena are expected to be the signature of the particular microscopic interaction driving the phase transition or to follow a more general behaviour. Indeed, despite the variety of microscopic interactions responsible for second order phase transitions of magnetic, structural, electronic, etc. nature, it is often found that completely different systems exhibit the same type of behaviour at the transition. Specifically, the temperature dependence of measurable quantities, such as the specific heat, the magnetic susceptibility, the compressibility, etc. near the critical point follow the same power law whose exponent is called *critical exponent*. The heuristic explanation of such common behaviour is given by the following argument, known as *universality principle*; the ensemble of systems characterised by the same behaviour is therefore called *universality class*. As the critical point is approached, the coherence length ξ of the order parameter begins to diverge. Sufficiently close to the critical point, ξ becomes much larger than the range of relevant interactions responsible for the transition. In particular, if ξ becomes much larger than the cutoff wavelength of the fluctuations

$$\xi \gg 1/\tilde{k} \tag{15}$$

it is expected that the specific nature of the above interactions is irrelevant in determining the critical behaviour of the system. The critical behaviour must then depend only on the symmetry of $\mathcal{H}_{eff}$, which implies, in particular, a given dimensionality of space and a given number of components of the order parameter. The Ginzburg–Landau $\mathcal{H}_{eff}$ in the free-field approximation and the $3d$XY $\mathcal{H}_{eff}$ mentioned above are two classic examples of universality class with one complex order parameter. These two classes are suitable for describing the weak and strong fluctuation regimes respectively. As previously mentioned, most conventional BCS superconductors, which all have long coherence lengths, are well described by the first class. On the other hand, superfluid ^{4}He, which has a short coherence length, exhibits strong fluctuations in agreement with the Ginzburg criterion (eq. (13)), and its fluctuation behaviour is well described by the $3d$XY Hamiltonian as reported, for example, in Stanley's book (1971). Other superfluids and superconductors belong to different universality classes, since their order parameter has more than two real components. This is the case of superfluid ^{3}He and of some heavy fermion superconductors. In the former, for example, it was established that the order parameter has p–wave symmetry, corresponding to an orbital momentum $l = 1$ and to a total spin $s = 1$ of the Cooper pair[1]. The number of real components of the order parameter is therefore 18. On the other hand, in cuprates, the available experimental data indicate that Cooper pairs form a singlet state ($s = 0$), as in conventional BCS superconductors and in superfluid ^{4}He. However, the question of the symmetry of the orbital part of the superconducting wave function is still controversial.

The predicted critical exponents of selected quantities are reported in Table 1 for the Ginzburg–Landau theory in the Gaussian approximation and for the $3d$XY model. In the next sections we present a comparison between these predictions and the experimental data on selected compounds of conventional and cuprate superconductors. We recall that the number of independent critical exponents is determined by applying scale invariance arguments to the length of the system (see, for example, Landau and Lifshitz, 1980). In the regime of weak fluctuations, there are two independent lengths, ξ and the dimension r_0 of the volume where the magnitude of the fluctuations is comparable to the magnitude of the equilibrium value of the order parameter. In this case, it can be shown that there are three independent exponents. In the regime of strong fluctuations, it is assumed that ξ is the only relevant length of the fluctuating system. This hypothesis is called *scaling hypothesis* and reduces to two the number of independent critical exponents.

The universality principle and the scaling hypothesis constitute the basis of the modern theory of critical phenomena, since they are expected to be generally valid, provided that the correlation length becomes sufficiently large at the transition, according to eq. (15). In real systems, however, it is not possible to study temperature regions closer than 0.1–1 mK to the critical point because of practical limitations. Therefore, in systems with sufficiently short ξ_0, the above inequality might be not fulfilled in the temperature

---

1. For an overview, we refer to the two lectures by R. Richardson and by D. Vollhardt published in this volume.

region accessible to experiment. In strongly anisotropic systems, it could occur that the condition for universality is satisfied along a given direction, but not along another. We shall apply the above considerations to amorphous alloys and to cuprates.

Table 1. Critical exponents of selected physical quantities in the weak (or Gaussian) and strong (or critical) fluctuation regimes according to respectively the Ginzburg–Landau and the 3dXY models for one complex order parameter.

Physical quantity		Critical exponent	Gaussian fluctuations Ginzburg–Landau theory	Critical fluctuations 3dXY model
Specific heat	$\Delta C \sim t^{-\alpha}$	α	0	Logarithmic divergence
Order parameter	$\psi \sim t^{\beta}$	β	1/2	1/3
Coherence length	$\xi \sim t^{-\nu}$	ν	1/2	2/3
Magnetic pen. depth	$\lambda \sim t^{-\beta}$	β	1/2	1/3
Fluctuation of the magnetic susceptibility	$\delta\chi \sim t^{-\upsilon}$	υ	1/2	2/3
Fluctuation of the conductivity	$\delta\sigma \sim t^{-\eta}$	η	$2-d/2$	2/3

4. EXPERIMENTAL ASPECTS OF FLUCTUATION STUDIES IN SUPERCONDUCTORS

Generally speaking, the effects of thermodynamic fluctuations near a second order phase transition appear as a rounding of the discontinuities or as a cutoff of the divergencies of physical quantities, such as the specific heat, the thermal expansion coefficient, the conductivity, etc. In conventional superconductors, these effects are too weak to be measured, because of the low T_c and of the long ξ. This can be seen from eqs. (1), (11) and (13). The situation is more favourable in amorphous alloys, since ξ is reduced by the amorphous structure down to 5–10 nm. In cuprates, fluctuation effects are particularly large, since T_c is large and ξ is extremely short.

An advantage of fluctuation studies is that fluctuating properties are determined by the condition of thermodynamic equilibrium even if the quantity to be measured is of kinetic origin, as in the case of the electrical conductivity. However, for some quantities, the signals to be measured are either small and/or the fluctuation contribution has to be subtracted from a large background. Moreover, the origin and temperature dependence of such background signal are unknown or difficult to determine. In superconductors, this is the case of the fluctuation contribution to the specific heat jump. A typical set of data reported by Roulin, Junod and Walker (1996) on the cuprate YBa$_2$Cu$_3$O$_{6+x}$ is shown in Fig. 3. In the case of other quantities, such as the magnetic susceptibility and the electrical conductivity, the fluctuation contribution is usually large and the background can be subtracted more easily, since it is well approximated by a constant or by a linear temperature dependence. An example of rounding of the superconducting transition of the resistivity in YBa$_2$Cu$_3$O$_{6+x}$ is shown in Fig. 4. Fluctuation measurements of the specific heat and of the conductivity have been reported by several groups on various compounds of cuprates and the results obtained exhibit good reproducibility. A comprehensive discussion on the results of specific heat measurements and on their interpretation was reported by Junod (1996), while a critical analysis of the experimental data on fluctuation conductivity was given by Gauzzi and Pavuna (1995).

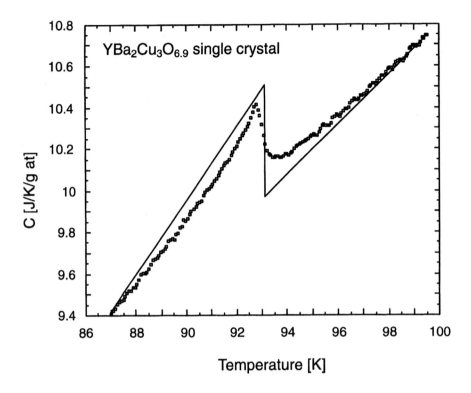

Figure 3. Experimental dependence of the specific heat of a single crystal of $YBa_2Cu_3O_{6.9}$ in the transition region. One notes the rounding of the jump at the transition attributed to the fluctuations of the order parameter. The solid line schematically represents the mean–field jump expected in the absence of fluctuations (after Roulin, Junod and Walker (1995)).

The analysis of experimental fluctuation data within the framework of the theories mentioned above is straightforward thanks to the simple universal predictions summarised in Table 1. The analytic derivation of these predictions can be found in any textbook. Here we just mention that the fluctuation of any physical property can be derived from the partition function of eq. (14) or from the incomplete thermodynamic potential of eq. (6). In the following, we give the result obtained in the regime of validity of the Gaussian approximation and in the region $T > T_c$ for 1) the fluctuation contribution $\delta\Delta C$ to the specific heat jump and 2) the fluctuation contribution $\delta\sigma$ to the electrical conductivity.

4.1. Specific heat jump

In first approximation, the fluctuating part $\delta\Phi$ of the incomplete thermodynamic potential is equal to the fluctuating part of the free–energy. Thus, by definition

$$\delta\Delta C = -\frac{1}{T_c}\frac{\partial^2 \delta\Phi}{\partial t^2} \qquad (16)$$

By inserting eq. (9) in the above expression and approximating the series with an integral, the result is (Levanyuk, 1963)

$$\delta\Delta C = \frac{2V^3\alpha^2}{k_B T_c^2} \int\limits_{|\mathbf{k}|\le|\tilde{\mathbf{k}}|} \langle|\delta\psi_\mathbf{k}|^2\rangle^2 \frac{d^3k}{(2\pi)^3} \tag{17}$$

The same result was obtained by Aslamazov and Larkin (1968) from the microscopic BCS theory. In the case of sufficiently large ξ_0, the product $\xi\tilde{k} \gg 1$ at all temperatures in the Ginzburg–Landau region $t \ll 1$, hence the limit of integration can be extended toward infinity. Levanyuk's *universal* result is obtained

$$\delta\Delta C = \frac{k_B}{16\pi} \frac{V}{\xi_{x,0}\xi_{y,0}\xi_{z,0}} \frac{1}{\sqrt{t}} \tag{18}$$

Similar results are obtained in other dimensions and in the region $T < T_c$. We note that the critical exponent depends only on the dimension of space d, while the amplitude of the fluctuations depends on the microscopic parameters of the Ginzburg–Landau Hamiltonian, which are characteristic of the system.

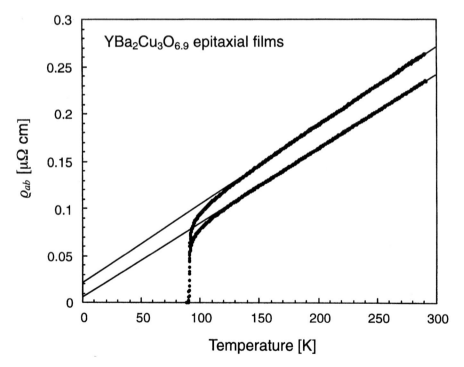

Figure 4. Experimental temperature dependence of the *ab*–plane resistivity of two epitaxial YBa$_2$Cu$_3$O$_{6.9}$ films. As in the preceding figure, the rounding of the jump at the transition is attributed to the fluctuations of the order parameter. In this case, the fluctuation contribution can be easily determined by extrapolating to low temperatures the linear temperature dependence at high temperatures where fluctuations are assumed to be absent. The fit of this linear dependence is represented by the solid lines. (Reproduced by permission of the American Physical Society from Gauzzi and Pavuna (1995)).

4.2. Electrical conductivity. Paraconductivity

As in the case of the specific heat, also the electrical conductivity is usually enhanced by a term $\delta\sigma$ due to thermodynamic fluctuations near a superconducting transition. This phenomenon is called *paraconductivity*, in analogy with paramagnetism. This enhancement appears as a rounding of the resistivity jump at T_c, which is visible in Fig. 4. This phenomenon was experimentally observed for the first time by Glover (1967) in stabilised amorphous bismuth films over a temperature range of ≈ 3 mK. The experiment was suggested by Ferrel and Schmidt (1967) to verify their predictions on critical phenomena near the superconducting transition. These predictions were based on scaling theory (Widom, 1965; Kadanoff, 1966). Hence, they are valid in the region of critical fluctuations like, for example, in the region near the lambda point of ^{4}He. Ferrel and Schmidt (1967) proposed that $\delta\sigma$ diverges near T_c as the correlation length $\delta\sigma \sim \xi \sim t^{-2/3}$. Later, Lobb (1987) proposed the same dependence and a cross–over of the critical exponent to $-1/3$ in the immediate vicinity of the transition on the basis of the full dynamic theory by Hohenberg and Halperin (1977). Glover's results motivated Aslamazov and Larkin (1968) to formulate predictions on the effects of the fluctuations in the framework of the BCS theory. They calculated the fluctuation contribution to the electrical conductivity, to the specific heat and to the sound absorption coefficient in first approximation of perturbation theory. Their results can be obtained also from the Ginzburg–Landau time–dependent equations, as showed for the first time by Abrahams and Woo (1968); the same results have indeed been partly obtained by Levanyuk (1963) with this method.

In what follows, only the results by Aslamazov and Larkin are recalled. They found that the most singular diagram describing the response function operator $\delta\hat{Q}_\omega \equiv i\omega\delta\hat{\sigma}_\omega$ corresponds to the excess conductivity of fluctuating Cooper pairs. The contribution of this diagram to the dc conductivity ($\omega = 0$) is found to have the following expression

$$\delta\sigma \propto -\int_{|\mathbf{k}| \leq |\tilde{\mathbf{k}}|} (\mathbf{k}\mathbf{u})\left(\mathbf{u}\frac{\partial}{\partial \mathbf{k}}\langle|\delta\psi_\mathbf{k}|^2\rangle^2\right)\frac{d^d k}{(2\pi)^d} \qquad (19)$$

where $\mathbf{u}$ denotes the unit vector along the electric field vector. As in the previous case, a *universal* result is obtained in the long–wavelength limit $\tilde{\xi}\tilde{k} \gg 1$. Within this approximation, by inserting eq. (9) into the above integral, Aslamazov and Larkin obtained the result reported in Table 1

$$\delta\sigma \sim t^{d/2-2} \qquad (20)$$

The proportionality factor is

$$\begin{cases} \dfrac{e^2}{32\hbar\xi_0} & d = 3 \\ \dfrac{e^2}{16\hbar s} & d = 2 \\ \dfrac{\pi e^2 \xi_0}{16\hbar S} & d = 1 \end{cases} \qquad (21)$$

where $s \ll \xi$ and $S \ll \xi^2$ are respectively the thickness of the film and the cross–sectional area of the whisker in the two– and one–dimensional cases respectively. As in the case of the specific heat, we note that the amplitude of the fluctuating term depends on the microscopic parameters of the system. Besides the diagram corresponding to eq. (19), the Aslamazov–Larkin theory predicts the existence of three other diagrams which are expected to give smaller contributions under ordinary conditions. Two of them give a negative contribution, which arises from the reduction of normal carrier density due to the appearance of fluctuating Cooper pairs. The third diagram gives an additional positive (paraconductivity) contribution, called Maki–Thompson or 'indirect' contribution, since it is associated with the interaction of fluctuating Cooper pairs with normal carriers. The reader can refer to Skocpol and Tinkham (1975) and to the original papers by Maki (1968; 1971) and by Thompson (1970) for further reading.

5. APPLICATION OF STANDARD FLUCTUATION MODELS TO CONVENTIONAL SUPERCONDUCTORS

Fluctuation effects in clean three–dimensional superconductors of the conventional BCS type are found to be too small to be observed, in agreement with eq. (11) and eq. (13). These effects are enhanced in dirty samples and in samples with reduced dimensionality and were indeed detected in stabilised amorphous films of simple metals, such as Bi, Pb and Ga, and alloys, such as Bi–Sb. The most studied properties were the fluctuation of specific heat, electrical conductivity and magnetic susceptibility (fluctuation diamagnetism), probably because the effects observed are larger more easily interpreted than in the case of other properties. In most reports, the experimental results were analysed within the framework of models based on the phenomenological Ginzburg–Landau theory or on the microscopic BCS theory. We recall that these two theories are equivalent in the transition region (Gor'kov 1959). The predictions of these theories, which are summarised above for the specific heat and for the electrical conductivity, are based on the simplest Gaussian approximation. Nevertheless, they are successful in accounting both qualitatively and quantitatively for the experimental data of most fluctuation properties, as reported in the review paper by Skocpol and Tinkham (1975). In particular, we mention the report by Glover (1969) showing the agreement between the universal Aslamazov–Larkin prediction in two–dimensions for the paraconductivity (see eq. (20) and eq. (21)) and the experimental behaviour observed in a series of amorphous films. More recently, it was reported by Xiang *et al.* (1993) that also Rb– and K–doped C_{60} follow the above predictions in three–dimensions, suggesting a rather conventional BCS behaviour in this new class of superconductors. On the other hand, experimental evidence for critical fluctuations of the conductivity according to the predictions by Ferrel and Schmidt (1967) was reported by Glover (1967) on amorphous Bi films in a temperature region a few mK above $T_c \approx 6$ K. Also in the case of the fluctuation specific heat, the universal prediction based on the Gaussian approximation (see eq. (18)) was found to account for the experimental data on conventional superconductors, such as Bi–Sb films (Zally and Mochel, 1971;1972). The same authors reported experimental evidence for a crossover to critical fluctuations in a temperature region close to the transition, as in the case of the fluctuation conductivity. On the other hand, a disagreement between predictions based on the Gaussian approximation and experimental data on conventional BCS superconductors was observed in the case of fluctuation diamagnetism. A universal prediction based on the Ginzburg–Landau theory was formulated by Prange (1970) after the first measurements reported by Gollub, Beasley, Newbower and Tinkham (1969) on several dirty metals. The departure of the experimental data from the above prediction was explained by taking into account a short wavelength cutoff in the fluctuation spectrum (Gollub, Beasley and Tinkham, 1970), thus suggesting the breakdown of the long–wavelength approximation. Similar conclusions were drawn by Johnson, Tsuei and Chaudhari (1978) to account for analogous deviations from the universal Aslamazov–Larkin predictions observed in the paraconductivity of amorphous alloys. We note that, in both cases, such cutoff effects were observed in samples where the correlation length was greatly reduced by the amorphous structure down to 5–10 nm.

6. APPLICATION OF STANDARD FLUCTUATION MODELS TO CUPRATES. CONTROVERSIAL RESULTS

In the preceding section, we have mentioned that some universal predictions based on the Gaussian approximation fail to account for the experimental behaviour of fluctuations in amorphous samples of conventional BCS superconductors with $\xi_0 \approx$ 5–10 nm. In this section, we discuss the same issue for cuprates. For these superconductors we expect an even larger departure of the experiments from the above universal predictions because of the extremely short correlation length. The short ξ follows directly from the high T_c through Heisenberg's uncertainty principle $\xi_0 \approx \hbar v_F/\Delta$, where v_F is the Fermi velocity and $\Delta \sim T_c$ is the superconducting energy gap. The expected small ξ is confirmed by susceptibility (Lee, Klemm and Johnston, 1989), specific heat (Lee and Ginsberg, 1992) and magnetoresistance (Matsuda, Hirai and Komiyama, 1988) measurements and is estimated to be ~1 nm in the *ab* plane of the CuO_2 planes and ~0.1 nm in the perpendicular (or *c*) direction. Under these conditions, *we expect that the universality principle and the scaling hypothesis are no longer valid*, at least in the *c* direction. In spite of this, with the exception of two studies (Freitas, Tsuei and Plaskett, 1988; Hopfengärtner, Hensel and Saemann–Ischenko, 1991), in previous reports, critical phenomena in cuprates have been extensively studied in several compounds by assuming the universal predictions of mean–field and scaling theories as valid. In these reports, the application of the models to the experimental data was aimed at studying the dimensionality *d* of the fluctuations by taking into account the dependence of the critical exponents on *d* (see Table 1) and at verifying whether the fluctuations are weak (Gaussian) or strong (critical). The motivation for these studies is that the electronic structure of cuprates in their normal state is quasi–two dimensional due to the presence of electronically active CuO_2 planes in their crystal structure. Therefore the role of the out–of–plane direction in determining the superconducting properties of these materials is considered as one of the fundamental open question related to high–temperature superconductivity in cuprates. Second, the short coherence length of cuprates combined with their small density of Cooper pairs has raised the question of the existence of critical fluctuations in the whole fluctuation region and, consequently, the breakdown of the mean–field regime. The experimental results of fluctuation measurements reported show a good reproducibility but the conclusions are highly controversial. The controversy concerns not only the dimensionality of the fluctuation spectrum, but also the number of components of the order parameter and the possible relevance of critical fluctuations. In what follows, we mention the most controversial points concerning 1) the specific heat in zero magnetic field and 2) in magnetic field; 3) the magnetic penetration depth and 4) the paraconductivity.

1) Specific heat in zero magnetic field

According to Table 1, in the case of Gaussian fluctuations, we would expect no divergence of the mean–field jump and a divergence of the fluctuation contribution with the critical exponent predicted by eq. (19). On the other hand, in the case of critical fluctuations described by the 3dXY model, a logarithmic divergence of the jump is expected. Inderhees *et al.* (1988) and Inderhees, Salamon, Rice and Ginsberg (1991) have reported experimental data on a $YBa_2Cu_3O_{6+x}$ single crystal in agreement with the first prediction, while the analysis of similar data on the same compound by Regan, Lowe and Howson (1991) favoured the second prediction. On the basis of a detailed analysis, Roulin, Junod and Walker (1995) conclude that both models provide an account of the experimental data on $YBa_2Cu_3O_{6+x}$.

2) Specific heat in magnetic field

Both theories mentioned above predict a universal scaling of the specific heat jump as a function of magnetic field. Also in this case, controversial interpretations were reported: the papers by Salamon, Shi, Overend and Howson (1993) and by Overend, Howson and Lawrie (1994) supported a 3dXY scaling picture, while Roulin, Junod and Muller (1995) and Junod (1996) showed that neither the Ginzburg–Landau nor the 3dXY scaling theory can be ruled out, as in the case of zero–field.

3) Magnetic penetration depth

Near the transition, the magnetic penetration depth is expected to diverge with a critical exponent $-1/2$ or $-1/3$ according to the Ginzburg–Landau and 3dXY scaling theories respectively (see Table 1). Measurements consistent with the former were reported on $YBa_2Cu_3O_{6+x}$ by Fiory, Hebard, Mankiewich and Howard (1988) and by Lin *et al.* (1995), while Kamal *et al.* (1994) reported experimental data supporting 3dXY behaviour.

4) Fluctuation conductivity. Paraconductivity

Several theoretical and experimental works on the paraconductivity in cuprates have been reported. Most experimental data were on $YBa_2Cu_3O_{6+x}$ but some data are available also on $La_{2-x}Ba_xCuO_4$ and on $Bi_2Sr_2CaCu_2O_{8+\delta}$. A summary on this subject goes beyond the scope of this work and can be found elsewhere (Gauzzi and Pavuna, 1995). Here we mention the most important and controversial conclusions reported in the literature. As to the nature of the fluctuations, no report supports the existence of critical fluctuations in zero magnetic field, while all reports are in favour of a predominant Aslamazov–Larkin contribution. In some cases, experimental evidence for an additional Maki–Thompson contribution was also reported. As to the dimensionality d of the fluctuation spectrum, all possible pictures have been supported in the case of $YBa_2Cu_3O_{6+x}$, according to the universal predictions by Aslamazov and Larkin: 1) 3d–fluctuations (Freitas, Tsuei and Plaskett, 1988; Xi *et al.*, 1989; Veira and Vidal, 1989; Sudhakar *et al.*, 1991); 2) 2d–fluctuations (Ausloos and Laurent, 1988; Hagen, Wang and Ong, 1988; Lang *et al.*, 1991); 1d–fluctuations (Ying and Kwok, 1990); 4) a cross–over from 2d to 3d associated with the layered structure of the cuprates (Oh *et al.*, 1988; Friedmann, Rice, Giapintzakis and Ginsberg, 1989; Gasparov, 1991; Baraduc *et al.*, 1992). We shall not mention the reports on the fluctuation conductivity in magnetic field nor the fluctuations of other quantities, such as the magnetic susceptibility, since this would add no substantially new elements to our discussion.

7. EVIDENCE FOR THE BREAKDOWN

OF THE UNIVERSALITY PRINCIPLE IN SHORT–ξ SYSTEMS

On the basis of the above summary of the state of the art on thermodynamic fluctuations in cuprates, we conclude that this field remains highly controversial. We argue that the controversy arises from three circumstances:
1) the different methods used to analyse the experimental data;
2) the extrinsic broadening of the transition due to local inhomogeneities and spatial distribution of T_c. In $YBa_2Cu_3O_{6+x}$, this can be easily produced by local deviations of the oxygen content x;
3) the assumption of validity of the universality principle to account for critical phenomena in cuprates, while this principle can not be applied to systems with sufficiently short ξ.

The first point is discussed in detail elsewhere (Gauzzi and Pavuna, 1995). The second point does not deserve any further comment. In this section, we draw our attention to the third point. In particular, we show that the controversial results obtained in cuprates can be explained by noting that both requirements $\xi\tilde{k} \gg 1$ and low anisotropy, which justify the universality principle and the scaling hypothesis, are not fulfilled in these materials. It follows that the critical behaviour of all fluctuating quantities generally depends on the cutoff of the fluctuation spectrum (Ivanchenko and Lisyanskii, 1992; Gauzzi, 1993). We shall discuss only the case of the fluctuation of the conductivity, since this is the only quantity for which the analysis of the experimental data is relatively simple. A numerical calculation of the Aslamazov–Larkin expression (see eq. (20) and eq. (21)) with the inclusion of such a cutoff was reported by Freitas, Tsuei and Plaskett (1988) and by Hopfengärtner, Hensel and Saemann–Ischenko (1991) to account for the deviation of their experimental data on $YBa_2Cu_3O_{6+x}$ from the universal Aslamazov–Larkin prediction in the long wavelength approximation $\xi\tilde{k} \gg 1$. This generalised approach did indeed improve the agreement between theory and experiment at high temperatures but not at low temperatures, i.e. in the region close to the transition. In this region, the above authors assumed that the universal behaviour predicted by the long–wavelength approximation is recovered. This discrepancy can be explained by an exact calculation of the Aslamazov–Larkin term (Gauzzi, 1993) which shows that, in the case of strongly anisotropic systems like cuprates, cutoff effects can be relevant not only at high temperatures, but also very close to the transition. In this case, a deviation from the universal predictions of eq. (20) and of eq. (21) is expected in the *whole* fluctuation region accessible to experiments. This possibility was indeed verified experimentally in $YBa_2Cu_3O_{6+x}$ (Gauzzi and Pavuna, 1995), as we can note from Fig. 5a. The same data are found to deviate also from the universal prediction of the scaling theory (see Table 1) in the whole fluctuation region down to 5 mK above T_c (see Fig. 5b). On the other hand, a quantitative account for the experimental data is given by the aforementioned exact calculation at both high and low temperatures down to ≈ 0.5 K above T_c (see Fig. 6a–b). We note that the results of the foregoing analysis are independent of the procedure of data analysis. In particular, the absence of a well–defined critical exponent is evident in the linear plot of the raw experimental data (see Fig. 5a). The usual procedure of determining the critical exponent from logarithmic plots can be misleading, since the slope of these plots is affected by the choice of T_c. Nevertheless, the use of logarithmic plots turns out to be useful *after* the data analysis to put into evidence the absence of a well–defined critical exponent (see Fig. 6b).

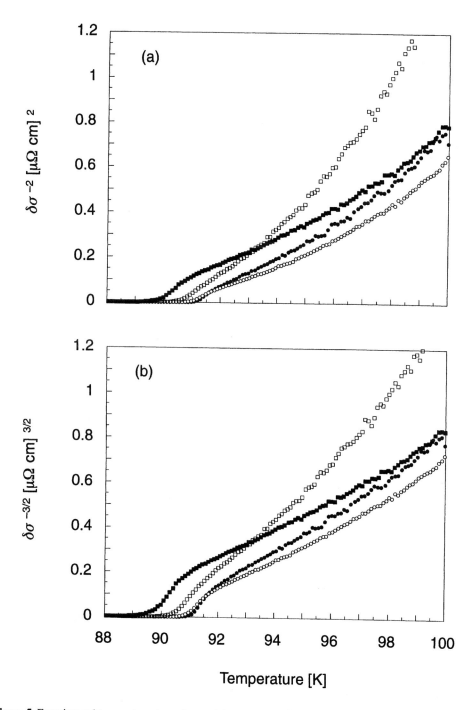

Figure 5. Experimental temperature dependence of the excess conductivity (paraconductivity) of four epitaxial $YBa_2Cu_3O_{6+x}$ films. **a)** The data are plotted with exponent 2 to verify a possible linear temperature dependence, which would support the universal prediction of mean–field theory in three dimensions (see Table 1, eq. (20) and eq. (21)). **b)** The same as in a) for the prediction of scaling theory $\delta\sigma \sim t^{-2/3}$ (see Table 1). Neither in a) nor in b) a linear dependence is observed, similarly to the case of the predictions of mean–field theory in one and in two dimensions, which are not reported here. This analysis suggests the failure of the universality principle in $YBa_2Cu_3O_{6.9}$. (Reproduced by permission of the American Physical Society from Gauzzi and Pavuna (1995)).

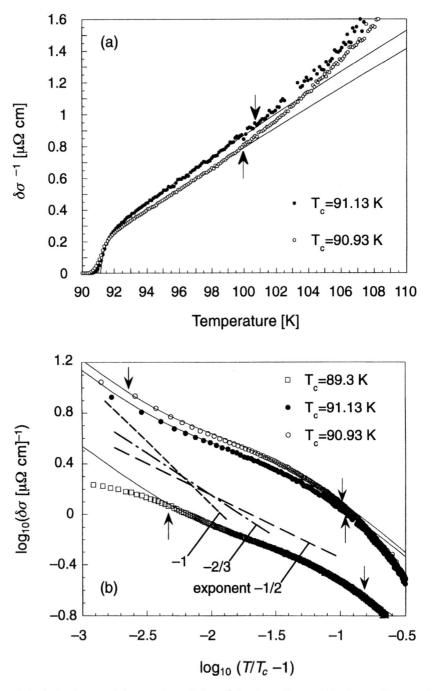

Figure 6. Analysis of some of the experimental data of the above figure within the cutoff approach. The result of the analysis is represented in both linear (a) and logarithmic (b) scales. In b), the deviations of the data from the universal predictions of mean–field and scaling theories are easily noted. These predictions are represented schematically by the broken lines with slopes −1/2, −2/3 and −1, corresponding to the critical exponents predicted by respectively mean–field theory in 3d, scaling theory in 3d and mean–field theory in 2d (see Table 1). The solid lines represent a fit of the data by using the exact calculation of the Aslamazov–Larkin term with cutoff included (Gauzzi 1993). The arrows indicate the range of temperatures where the data have been analysed. The values of the critical temperatures of the samples are indicated by the legend. (Reproduced by permission of the American Physical Society from Gauzzi and Pavuna (1995)).

The above results indicate that, in the short–ξ superconductor YBa$_2$Cu$_3$O$_{6+x}$, the correlation length ξ of the fluctuations never becomes sufficiently large with respect to the interatomic distance to make the critical behaviour universal, i.e. independent of the microscopic interaction driving the transition. An experimental support for this picture could be provided by a study of the disorder–induced reduction of T_c in YBa$_2$Cu$_3$O$_{6+x}$ films (Gauzzi and Pavuna, 1997; Gauzzi, Jönsson, Clerc–Dubois and Pavuna, 1997). In this report, we found that the minimum size r_c of structurally ordered domains necessary for obtaining the maximum value of T_c ($\approx$ 91 K) is $r_c \approx$ 14 nm (see Fig. 7). In more disordered samples, r_c is reduced and a rapid reduction of T_c is observed. The interpretation of these data is that the correlation length ξ of the superconducting order parameter in copper–oxide metals is limited by the value of the structural correlation length r_c. This is due to the strong directionality of the 3d orbitals of the copper in the CuO$_2$ planes forming the conduction band, while in conventional metals the conduction band is formed by the isotropic s–orbitals or by the less directional p–orbitals. As a result, the divergence of the superconducting correlation length ξ at the transition (see eq. (12)) is cut off by the value of the structural correlation length r_c, thus explaining the observed reduction of T_c in disordered samples. By assuming that this interpretation is valid, we conclude that r_c is a probe of the divergence of ξ at the transition. From the diagram of Fig. 7, we estimate that ξ does not become larger than $\approx$ 14 nm at the transition in well–ordered YBa$_2$Cu$_3$O$_{6+x}$ with $T_c \approx$ 91 K. This is a surprisingly small value, which suggests the breakdown of the universality principle and of the scaling hypothesis in cuprates, in agreement with the experimental observations discussed above.

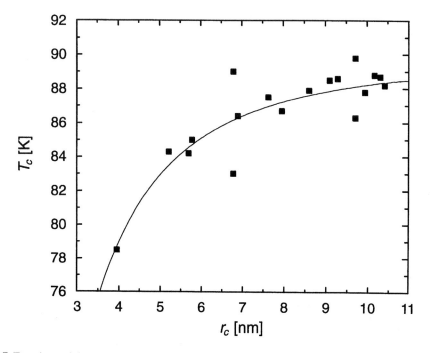

Figure 7. Experimental dependence of the disorder–induced reduction of critical temperature as a function of structural correlation length r_c in epitaxial YBa$_2$Cu$_3$O$_{6.9}$ films. The solid line represents the predicted behaviour by assuming that the square–root divergence of the Ginzburg–Landau superconducting coherence length ξ at the transition is limited by r_c. According to this interpretation, the above data indicates that, in YBa$_2$Cu$_3$O$_{6.9}$, ξ increases up to only $\approx$ 14 nm at the transition in well–ordered samples (Gauzzi, Jönsson, Clerc–Dubois and Pavuna (1997)).

8. SUMMARY. DISCUSSION AND CONCLUSIONS

In this paper we have presented experimental evidence for the deviation of the fluctuation behaviour of the cuprate superconductor $YBa_2Cu_3O_{6+x}$ from the universal predictions derived from both mean–field and scaling theories. Similar deviations were observed in other cuprates, such as $La_{2-x}Ba_xCuO_4$ and $Bi_2Sr_2CaCu_2O_{8+\delta}$ and also in amorphous alloys. In particular, no well–defined critical exponents of the fluctuation conductivity (paraconductivity) were found in the whole fluctuation region accessible to experiments by using a method of data analysis which is independent of any arbitrary choice of T_c. We have explained such a breakdown of universal predictions by applying an exact cutoff scheme to the mean–field fluctuation spectrum. The relevance of cutoff effects in cuprates is due to the extremely short correlation length ξ of the order parameter in these materials. This implies that ξ remains too small at the transition to allow the application of the universality principle to the critical behaviour at the transition. This circumstance is supported by an independent study carried out on the disorder–induced reduction of T_c in $YBa_2Cu_3O_{6+x}$ films. This study suggests that ξ increases up to only ≈ 14 nm at the transition in well–ordered $YBa_2Cu_3O_{6+x}$. This microscopic size is unusual for a coherent quantum state such as the superconducting state. Therefore, the expression 'divergence of the correlation length at the transition', which is commonly used in the field of second order phase transitions, does not seem to be appropriate for cuprates.

The loss of universality induced by the short ξ is not a negative statement as it might seem at first glance. Although the universality class of the effective Hamiltonian can no longer be recognised because of the non–universal behaviour at the transition, this behaviour is interesting, since it is distinctive of the type and of the parameters of the microscopic interaction driving the transition. For example, our study of the fluctuation spectrum within the cutoff approach leads to the *qualitative* conclusion that the fluctuations are rapidly attenuated with decreasing wavelength as compared to the case of conventional superconductors with long ξ (see Fig. 2). This conclusion is in agreement with previous strong–coupling calculations of the BCS–Eliashberg equations (Bulaevskii and Dolgov, 1988). Hence, the microscopic picture emerging from our fluctuation data is that of pair correlations between localised states, such as those predicted by the bipolaron theory of superconductivity of Alexandrov and Mott (1994).

We recall that our fluctuation measurements probe only the integral of the fluctuation spectrum over the entire momentum space (see eq. (17) and eq. (19)). It would therefore be essential to determine more precisely the momentum dependence of the spectrum of cuprates and compare it with the Lorentzian dependence of the conventional mean–field spectrum (see eq. (11)). This would enable us to determine the corrective terms of the conventional BCS Hamiltonian which are expected to give a better description of the superconducting state of cuprates. In this respect, a systematic study of other fluctuation properties, such as the fluctuation diamagnetism, would be important. Fluctuation studies on other short–ξ superconductors, such as Bi–based oxides, Chevrel phases and alkali–doped C_{60} would also be useful for exploring possible new aspects of the physics of phase transitions in short–ξ systems.

The author acknowledges A. Junod for providing the experimental data of Fig. 3, M. Bauer, V. A. Gasparov, M. Howson, D. Pavuna, A. Rigamonti and A. A. Varlamov for useful discussions and J. Berrocosa for his experimental help.

REFERENCES

Abrahams, E. and Woo, 1968, J., Phys. Lett. **27A**, 117.
Alexandrov, A. S. and Mott, 1994, N., Rep. Prog. Phys. **57**, 1197.
Aslamazov, L. G. and Larkin, A.I., 1968, Fiz. Tverd. Tela **10**, 1104 [Sov. Phys. Solid State **10**, 875 (1968)].
Ausloos, M. and Laurent, Ch., 1988, Phys. Rev. B **37**, 611.
Baraduc, C., Pagnon, V., Buzdin, A., Henry, J. Y. and Ayache, C., 1992, Phys. Lett. A **166**, 267.
Bulaevskii, L. N. and Dolgov, O. V., 1988, Solid State Commun. **67**, 63.
Coleman, S. and Weinberg, E., 1973, Phys. Rev. D **7**, 1988.
de Gennes, P. G., 1972, Solid State Commun. **10**, 753.
Ferrel, R. A. and Schmidt, H., 1967, Phys. Lett. **25A**, 544.
Fiory, A. T., Hebard, A. F., Mankiewich, P. M. and Howard, R. E., 1988, Phys. Rev. Lett. **61**, 1419.
Freitas, P. P., Tsuei, C. C. and Plaskett, T. S., 1988, Phys. Rev. B **36**, 833.
Friedmann, T. A., Rice, J. P., Giapintzakis, J. and Ginsberg, D. M., 1989, Phys. Rev B **39**, 4258.
Gasparov, V. A., 1991, Physica C **178**, 445.
Gauzzi, A., 1993, Europhys. Lett. **21**, 207.
Gauzzi, A. and Pavuna, D., 1995, Phys. Rev. B **51**, 15420.
Gauzzi, A. and Pavuna, D., 1997, J. of Superconductivity, in press.
Gauzzi, A., Jönsson, B. J., Clerc–Dubois, A. and Pavuna, D., 1997, preprint.
Ginzburg, V. L., 1960, Fiz. Tverd. Tela **2**, 2031 [Sov. Phys. Solid State **2**, 1824 (1960)].
Glover, R. E., 1967, Phys. Lett. **25A**, 542.
Glover, R. E., 1969, *Proc. 11th Int. Conf. on Low Temperature Physics, St. Andrews, 1968,* Allen, J. F., ed.
Gollub, J. P., Beasley, M. R., Newbower, R. S. and Tinkham, M., 1969, Phys. Rev. Lett. **22**, 1288.
Gollub, J. P., Beasley, M. R. and Tinkham, M., 1970, Phys. Rev. Lett. **25**, 1646.
Gor'kov, L. P., 1959, Z. Eksp. Theor. Fiz. **36**, 1918 [Sov. Phys. JETP **36**, 1364 (1959)].
Hagen, S. J., Wang, Z. Z. and Ong, N. P., 1988, Phys. Rev. B **38**, 7137.
Hohenberg, P. C. and Halperin, B. I., 1977, Rev. Mod. Phys. **49**, 435.
Hopfengärtner, R., Hensel, B. and Saemann–Ischenko, G., 1991, Phys. Rev. B **44**, 741.
Inderhees, S. E., Salamon, M. B., Goldenfeld, N., Rice, J. P., Pazol, B. G., Ginsberg, D. M., Liu, J. Z. and Crabtree, G. W., 1988, Phys. Rev. Lett. **60**, 1178.
Inderhees, S. E., Salamon, M. B., Rice, J. P. and Ginsberg, D. M., 1991, Phys. Rev. Lett. **66**, 232.
Ivanchenko, Yu. M. and Lisyanskii, A. A., 1992, Phys. Rev. A **45**, 8525.
Johnson, W. L., Tsuei, C. C. and Chaudhari, P., 1978, Phys. Rev. B **17**, 2884.
Junod, A., 1990, in *Physical Properties of High Temperature Superconductors II*, Ginsberg, D. M. ed. (World Scientific, Singapore).
Junod, A., 1996, in *Studies of High Temperature Superconductors Vol. 19*, Narlikar, A. V., ed. (Nova Science, Commack New York).
Kadanoff, L. P., 1966, Physics **2**, 263.
Kamal, S., Bonn, D. A., Goldenfeld, N., Hirschfeld, P. J., Liang, R. and Hardy, W. N., 1994, Phys. Rev. Lett. **73**, 1845.
Kapitulnik, A., Beasley, M. R., Di Castro, C. and Castellani, C., 1988, Phys. Rev. B **37**, 537.
Landau, L. D., 1937, Zh. Eksp. Theor. Fiz. **7**, 19 [Phys. Z. der Sowjet Union **11**, 26 (1937).
Landau, L. D. and Lifshitz, E. M., 1980, *Statistical Physics, part I* (Pergamon, Oxford).
Lang, W., Heine, G., Jodlbauer, H., Schlosser, V., Markowitsch W., Schwab, P., Wang, X.Z. and Bäuerle, D., 1991, Physica C **185–189**, 1315.
Lee, W. C. and Ginsberg, D. M., 1992, Phys. Rev. B **45**, 7402.
Lee, W. C., Klemm, R. A. and Johnston, D. C., 1989, Phys. Rev. Lett. **63**, 1012.
Levanyuk, A. P., 1959, Zh. Eksp. Teor. Fiz. **36**, 810 [Sov. Phys. JETP **36**, 571 (1959)].
Levanyuk, A. P., 1963, Fiz. Tverd. Tela **5**, 1776 [Sov. Phys. Solid State **5**, 1294 (1964)].
Lin, Z.-H, Spalding, G. C., Goldman, A. M., Bayman, B. F. and Valls, O.T., 1995, Europhys. Lett. **32**, 573.
Lobb, C. J., 1987, Phys. Rev. B **36**, 3930.
Ma, S.-K., 1976, *Modern Theory of Critical Phenomena* (Benjamin, New York).
Maki, K., 1968, Prog. Theor. Phys. **40**, 193.
Maki, K., 1971, Prog. Theor. Phys. **45**, 1016.
Matsuda, Y., Hirai, T. and Komiyama, S., 1988, Solid State Commun. **68**, 103.

Oh, B., Char K., Kent, A. D., Naito, M., Beasley, M. R., Geballe, T. H., Hammond, R. H., Kapitulnik, A. and Graybeal, J. M., 1988, Phys. Rev. B **37**, 7861.
Overend, N., Howson, M. A. and Lawrie, I. D., 1994, Phys. Rev. Lett. **72**, 3238.
Patashinskii, A. Z. and Pokrowskii, V. L., 1979, *Fluctuational Theory of Phase Transitions* (Pergamon, Oxford).
Prange, R. E., 1970, Phys. Rev. B **1**, 2349.
Regan, S., Lowe, A. J. and Howson, M. A., 1991, J. Phys.: Condens. Matter **3**, 9245.
Roulin, M., Junod, A. and Muller, J., 1995, Phys. Rev. Lett. **75**, 1869.
Roulin, M., Junod, A. and Walker, E., 1996, Physica C **260**, 257.
Salamon, M. B., Shi, J., Overend, N. and Howson, M. A., 1993, Phys. Rev. B **47**, 5520.
Skocpol, W. J. and Tinkham M., 1975, Rep. Prog. Phys. **38**, 1049.
Stanley, H. E., 1971, *Introduction to Phase Transitions and Critical Phenomena* (Oxford University Press, Oxford).
Sudhakar, N., Pillai, M. K., Banerjee, A., Bahadur, D., Das A., Gupta, K. P., Sharma, S. V. and Majumdar, A. K., 1991, Solid State Commun. **77**, 529.
Thompson, R. S., 1970, Phys. Rev. B **1**, 327.
Veira, J. A. and Vidal, F., 1989, Physica C **159**, 468.
Widom, B., 1965, J. Chem. Phys. **43**, 3892.
Wilson, K. G., 1971, Phys. Rev. B **4**, 3174.
Xi, X. X., Geerk, J., Linker, G., Li, Q. and Meyer, O., 1989, Appl. Phys. Lett. **54**, 2367.
Xiang, X. D., Hou, J. G., Crespi, V. H., Zettl, A. and Cohen, M., 1993, Nature **361**, 54.
Ying, Q. Y. and Kwok, H. S., 1990, Phys. Rev. B **42**, 2242.
Zally, G. D. and Mochel, J. M., 1971, Phys. Rev. Lett. **27**, 1710.
Zally, G. D. and Mochel, J. M., 1972, Phys. Rev. B **6**, 4142.

MICROWAVE RESPONSE OF HIGH-T_C SUPERCONDUCTORS

A. Agliolo Gallitto, I. Ciccarello, M. Guccione, M. Li Vigni, and
D. Persano Adorno

Istituto Nazionale di Fisica della Materia, Unità di Palermo and Istituto di Fisica dell'Università, Via Archirafi 36, I-90123 Palermo, Italy

INTRODUCTION

The study of the microwave response by high-T_c superconductors (HTSC) is of interest to the understanding of the processes responsible for microwave energy losses in such materials. The investigation is commonly performed by measuring the power absorption with standard EPR spectrometers or, alternatively, the surface impedance. A further method concerns the study of the nonlinear response; in this case the samples are exposed to an intense microwave (mw) field oscillating at the fundamental frequency ω and signals radiated at harmonic frequencies of the driving field are detected. Measurements of harmonic generation allow to investigate the mechanisms responsible for energy dissipation in HTSC, yelding complementary information to the direct power absorption method. In this paper we give a review of the results reported in the literature about the power absorption, the nonlinear response and the surface impedance of HTSC exposed to mw fields.

1. POWER ABSORPTION

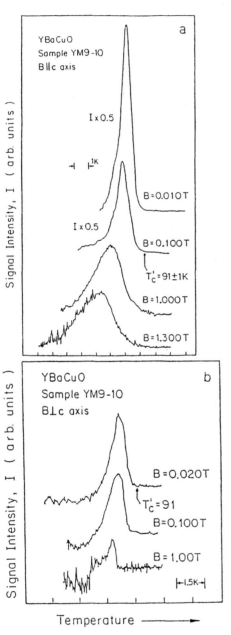

Figure 1. Intensity of the modulated absorption signal as a function of temperature in a YBCO single crystal for different magnetic fields, (a) parallel and (b) perpendicular to the c direction [after Ref. 3].

Microwave losses can be detected by measuring either the power absorption or the resistive component of the surface impedance. In the literature the study has been mainly devoted to investigation of field-induced losses[1-8]. Power absorption measurements are performed by using EPR techniques with, or without, field modulation. The intensity of the unmodulated signal is proportional to the real component of the microwave surface impedance. EPR spectrometers with magnetic field modulation allow to detect the magnetic field derivative of the absorbed power. Modulated and unmodulated signals have been studied in both ceramic[1, 2, 8] and single crystal[3-6] HTSC. Measurements in ceramic samples are mainly performed at low applied fields, where mw losses by Josephson junctions play an important role. On the contrary, measurements in single crystals are generally performed in the mixed state, where dissipation is ascribed to fluxon dynamics. Here we particularly address our attention to results on crystalline samples.

The modulated power absorption has been studied as a function of the temperature[3] and the applied field[4]. For

$Bi_2Sr_2CaCu_2O_{8+\delta}$ (BSCCO) and $YBa_2Cu_3O_{7-\delta}$ (YBCO) single crystals the temperature dependence of the absorption signal shows a sharp peak just below T_c.[3] On increasing the external applied field H_0 the peak broadens, shifts toward lower temperatures and decreases in intensity. Figures 1a and 1b show the mw absorption signals detected by Shaltiel et al.[3] in a sample of YBCO single crystal, for different values of the applied magnetic field, (a) parallel and (b) perpendicular to the crystal c-axis. The field variation of the peak intensity is more enhanced for H_0 parallel than for H_0 perpendicular to the crystal c-axis. A peak with similar features is detected in ceramic and powdered YBCO samples[8]. However, in this materials another more intense peak, extending in a wide range of temperatures below T_c, is also observed[8]. Very likely it is connected to dissipative processes localized in weak-link boundary regions of the sample, where a Josephson coupling may be hypothesized.

The unmodulated power absorption has been studied in YBCO samples of different nature: powders[7], crystals[6] and films[5]. The signal intensity is observed to increase on increasing the temperature and the external field amplitude. The temperature variations occur in a very small range of temperatures below T_c. This behavior has been ascribed to the fast increase of the sample resistivity when the temperature approaches T_c. The field dependence has been accounted for as due to the growing contribution of the vortex motion to energy dissipation at increasing H_0. By way of example, we show in Fig.2 the results obtained by Blackstead et al.[5] in a sample of YBCO film exposed to a mw field oscillating at 12.92 GHz. Their measure-

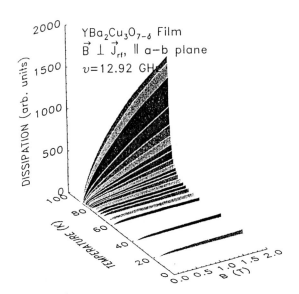

Figure 2. Temperature and field dependencies of the unmodulated absorption signal in a film of YBCO [after Ref. 5].

ments have been performed with the static magnetic field H_0 parallel to $H(\omega)$ and perpendicular to the mw current $J(\omega)$. In this field geometry the Lorentz force between fluxons and $J(\omega)$ is of maximal extent, so that relevant dissipative effects due to the vortex motion are expected.

Results relative to both modulated and unmodulated absorption have been discussed in the framework of different models. The most widely considered models are those elaborated by Tinkham[9] and Coffey and Clem[10]. Tinkham has calculated the dc resistivity of HTSC taking into account flux-flow and flux-creep effects[9]:

$$R(B,T)=R_n[I_0(\gamma)]^{-2}, \qquad (1)$$

where R_n is the normal state resistivity, I_0 is the zero-order Bessel function and

$$\gamma = \frac{A(1-T/T_c)^{3/2}}{2B}, \quad \text{with} \quad A = \frac{UB_{c2}(0)}{KT}; \qquad (2)$$

here U is the pinning barrier energy.

Several authors have used Eq.(1) to account for the observed data of $R(B, T)$ at microwave frequencies[3, 4].

Coffey and Clem (CC) have developed a comprehensive theory for the electromagnetic (em) response of type II superconductors in the mixed state, taking into account flux-creep, flux-flow and pinning effects in the two-fluid model of the superconductivity[10]. Their theory applies for $H_0 > 2H_{c1}$, when the static magnetic field inside the crystal can be supposed as generated by a uniform density of fluxons n_0; in this case $H_0 \approx B_0 = n_0\phi_0$. The mw field inside the sample is characterized by a complex penetration depth, $\tilde{\lambda}$, influenced by the fluxon motion and the very presence of vortices which bring along normal material in their cores. In the linear approximation $[H(\omega) \ll H_0]$ the following expression of $\tilde{\lambda}$ has been obtained[10]:

$$\tilde{\lambda}^2 = \frac{\lambda^2 + i\delta_v^2/2}{1 - 2i\lambda^2/\delta^2}, \qquad (3)$$

where λ is the London penetration depth, δ is the normal fluid skin depth and δ_v is the complex effective skin depth arising from the vortex motion.

Both λ and δ depend on the magnetic field as

$$\lambda = \lambda_0 \{(1-w_0)[1-B_0/B_{c2}(T)]\}^{-1/2} \qquad (4)$$

$$\delta = \delta_0 \{1-(1-w_0)[1-B_0/B_{c2}(T)]\}^{-1/2}, \qquad (5)$$

where $\lambda_0=(mc^2/4\pi ne^2)^{1/2}$ is the London penetration depth at $T=0$ K; $\delta_0=(c^2/2\pi\omega\sigma_0)^{1/2}$ is the normal metal skin depth at $T=T_c$; w_0 and $(1-w_0)$ are the fractions of normal and superconducting electrons at $H_0=0$, with $w_0=(T/T_c)^4$.

At microwave frequencies and for temperatures close to T_c one may assume that vortices are in the flux-flow regime[5, 11]. In this case δ_v is given by

$$\delta_v = \frac{2B_0\phi_0}{4\pi\eta\omega}, \qquad (6)$$

where $\eta=\phi_0 B_{c2}(T)\sigma_0/c^2$ is the viscous drag coefficient of the fluxon motion.

In the London local limit[12] the surface impedance is given by

$$Z_s = R_s + iX_s = i\omega\tilde{\lambda}; \qquad (7)$$

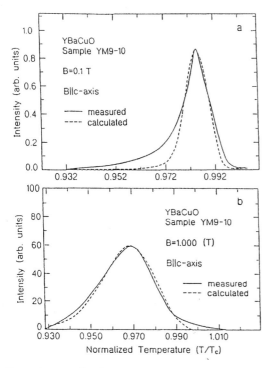

Figure 3. Comparison between the observed modulated absorption signal and the calculated spectrum using Tinkham model, as discussed in the text [after Ref. 3].

so, in the CC model, the field dependence of Z_s and, therefore, of the absorbed power, arises from changes of the complex penetration depth induced by the presence of fluxons and their motion.

Data relative to modulated power absorption in HTSC single crystals has been mainly discussed in the framework of the Tinkham model[3, 4]. On the contrary, results on the field dependence of unmodulated power absorption and surface resistance near T_c have been ascribed to changes of the flux-flow resistance[7, 11] or, alternatively, to changes of the complex penetration depth[5,13] deduced by Coffey and Clem, on the hypothesis that fluxons be in the flux-flow regime.

Shaltiel et al.[3] have discussed their data using the Tinkham model. Figures 3a and 3b show the comparison between the observed microwave signals and the expected curves, for two different orientations of the applied field with respect to the crystal c-axis. The dashed line in Fig.3a has been obtained by calculating the field derivative of Eq.(1) with A=200 T. The fit of Fig.3b has been obtained in a similar way, with A=500 T.

Blackstead et al.[5] have discussed their data in the framework of the CC model. Figure 4 shows the comparison between experimental and expected results for the temperature dependence of the power absorption at a fixed field of 1.5 T, for two different field geometries. The authors have obtained a very good agreement for the configuration $B \perp J(\omega)$. However, the model does not account for the data obtained when $B \| J(\omega)$. This incongruity has also been reported by several authors[6,14]. When the applied field is parallel to the mw current, the Lorentz force between the fluxons and the mw current is nominally zero; therefore, the field variation of R_s should be only due to the effect of the normal fluid[10]. However, the observed variation is larger than the expected one. The effect has been ascribed to two possible causes: a few authors[5,14] explain their data by assuming Josephson junctions to be present between the superconducting planes; others[6] ascribe the result to the fact that, at temperatures close enough to T_c, thermally induced vortex bending may occur, so that the Lorentz force can be locally different from zero, as suggested by Coffey and Clem[10].

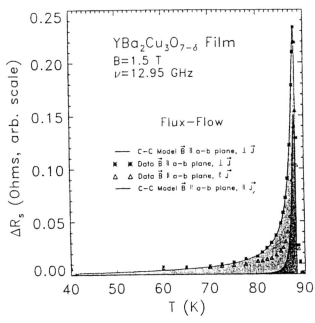

Figure 4. Comparison between the detected unmodulated signal (symbols) and the expected one using the CC model (lines), at different field geometries [after Ref.5].

2. NONLINEAR RESPONSE

When an electromagnetic wave propagates through a material, it induces magnetic (and/or electric) polarization related to the driving field by a magnetic (and/or electric) susceptibility which itself may depend on the em field. Therefore, in the most general case, the induced polarization is expected to be not just linearly dependent on the driving field, with quadratic, cubic, etc., components also present, giving rise to harmonic emission.

High-T_c superconductors exposed to intense em fields are characterized by markedly nonlinear properties; this means that a not negligible fraction of the absorbed power is conveyed into harmonic signals. Harmonic microwave emission has been studied in both ceramic[15-18] and crystalline[19-22] HTSC. It has been shown that there are several mechanisms responsible for the nonlinear response; they depend on the temperature, magnetic field and type of superconductor. Nonlinearity is a peculiar property of the superconducting state: harmonic emission vanishes whenever the superconductors go into the normal state for an increase of either the temperature or the applied magnetic field[15,22]. Furthermore, it has been shown that only the magnetic polarization is responsible for nonlinear emission by HTSC.

In this section we review the main results on nonlinear response of YBCO samples exposed to pulsed microwave fields. In addition, we report preliminary results of third-harmonic emission by $Ba_{0.6}K_{0.4}BiO_3$ (BKBO) crystals.

The experimental apparatus is shown in Fig.5. Basilar element of the nonlinear spectrometer is a bimodal cavity oscillating at the two frequencies ω and $n\omega$, with n=2 and n=3 for the second-harmonic (SH) and third-harmonic (TH) detection, respectively. The samples are located in a region of the cavity where both $H(\omega)$ and $H(n\omega)$ are maximal and parallel to each other. The ω-mode of the cavity is fed by a triode pulse oscillator, with a pulse repetition rate of 100÷1000 pps and a pulse width of 0.5÷5 μsec, giving a maximum peak power of the order of 1kW. A low pass filter at the input of the cavity cuts any harmonic content of the oscillator by more than 60 dB. The detection mode of the cavity is the TE_{102} mode, resonating at 6 GHz, for both n=2 and n=3. The harmonic signals generated by the sample are filtered by a bandpass filter, with more than 60 dB rejection at the fundamental frequency, and are detected by a superheterodyne receiver. The dc magnetic field H_0 can be rotated in a plane containing the magnetic and electric mw fields over the full range (0-360)°. Two additional coils, parallel to the main coils of the magnet and independently fed, are used to reduce the residual field within 0.2 Oe and generate dc

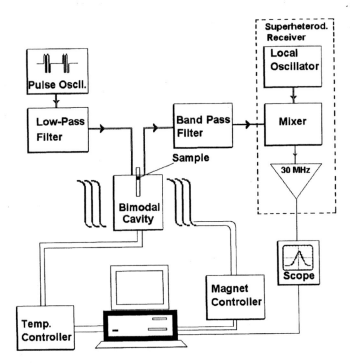

Figure 5. Experimental apparatus for the detection of harmonic emission at microwave frequencies.

fields of a few Oersted. A temperature controller allows to work either at a constant temperature or at increasing (or decreasing) temperature with a constant rate of variation.

2.1 Results at Temperatures far from T_c

Experiments on harmonic emission in ceramic YBCO have shown two different mechanisms responsible for harmonic generation at microwave frequency. For weak external magnetic fields the harmonic response has been ascribed to nonlinearity of the Josephson currents[15-16]. At high fields, where the Josephson junctions are decoupled, harmonic generation has been ascribed to intragranular fluxon dynamics with fluxons in the Bean critical state[18].

Figure 6 shows the SH signal intensity as a function of the temperature in a zero-field-cooled (ZFC) sample of ceramic YBCO in a dc field of 3 Oe and a mw magnetic field of about 5 Oe. At temperatures far from T_c the signal decreases on increasing the temperature, while near T_c the signal shows a maximum just below T_c. A theory which accounts quite well for the experimental results far from T_c has been proposed[16]. However, it does not explain the presence of the peak in the SH emission near T_c. In this theory[16] it is assumed that currents induced in the superconductors form loops involving Josephson junctions. The equation for the loop current component oscillating at the SH frequency of the driving field is solved analytically by considering the self-induced magnetic flux. In this model the temperature dependence of the SH signal intensity is determined by the temperature

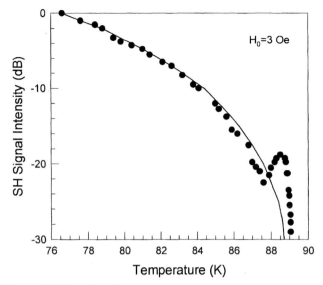

Figure 6. SH signal intensity as a function of temperature in a sample of ceramic YBCO. Input mw field ~5 Oe. Symbols: experimental points; continuous line: expected results using the theory developed in Ref.16.

dependence of the Josephson critical current. The continuous line in Fig.6 is a plot of the expected results: the temperature dependence of the SH signal intensity is well explained except for the peak near T_c.

As regards the field dependence of the SH signal far from T_c, it has been observed that, for ZFC samples, the signal increases sharply with the field, reaches the maximum value at fields of the order of a few Oe and decreases irreversibly when the field intensity is of the order of a hundred Oe[16]. The SH vs H_0 curves obtained at low fields are well accounted for by the model discussed in the Ref.16. At high fields a weak "remanent" signal is detected. This signal is also observed in field-cooled (FC) samples. It has to be investigated at input power levels higher than those used for ZFC low-field signals[18]. Figure 7 shows the intensity of the remanent signal as a function of the magnetic field H_0, at T=4.2 K, for the same sample of Fig.6. The intensity of the signal is field independent in all the range of fields investigated as long as H_0 is increased (or decreased) at constant rate. Sharp minima are observed whenever the magnetic field sweep is reversed, independently of the value of H_0 at which the inversion is operated. It has been shown that the microwave response of a superconductor in a critical state "à la Bean" is expected to be asymmetrical during the period of the em wave[18]. In fact, because of the rigidity of the fluxon lattice, when the sample is in a critical state developed by increasing fields, the induction flux is essentially influenced only during the negative semi-period of the mw field, and *vice versa* for a critical state developed by decreasing fields; so, a magnetization with both odd and even Fourier components comes out. According to this hypothesis, a theory that accounts quite well for the SH signals observed in ceramic YBCO at high fields and low temperatures has been developed[18]: a SH signal displaying the features shown in Fig.7 is expected.

Measurements of SH and TH emission in YBCO single crystals at low fields have shown that only weak harmonic signals are present for $T \ll T_c$. On the other hand, a high-field SH signal with peculiarities similar to the ones of Fig.7 has been detected, showing that a "rectification" process of the mw field occurs also in crystalline samples in the critical state[22]. This signal has been investigated as a function of the temperature for different cooling conditions[22]. It has been ascertained that the SH signal intensity does depend on the cooling conditions; in particular, the SH signal is more intense in ZFC samples than in FC samples. Concerning the field dependence of the SH signal, it has been observed that at T=4.2 K the SH vs H_0 curve is similar to the one of Fig.7 for FC samples as well as for ZFC samples after the first run to high fields has been performed. Nevertheless, on increasing the temperature, the minima are less and less enhanced. Moreover, at temperatures of the order of 10 K, the SH signal detected in ZFC samples is unstable: it decays to a stationary value after a proper time elapses from the instant in which H_0 has been settled at a given value. The decay rate depends on the dc field and increases with the temperature. Finally, above the irreversibility line all signals are stable and reversible, independently of the cooling conditions: these features are peculiar of the peak near T_c of Fig.6. Clearly, a mechanism different from those discussed above is responsible for the harmonic emission at temperatures close to T_c.

The mechanisms discussed up to now are responsible for SH as well as TH emission in HTSC. Though we do not report here the results relative to the TH microwave response in ceramic samples, we remark that the temperature dependence of the TH signal in ceramic

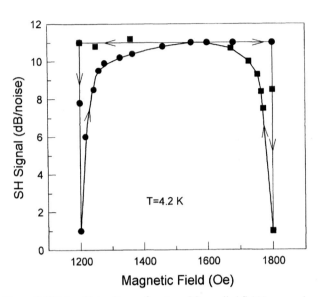

Figure 7. SH signal intensity as a function of the applied field in a sample of ceramic YBCO after the first run at high fields has been performed. Input mw field ~ 10 Oe. Measurements performed at increasing field (●) and decreasing field (■).

YBCO is similar to the one shown in Fig.6.

2.2 Results near T_c

The nonlinear microwave emission at temperatures close to T_c has been investigated to a degree in YBCO single crystals. The temperature dependence of the harmonic signals exhibits an enhanced peak just below T_c for both SH and TH emission. The features of the peak depend on the input power level[21], the value of H_0 and the orientations of $H(\omega)$ and H_0 with respect to the crystal c-axis[23]. In particular, on increasing H_0 the peak broadens and shifts toward lower temperatures[23]. The temperature and field dependencies of the SH and TH signals have similar behavior, except for the fact that the peak in the TH vs temperature curve has maximal intensity at $H_0=0$[21], where, as expected, no SH emission is observed. In order to study the mechanisms responsible for the nonlinear response near T_c, the TH response of YBCO single crystals has been thoroughly investigated in the absence of dc fields[21]. In this case, the signal is for sure not mixed with others arising from Josephson junction nonlinearity or fluxon dynamics. It has been found that the TH signal has maximal intensity when the mw field is parallel to the crystal c-axis and minimal intensity when $c \perp H(\omega)$[21]. Figure 8 shows the temperature dependence of the TH signal intensity for a YBCO single crystal with $c \| H(\omega)$, at $H_0=0$. Features to be remarked in Fig.8 are the enhanced peak just below T_c and the weak signal extending over all the range of temperatures investigated. It was also measured the input power dependence of the

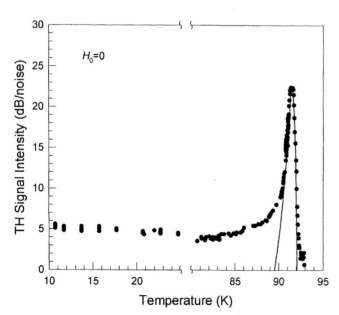

Figure 8. TH signal intensity as a function of temperature in a YBCO single crystal with $c\|H(\omega)$. Input mw field 6.5 Oe. Symbols: experimental points; continuous lines: plots of Eq.(14) using $\lambda_0/\delta_0=0.083$ and $\alpha_1=6.5\times10^{-3}$ Oe^{-1}.

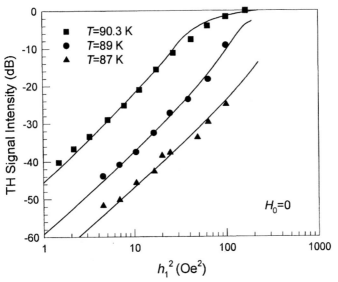

Figure 9. TH signal intensity as a function of input power for a sample of YBCO crystal with $\mathbf{c} \| H(\omega)$, at different values of temperature. Symbols: experimental points; continuous lines: plots of Eq.(14) using the same values of λ_0/δ_0 and α_1 as for Fig.8.

TH signal near T_c at different values of the temperature; the results are shown in Fig.9. As one can see, the power dependence of the TH signal is different at different values of the temperature. Furthermore, the power dependence cannot be described by a mere n-order power law: at a given temperature the slope of the TH vs P_{input} line is not the same in all the range of power levels investigated[21]. In all the cases, a less than cubic power dependence has been found. Further investigation[23] has shown that the intensity of the TH signal near T_c depends on the value of the applied field H_0[23]. On the contrary, the signal detected at temperatures far from T_c is nearly independent of H_0; it varies roughly quadratically with the input power level. Its origin has been not yet fully understood; the signal is too weak to be studied in detail.

The origin of the enhanced peak near T_c is not accounted for by the standard BCS theory in the limit of weak electron-phonon interaction. At present, three models have been put forward concerning the nonlinear microwave emission in HTSC near T_c[19-21, 24], but only two of these predict harmonic emission at zero applied field. Coffey[24] generalized the theory of the em response described in Section 1 to the nonlinear response regime. In the framework of this model harmonic generation is expected arising from the vortex motion, so no harmonic emission is expected at zero dc field. The TH response near T_c in the absence of dc magnetic field has been studied by Leviev and co-workers[19,20] in YBCO single crystals with the c-axis perpendicular to the mw magnetic field. The data has been explained by a model based on the Eliashberg equations[25], in which strong-coupling effects are taken into account. The authors report a cubic dependence of the TH signal on the input power level; this behavior is well accounted for by their model. However, the TH signal detected in YBCO crystals with c parallel to the mw magnetic field[21] shows temperature and input power dependencies which cannot be accounted for by the model proposed by Leviev et al.

A phenomenological theory has been elaborated that well accounts for the TH emission near T_c in the absence of applied field, in all the range of temperatures and input power levels investigated[21]. The theory is based on the two-fluid model, with the additional hypothesis that the mw field modulates the partial concentrations of both normal and superconducting electrons. By assuming that strong coupling is effective in HTSC, when an em field is traveling inside a crystal it perturbs the order parameter, so that electrons are locally out of thermal equilibrium. If the frequency is such that $\omega\tau \ll 1$ (τ is the electron-phonon relaxation time), the partial concentrations of normal and superconducting electrons adjust themselves through thermally activated processes, following adiabatically the em field variations[21]. These processes occur in a surface layer of the order of the field penetration depth; they are effective only at temperatures close to T_c, where the energy gap is not large and the penetration depth is of maximal extent.

In the framework of the two fluid model, when $H_0=0$, the complex penetration depth of the mw field inside a superconductor can be easily inferred from Eq.(3) by setting $B_0=0$. One obtains

$$\tilde{\lambda}^2(B_0=0) = \frac{\lambda^2(B_0=0)}{1-2i\lambda^2(B_0=0)/\delta^2(B_0=0)}, \qquad (8)$$

with:

$$\lambda(B_0=0) = \lambda_0/\sqrt{1-w_0}; \qquad \delta(B_0=0) = \delta_0/\sqrt{w_0}. \qquad (9)$$

The mw field induction obtained using expression (8) does not contain harmonic Fourier components. Nonlinearity comes out if one assumes that the mw field modulates the partial concentrations of the normal and superconducting fluids. Following this idea, it has been set[21]

$$w(t) = w_0 + |\alpha_1 h_1 \cos(\omega t)|, \qquad (10)$$

where h_1 is the amplitude of the mw field and α_1 is a phenomenological parameter. By replacing Eq.(10) into Eqs. (8) and (9) the average mw magnetic induction inside the crystal can be calculated. For a sample of thickness D much greater than the characteristic penetration depths, the mw magnetic induction averaged over the volume of the crystal is given by

$$<B> = \frac{2}{D} Re \int_0^\infty h_1 e^{-i\omega t} e^{-x/\tilde{\lambda}(B_0=0)} dx =$$

$$= -\frac{h_1}{D} \left[Im\{\tilde{\lambda}(B_0=0)\} \cos(\omega t) + Re\{\tilde{\lambda}(B_0=0)\} \sin(\omega t) \right],$$

(11)

where $\tilde{\lambda}(B_0=0)$ is that of Eq.(8) with w_0 of Eqs. (9) replaced by $w(t)$ of Eq.(10). The Fourier coefficients of $<B>$ oscillating at 3ω are given by

$$a_3 = \frac{1}{\pi} \int_0^{2\pi} <B> \cos(3\omega t) d(\omega t),$$

(12)

$$b_3 = \frac{1}{\pi} \int_0^{2\pi} <B> \sin(3\omega t) d(\omega t).$$

(13)

From a_3 and b_3 one can easily calculate the induced magnetization oscillating at 3ω and, eventually, the intensity of the TH signal. This is proportional to the power emitted by the sample at 3ω:

$$P(3\omega) \propto [M(3\omega)]^2 = (a_3^2 + b_3^2) \cos^2(3\omega t + \phi_3),$$

(14)

where

$$\tan\phi_3 = b_3/a_3.$$

The continuous lines in Figs.8 and 9 are plots of Eq.(14) with $\lambda_0/\delta_0 = 0.083$ and $\alpha_1 = 6.5 \times 10^{-3} Oe^{-1}$.

As one can see, the model accounts quite well for both the temperature and the power dependencies of the TH signal detected at temperatures close to T_c[21].

For a comparison, we report in Fig.10 preliminary results on TH emission in a crystal of BKBO, at zero dc field. The TH signal shows a temperature dependence different from that observed in YBCO crystals: besides the peak near T_c, an intense signal is also detected at temperatures far from T_c. We have measured the power dependence of this signal in a large range of temperatures. Near T_c we have essentially observed the same behavior as for YBCO; far from T_c the slope of the TH vs P_{input} curve decreases on decreasing the temperature, showing a nearly linear dependence at temperatures of the order of a few K. We believe that the origin of the peak near T_c is the same as for YBCO; the different features of the signal in BKBO can be ascribed to the different quality of crystal. Concerning the signal detected at temperatures far from T_c, it clearly arises from a different mechanism. However, a more detailed investigation is need to better understand its origin.

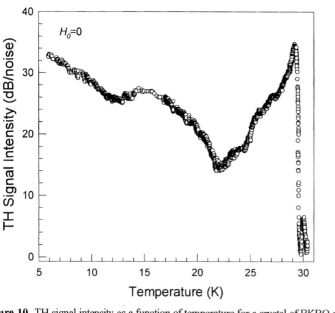

Figure 10. TH signal intensity as a function of temperature for a crystal of BKBO at $H_0=0$. Input mw field ~10 Oe.

3. SURFACE IMPEDANCE

The complex surface impedance, $Z_s = R_s + iX_s$, accounts for the absorption and reflection of high-frequency em waves at the surface of conductors. Measurements of both the resistive, R_s, and the reactive, X_s, components of Z_s in HTSC allow to investigate important properties of the superconducting state, such as gap parameter, quasiparticle density and nature of scattering. Measurements of the resistive component R_s yield information about dissipative processes, while those of the reactive component X_s provide a convenient method for determining the temperature dependence of the field penetration depth λ(T). The most common technique for measuring the mw surface impedance of HTSC is the resonant cavity perturbation technique. The sample is introduced into a resonant cavity with a high quality factor Q and the variations of the cavity characteristics induced by the sample are measured. In particular, changes of R_s induce changes of Q, while changes of X_s shift the resonance frequency of the cavity. In HTSC Z_s has been measured in a large range of temperatures. It has been shown that measurements of mw surface impedance at low temperatures may contribute to clarify the nature of the superconducting state.

Since the discovery of HTSC a lot of papers has been devoted to the understanding of the effective pairing mechanism in the growing up the superconducting state in these materials. Several pairing mechanisms have been proposed, with different order parameter symmetries. In the framework of the BCS theory an isotropic s-wave order parameter is expected, leading to an exponential decrease of both R_s and X_s on decreasing the temperature. However, high quality single crystal HTSC show a different temperature dependence of Z_s. From measurements of X_s it has been inferred that, in the most general case, the low temperature penetration depth in the superconducting planes varies according to a polinomial law with a leading linear term[26, 27]. This linear term can be accounted for by assuming a d-wave or, alternatively, an anisotropic s-wave order parameter symmetry[26]. However, a theory that fully explains all the experimental data in HTSC has not yet been elaborated.

Measurements of the mw surface impedance at temperatures close to T_c have been performed to investigate processes responsible for the mw energy losses induced by the external magnetic field[11, 13, 28, 29]. As discussed in Section 1, measurements of the resistive component of Z_s give the same information as those provided by EPR measurements; they have been mainly performed in samples in the mixed state at high fields. Furthermore, using the models discussed in the literature[10, 30], it is expected that the reactive component of Z_s does not change appreciably with the applied field. Recently, we have elaborated a model which satisfactorily accounts for the field-induced changes of Z_s at low fields[31]. In the framework of this model variations of both R_s and X_s, of about the same extent, are expected. In the following we discuss into detail these results.

We have investigated the field-induced variations of the mw surface impedance Z_s of a YBCO single crystal at temperatures close to T_c[31]. The sample is the very one used to detect TH emission data of Fig.8; it is located inside a rectangular cavity, resonating at 6

GHz and tuned in the TE_{102} mode, in a region in which the magnetic field $H(\omega)$ is of maximal intensity. The input power is of the order of 1 mW. Output signals are detected by a superheterodyne receiver. Measurements refer to the amplitude of the wave transmitted by the cavity as a function of the applied field H_0; at $H_0=0$ the wave amplitude is reduced to zero through interference with a reference signal. Since we do not make use of automatic frequency control of the mw oscillator, the output signal is influenced by both the resistive and the reactive components of the surface impedance. Indeed, when the input power to the cavity is fixed at a constant level, the detected signal is proportional to the field-induced changes of Z_s

$$\Delta Z_s (H_0, T) = |Z_s(H_0, T) - Z_s(0, T)|.$$

All measurements have been performed in the ZFC sample with both the static magnetic field H_0 and the microwave field $H(\omega)$ parallel to the crystal c-axis, at different temperatures close to T_c.

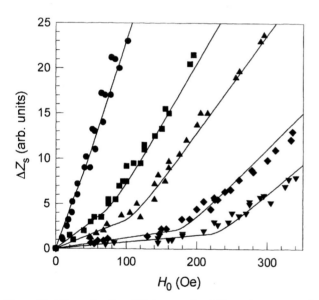

Figure 11. ΔZ_s as a function of H_0 in a sample of YBCO single crystal with $c\|H(\omega)\|H_0$, at different values of temperature. Symbols: experimental points; continuous lines: expected curves using $\alpha_0=2\times10^{-5}$ Oe^{-1}, $\lambda_0=1400$ Å and $\lambda_0/\delta_0=0.083$ [after Ref. 31].

Figure 11 shows ΔZ_s as a function of H_0 at different values of the temperature: $T=91.5$ K (●), $T=90.4$ K (■), $T=89.4$ K (▲), $T=87.2$ K (♦) and $T=85.5$ K (▼). Each ΔZ_s vs H_0 curve exhibits an "elbow" at a particular value of the dc magnetic field, which we call H^*. Except very near the elbow, the experimental points relative to a fixed temperature fall along two straight lines of different slopes intersecting at H^*; on increasing the temperature, H^* decreases and both slopes increase. From Fig.11 one can notice that ΔZ_s varies with the applied field even at low fields, when the sample is for sure in the full Meissner state. We have ascribed the increased rate of variation above H^* to the transition from the Meissner to the mixed state.

127

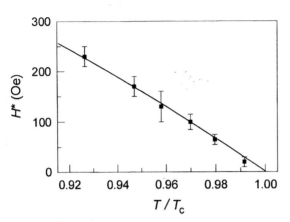

Figure 12. H^* as a function of temperature. Symbols are the values of H^* deduced from ΔZ_s measurements. Continuous line: expected temperature dependence of the lower critical field [after Ref.31].

In Fig.12 is shown H^* as a function of the reduced temperature. H^* data has been fitted by the law expected for the temperature dependence of the lower critical field, $H_{c1}(T)=H_{c1}(0)[1-(T/T_c)^4]$. The continuous line of the figure corresponds to the best fit obtained with $H_{c1}(0)=840$ Oe; this value is consistent with the ones reported for YBCO single crystals[32].

The experimental data at relatively high magnetic fields, $H_0 > 2H_{c1}$, has been fitted by Eq.(7) using λ of Eq.(3); we have ascertained that the CC model[10] accounts quite well for these data. However, the experimental data at low fields, $H_0 < H^*$, cannot be explained by anyone of the models discussed in the literature, including the CC model. The surface impedance depends on H_0 even when $H_0 \leq H_{c1}$ and no fluxons are present inside the crystal. To account for this finding we consider processes which take place near the sample surface in a layer of thickness of the order of λ. As it has been shown in Section 2, at temperatures close to T_c a mw field of few Oe, decaying inside a YBCO crystal, modulates to a detectable extent the partial concentrations of both the normal and the condensate fluids, giving rise to nonlinear emission[21]. On the other hand, it is well known that a magnetic field smaller than H_{c1} decays exponentially inside a superconductor within λ. We assume that the dc magnetic field which decays in the surface layers perturbs in a similar way the electron concentrations: on increasing H_0 the normal fluid density increases and that of the condensate decreases. Since the measurements have been performed using microwave magnetic field intensities of the order of a few mOe, we can now neglect the perturbation due to the mw field and we set[31]

$$w(H_0) = w_0(1 + \alpha_0 H_0'), \qquad (16)$$

where α_0 is a phenomenological parameter and H_0' is that part of the applied field which decays in the surface layers. Since the magnetic field decreases exponentially inside the superconductor, the perturbation term decreases as well. However, we consider the perturbation to be uniform, so that α_0 accounts for an average coupling between H_0 and the electron fluids within the penetration length. The perturbation term increases linearly with H_0 as far as H_0 is equal to the penetration field H^*, when the whole superconductor goes into the mixed state. On further increasing H_0 above H^*, the perturbation mechanism becomes less and less efficient because of the decreasing magnetization.

We have calculated[31] the surface impedance [Eq. (7)] by using the expression of $\tilde{\lambda}$ expected from the CC model [Eq. (3)], with w_0 of Eqs. (4) and (5) replaced by $w(H_0)$ of Eq. (16). Calculations have been carried out numerically in the following approximation: H_0' of Eq. (16) is set equal to H_0 for $H_0 \leq H^*(T)$; H_0' takes on values linearly decreasing from $H^*(T)$ to zero for $H^*(T) \leq H_0 \leq H_{c2}(T)$. Moreover, we have substituted B_0 in Eqs. (4) and (5) with $H_0 - H_0'$.

The continuous lines of Fig.11 are expected results from this model. We have used for λ_0/δ_0 the same value as that used in the fitting of data reported in Figs. 8 and 9 and $\lambda_0 = 1400$ Å, which is the value reported in the literature for YBCO single crystals. The only parameter we have adjusted is α_0; the best fitting has been obtained using $\alpha_0 = 2 \times 10^{-5}$ Oe^{-1}. The model accounts satisfactorily for both the field dependence of the surface impedance in the Meissner state and the slope variation observed in the ΔZ_s vs H_0 curves at low fields. Separate calculations of the resistive R_s and the reactive X_s components of Z_s lead to the conclusion that, at low fields, both R_s and X_s contribute to the field-induced changes of Z_s, while at high fields R_s plays the most important role.

In the field geometry in which all measurements have been performed the value of the magnetic field at which fluxons start penetrating the sample is expected to be smaller than H_{c1}, because of demagnetization effects. However, Zeldov et al.[33] suggested that at temperatures close to T_c, when pinning effects are negligible, the Meissner currents drive the entering vortices to the center of the sample with the result that, near the sample edge, demagnetization and geometrical-barrier effects may offset each other. The effective penetration field at the sample edge may coincide with H_{c1}. Indeed, the measured values of $H^*(T)$ are close to $H_{c1}(T)$. We suggest that the values of $H^*(T)$ are those at which vortices are present over the whole sample. Measurements of the microwave surface impedance at low fields may afford a convenient way for determining geometrical barriers in high-T_c superconductors.

REFERENCES

1. M. Giura, R. Marcon and R. Fastampa, Phys. Rev. **B 40**, 4437 (1989); E. Silva, R. Marcon and F. C. Matacotta, Physica **C 218**, 109 (1993).
2. A. Dulcic, B. Rakvin and M. Pozek, Europhys. Lett. **10**, 593 (1989).
3. D. Shaltiel, V. Ginodman, M. Golosovky, U. Katz, H. Boasson, W. Gerhouser and P. Fisher, Physica **C 202**, 303 (1992); M. Golosovky, V. Ginodman, D. Shaltiel, W. Gerhouser and P. Fisher, Phys. Rev. **B 47**, 9010 (1993).
4. N. H. Tea, M. B. Salamon, T. Datta, H. M. Duan and A. M. Hermann, Phys. Rev. **B 45**, 5628 (1992); F. Zuo, M. B. Salamon, E. D. Bukowski, J. P. Rice and D. M. Ginsberg, Phys. Rev. **B 41**, 6600 (1990).
5. H. A. Blackstead, D. B. Pulling, J. S. Horwitz and D. B. Chrisey, Phys. Rev. **B 49**, 15335 (1994).
6. E. K. Moser, W. J. Tomasch, P. J. McGinn and J. Z. Liu, Physica **C 176**, 235 (1991).
7. W.J.Tomasch, H. A. Blackstead, S. T. Ruggiero, P. J. McGinn, J. R. Clem, K. Shen, J. W. Weber, D. Boyne, Phys. Rev. **B 37**, 9864 (1988).
8. E. Buluggiu and A. Vera, Appl. Mag. Reson. 8, 77 (1995); J. Low Temp. Phys., in press.
9. M.Tinkham, Phys. Rev. Lett. **61**, 1658 (1988).
10. M. W. Coffey and J. R. Clem, Phys. Rev. Lett. **67**, 386 (1991); Phys. Rev. **B 45**, 9872 (1992).
11. M. Golosovky, M. Tsindlekht, H. Chayet and D. Davidov, Phys. Rev. **B 50**, 470 (1994).
12. F. London, *Superfluids*, vol.I (Dover Publications, New York 1961).
13. Dong-Ho Wu, S. Sridhar, W. Kennedy, Phys. Rev. Lett. **63**, 1873 (1989).
14. K.H.Lee and D.Stroud, Phys.Rev.**B 46**, 5699 (1992).
15. C. D. Jeffries, Q. H. Lam, Y. Kim, L. C. Bourne and A. Zettl, Phys. Rev. **B 37**, 9840 (1988); Q. H. Lam and C. D. Jeffries, Phys. Rev. **B 39**, 4772 (1989).
16. I. Ciccarello, M. Guccione and M. Li Vigni, Physica **C 161**, 39 (1989).
17. L. Ji, R. H. Sohn, G. C. Spalding, C. J. Lobb and M. Tinkham, Phys. Rev. **B 40**, 10936 (1989); K. H. Muller, J. C. MacFarlane and R. Driver, Physica **C 158**, 69 (1989).
18. I. Ciccarello, C. Fazio, M. Guccione and M. Li Vigni, Physica **C 159**, 769 (1989).
19. G. I. Leviev, A. V. Rylyakov and M. R. Trunin, Physica **C 162-164**, 1595 (1989).

20. M. R. Trunin and G. I. Leviev, J. Phys. III France **2**, 355 (1992).
21. I. Ciccarello, C. Fazio, M. Guccione, M. Li Vigni and M. R. Trunin, Phys. Rev. **B 49**, 6280 (1994).
22. I. Ciccarello, C. Fazio, M. Guccione and M. Li Vigni, Il Nuovo Cimento **D 15**, 429 (1993).
23. A. Agliolo Gallitto and M. Li Vigni, Physica **C 259**, 365 (1996).
24. M. W. Coffey, Phys. Rev. **B 46**, 567 (1992).
25. G. M. Eliashberg, Sov. Phys. JETP **34**, 668 (1972); **11**, 696 (1960).
26. J. Buan, A. M. Goldman, C. C. Huang, O. T. Valls, T. Jacobs, N. Israeloff, S. Sridhar, B. P. Stojkovic, J. Z. Liu, R. Shelton, C. R. Shih and H. D. Yang, Tr. J. of Physics **20**, 655 (1996).
27. W. N. Hardy, D. A. Bonn, D. C. Morgan, R. Liang and K. Zhang, Phys. Rev. Lett. **70**, 3999 (1993); J. Mao, D. H. Wu, J. Peng, R. L. Greene and S. M. Anlage, Phys. Rev. **B 51**, 3316 (1995).
28. J. Owliaeie, S. Sridhar and J. Talvacchio, Phys. Rev. Lett. **69**, 3366 (1992).
29. T. Jacobs, S. Sridhar, Q. Li, G. D. Gu and N. Koshizuka, Phys. Rev. Lett. **75**, 4516 (1995).
30. A. Dulcic, M. Pozek, Physica **C 218**, 449 (1993).
31. A. Agliolo Gallitto, I. Ciccarello, M. Guccione, M. Li Vigni and D. Persano Adorno, Phys. Rev. **B 56**, 1 (1997).
32. Dong-Ho Wu and S. Sridhar, Phys. Rev. Lett. **65**, 2074 (1990).
33. E. Zeldov, A. I. Larkin, V. B. Geshkenbein, M. Konczykowski, D. Majer, B. Khaykovich, V. M. Vinokur and H. Shtrikman, Phys. Rev. Lett. **73**, 1428 (1994).

II. ORGANIC SUPERCONDUCTIVITY. FULLERENES.

STRUCTURE AND PHASE DIAGRAM OF ORGANIC SUPERCONDUCTORS

Takehiko Ishiguro and Hiroshi Ito

Department of Physics,
Kyoto University,
Kyoto 606-01, Japan

INTRODUCTION

The organic superconductors or molecule-based superconductors have been realized with charge transfer complexes in which donor and acceptor molecules are paired. In most organic superconductors, the electronically dominant molecule has TTF-type skeleton acting as donor as illustrated in Fig. 1. M(dmit)$_2$ type molecule is exceptional and acts as acceptor. Tables 1 and 2 lists the organic superconductors consisting of donor molecules of TTF-derivative. The salts with symmetrical molecules are given in Table 1 while the salts with nonsymmetrical TTF-derivatives are listed in Table 2. The molecule-based superconductors with M(dmit)$_2$ are given in Table 3. The fullerene salts are listed in Table 4.

The first organic material exhibited superconductivity was (TMTSF)$_2$PF$_6$. The salt was nonmetallic below 12 K at zero pressure but application of pressure brought about superconductivity. Further increase of the pressure, however. suppressed the superconductivity. The salt was followed by (TMTSF)$_2$ClO$_4$ exhibiting the superconductivity at zero pressure. It is found that the properties of (TMTSF)$_2$ClO$_4$ are scaled more or less to those of (TMTSF)$_2$PF$_6$ if we take into account the intermolecular spacing, which can be varied by pressure: the TMTSF salts with smaller anion correspond to salts with larger anion under pressure. The TMTSF salts produced with counter anions AsF$_6$, SbF$_6$, TaF$_6$, ReO$_4$ and FSO$_3$ are of isomorphous crystalline structure. Their properties are also influenced by the symmetry of counter anions. The non-centrosymmetric anion induces an ordering phase transition associated with changes in the electronic properties. This contrasts to pressure with which the change in inter-molecular spacing can be a measure. Meanwhile, the replacement of counter molecules can be regarded as chemical pressure or lattice pressure.

The second family of the organic superconductors was formed with ET molecule. β-(ET)$_2$I$_3$ exhibiting superconductivity at ambient pressure is the representative. We get its family with isomorphous structure by replacing I$_3$ anion to IBr$_2$ anion and so on.

Table 1 List of molecule-based superconductors, in which symmetric donors are dominant. P_c is critical pressure and T_c is critical temperature.

Material	Symmetry or type of counter molecule	P_c [kbar]	T_c [K]
(TMTSF)$_2$PF$_6$	Octahedral	12	0.9
(TMTSF)$_2$SbF$_6$	Octahedral	10.5	0.38
(TMTSF)$_2$TaF$_6$	Octahedral	11	1.35
(TMTSF)$_2$AsF$_6$	Octahedral	9.5	1.4
(TMTSF)$_2$ClO$_4$	Tetrahedral	0	1.4
(TMTSF)$_2$ReO$_4$	Tetrahedral	9.5	1.2
(TMTSF)$_2$FSO$_3$	Tetrahedral-like	5	≈ 3
(TMTTF)$_2$Br	Spherical	26	0.8
(ET)$_2$ReO$_4$	Tetrahedral	4.0	2.0
β_L-(ET)$_2$I$_3$	Linear	0	1.5
β_H-(ET)$_2$I$_3$	Linear	0	8.1
β-(ET)$_2$IBr$_2$	Linear	0	2.7
β-(ET)$_2$AuI$_2$	Linear	0	4.9
θ-(ET)$_2$I$_3$	Linear	0	3.6
γ-(ET)$_3$I$_{2.5}$	Linear	0	2.5
κ-(ET)$_2$I$_3$	Linear	0	3.6
κ-(ET)$_4$Hg$_{2.89}$Cl$_8$	Polymeric	12	1.8
κ-(ET)$_4$Hg$_{2.89}$Br$_8$	Polymeric	0	4.3
κ-(ET)$_2$Cu(NCS)$_2$	Polymeric	0	10.4 (8.7)[a]
κ-(ET)$_2$Cu(NCS)$_2$ deuterated	Polymeric	0	11.2 (9.0)[a]
κ-(ET)$_2$Cu[N(CN)$_2$]Br	Polymeric	0	11.8 (10.9)[a]
κ-(ET)$_2$Cu[N(CN)$_2$]Br deuterated	Polymeric	0	11.2 (10.6)[a]
κ-(ET)$_2$Cu[N(CN)$_2$]Cl	Polymeric	0.3	12.8
κ-(ET)$_2$Cu[N(CN)$_2$]Cl deuterated	Polymeric	0.3	13.1
κ-(ET)$_2$Cu[N(CN)$_2$]Cl$_{0.5}$Br$_{0.5}$	Polymeric	0	11.3
κ-(ET)$_2$Cu[N(CN)$_2$]Cl$_{0.25}$Br$_{0.75}$	Polymeric	0	11.5
κ-(ET)$_2$Cu[NC(N)$_2$]Cl$_{0.15}$Br$_{0.85}$	Polymeric	0	10
κ-(ET)$_2$Cu[N(CN)$_2$]Br$_{0.9}$I$_{0.1}$	Polymeric	3	5.9
κ-(ET)$_2$Cu(CN)[N(CN)$_2$]	Polymeric	0	11.2
κ-(ET)$_2$Cu(CN)[N(CN)$_2$] deuterated	Polymeric	0	12.3
κ-(ET)$_2$Cu$_2$(CN)$_3$	Polymeric	1.5	2.8
κ'-(ET)$_2$Cu$_2$(CN)$_3$	Polymeric	0	4.1
α-(ET)$_2$NH$_4$Hg(SCN)$_4$	Polymeric	0	0.8 ~ 1.7
α-(ET)$_2$KHg(SCN)$_4$	Polymeric	0	0.3
		1.2[b]	1.2
α-(ET)$_2$RbHg(SCN)$_4$	Polymeric	0	0.5
α-(ET)$_2$TlHg(SCN)$_4$	Polymeric	0	0.1
κ-(ET)$_2$Ag(CN)$_2$H$_2$O	Cluster	0	5.0
(ET)$_3$Cl$_2\cdot$(H$_2$O)$_2$	Cluster	16	2
(ET)$_4$Pt(CN)$_4$H$_2$O	Cluster	6.5	2
(ET)$_4$Pd(CN)$_4$H$_2$O	Cluster	7	1.2

continued

Material	Symmetry or type of counter molecule	P_c [kbar]	T_c [K]
κ_L-(ET)$_2$Cu(CF$_3$)$_4$·TCE	Planar	0	4.0
κ_L-(ET)$_2$Cu(CF$_3$)$_4$·TBE	Planar	0	5.2
κ_L-(ET)$_2$Au(CF$_3$)$_4$·TBE	Planar	0	5.8
κ_L-(ET)$_2$Ag(CF$_3$)$_4$·TBE	Planar	0	4.8
κ_L-(ET)$_2$Ag(CF$_3$)$_4$·121DBCE	Planar	0	4.5
κ_L-(ET)$_2$Ag(CF$_3$)$_4$·121DCBE	Planar	0	3.8
κ_L-(ET)$_2$Cu(CF$_3$)$_4$·112DCBE	Planar	0	4.9
κ_L-(ET)$_2$Ag(CF$_3$)$_4$·112DCBE	Planar	0	4.1
κ_L-(ET)$_2$Cu(CF$_3$)$_4$·121DBCE	Planar	0	5.5
κ_L-(ET)$_2$Au(CF$_3$)$_4$·112DCBE	Planar	0	5.0
κ_L-(ET)$_2$Au(CF$_3$)$_4$·121DBCE	Planar	0	5.0
κ_L-(ET)$_2$Ag(CF$_3$)$_4$·121DBCE	Planar	0	4.5
κ_L-(ET)$_2$Cu(CF$_3$)$_4$·121DCBE	Planar	0	3.5
κ_L-(ET)$_2$Au(CF$_3$)$_4$·121DCBE	Planar	0	3.2
κ_L-(ET)$_2$Ag(CF$_3$)$_4$·TCE	Planar	0	2.4
κ_L-(ET)$_2$Au(CF$_3$)$_4$·TCE	Planar	0	2.1
κ_H-(ET)$_2$Cu(CF$_3$)$_4$·TCE	Planar	0	9.2
κ_H-(ET)$_2$Ag(CF$_3$)$_4$·TCE	Planar	0	11.1
κ_H-(ET)$_2$Ag(CF$_3$)$_4$·TBE	Planar	0	7.2
κ_H-(ET)$_2$Au(CF$_3$)$_4$·TCE	Planar	0	10.5
κ_H-(ET)$_2$Ag(CF$_3$)$_4$·112DCBE	Planar	0	10.2
κ_H-(ET)$_2$Ag(CF$_3$)$_4$·121DCBE	Planar	0	7.3
β''-(ET)$_4$Fe(C$_2$O$_4$)$_3$·H$_2$O·PhCN	Octahedral	0	6.5 ~ 7.7
β''-(ET)$_2$SF$_5$CH$_2$CF$_2$SO$_3$		0	5.3
(BEDO)$_3$Cu$_2$(NCS)$_3$	Polymeric	0	1.06
(BEDO)$_2$ReO$_4$·H$_2$O	Tetrahedral	0	1.5
λ-(BETS)$_2$GaCl$_4$	Tetrahedral	0	8
λ-(BETS)$_2$GaBr$_x$Cl$_y$	Tetrahedral	0	7~ 8
λ-(BETS)$_2$GaCl$_3$F	Tetrahedral	0	3.5

[a] The parenthesized values are given through an analysis taking account of dimensionality and thermal fluctuations.

[b] Uniaxial

TCE	: 1, 1, 2-trichloroethane, TBE : 1, 1, 2-tribromoethane
121DBCE	: 1, 2-dibromo-1-chloroethane
121DCBE	: 1, 2-dichloro-1-bromoethane
112DCBE	: 1, 1-dichloro-2-bromoethane
PhCN	: benzonitrile

In these cases the ET molecules are aligned to form layers, which act as conductors, separated by anion molecules. However, the salt with chemical composition (ET)$_2$I$_3$ has different kinds of stacking structures of ET molecules discriminated as α-, κ-, and θ-types in addition to β-(ET)$_2$I$_3$. Further, by replacing the anion molecule, a variety of crystal structures are formed even if the charge transfer ratio between ET molecule and

Table 2 List of molecule-based superconductors in which hybrid donors are dominant. P_c is critical pressure and T_c is critical temperature.

Material	Symmetry or type of counter molecule	P_c [kbar]	T_c [K]
(DMET)$_2$Au(CN)$_2$	Linear	3.5	0.8
(DMET)$_2$AuCl$_2$	Linear	0	0.83
(DMET)$_2$AuBr$_2$	Linear	1.5	1.6
(DMET)$_2$AuI$_2$	Linear	5	0.55
(DMET)$_2$I$_3$	Linear	0	0.47
(DMET)$_2$IBr$_2$	Linear	0	0.58
κ-(DMET)$_2$AuBr$_2$	Linear	0	1.9
κ-(MDT-TTF)$_2$AuI$_2$	Linear	0	4.1
κ-(S,S-DMBEDT-TTF)$_2$ClO$_4$	Tetrahedral	5.8	2.6
(DMET-TSF)$_2$AuI$_2$	Linear	0	0.58
(DMET-TSF)$_2$I$_3$	Linear	0	0.4
(DTEDT)$_3$Au(CN)$_2$	Linear	0	4
(DTEDT)$_3$SbF$_6$	Octahedral	0	0.3
(TMET-STF)$_2$BF$_4$	Tetrahedral	0	3.8

Table 3 List of molecule-based superconductors in which metal(dmit)$_2$ are dominant. P_c is critical pressure and T_c is critical temperature.

Material	Symmetry of counter molecule	P_c [kbar]	T_c [K]
TTF[Ni(dmit)$_2$]$_2$		7	1.6
N(Me)$_4$[Ni(dmit)$_2$]$_2$		7	5.0
α-(EDT-TTF)[Ni(dmit)$_2$]$_2$		0	1.3
α-TTF[Pd(dmit)$_2$]$_2$		20	6.5
α'-TTF[Pd(dmit)$_2$]$_2$		22	1.7
β-N(Me)$_4$[Pd(dmit)$_2$]$_2$		6.5	6.2
α-NMe$_2$Et$_2$[Pd(dmit)$_2$]$_2$		2.4	4

Table 4 List of molecule-based superconductors in which fullerene is dominant. P_c is critical pressure and T_c is critical temperature.

Material	Symmetry of salt	P_c [kbar]	T_c [K]
K_3C_{60}	fcc	0	19.3
Rb_3C_{60}	fcc	0	29.6
Cs_2RbC_{60}	fcc	0	33
$CsRb_2C_{60}$	fcc	0	31
Rb_2KC_{60}	fcc	0	27
RbK_2C_{60}	fcc	0	23
CsK_2C_{60}	fcc	0	24
Cs_3C_{60}	bct/bcc	14.3	40
$(NH_3)_4Na_2CsC_{60}$	fcc	0	29.6
$(NH_3)_{0.5-1}NaRb_2C_{60}$	fcc	0	8 ~ 12
$(NH_3)_{0.5-1}NaK_2C_{60}$	fcc	0	8.5 ~ 17
$NH_3K_3C_{60}$	Orthorhombic	14.8	28
$Na_2Rb_{0.25}Cs_{0.75}C_{60}$		0	9.4
$Na_2Rb_{0.5}Cs_{0.5}C_{60}$		0	8.4
$Na_2Rb_{0.75}Cs_{0.25}C_{60}$		0	5.2
$RbNa_2C_{60}$	sc	0	3.5
Na_2CsC_{60}	sc	0	10.5
$RbTl_{1.5}C_{60}$		0	27.5
$Na_3N_3C_{60}$	sc	0	15
Sr_6C_{60}	bcc	0	4
Ba_6C_{60}	bcc	0	7
$K_3Ba_3C_{60}$	bcc	0	5.6
$K_x(OMTTF)C_{60}$ (benzene)		0	18.8
Ca_5C_{60}	sc	0	8.4
$Yb_{2.75}C_{60}$	Orthorhombic	0	6
Sm_xC_{60} ($x \simeq 3$)		0	8
$Rb_x(OMTTF)C_{60}$ (benzene)		0	26

fcc : Face-centered cubic
sc   : Simple cubic
bcc : Body-centered cubic
bct : Body-centered tetragonal
OMTTF : octamethylenetetrathiafulvalene

the anion the same, that is 0.5e per ET molecule. Consequently the phase diagram of ET salts has rich variety corresponding to the variety of the molecular arrangement. It is noteworthy that ET molecules has more flexible skeleton compared to the TMTSF molecule and the role of on-site Coulomb interaction is varied with the structure. This is observed in the normal state and gives influence to the superconductivity.

The fullerene molecule with spherical structure aggregates to form, typically, cubic crystal. By doping alkali or alkali earth metal atoms, superconductivity is achieved. In this case the structure is three-dimensional in contrast to the TMTSF and ET salts exhibiting the restricted dimensionality. The π-electrons, moving among C_{60} molecules dominate the conducting properties and T_c of the superconductivity is scaled by the inter-molecular distance as for TMTSF and ET salts. The phase diagrams are rather simple.

In this article, the specific features in structures and phase diagrams of TMTSF and ET superconductors with respect to the superconductivity and metal-nonmetal transition, which is related to the electronic structure and hence the superconducting mechanism, are briefly described.

STRUCTURE, ELECTRONIC STRUCTURE AND DIMENSIONALITY

The principal molecules shown in Fig. 1 are planer except for the case of C_{60} molecule, which is spherical. The planer molecules are either stacked in face-to-face array or aligned with face-to-face pairs. In the face-to-face stacked case the π electrons in the adjacent molecules are overlapping along the stacking axis, resulting in quasi one-dimensional conductor. This is the case for TMTSF salts exhibiting superconductivity. In contrast to TMTSF salts, for ET compounds the steric confomation of the end-ethylene groups prevents close face-to-face stacking of the molecules. Thus the side-by-side interaction becomes comparable to that of the face-to-face interaction. These interactions make the system two-dimensional. In this case the electrons transfers between columns formed by the stacking and conductance is of two-dimensional nature. The conducting plane can be formed also by orthogonally arranged face-to-face paired dimers, which is called as κ type arrangement. The sheets formed by stacking of planer molecules are alternatively separated by insulating layers formed by counter anions as shown in Fig. 2. The above-mentioned face-to-face stackings are classified into α, β and θ stackings, depending on the relative phase relationship between adjacent columns.

The electronic structure responsible for the superconductivity is constructed by π-electrons in the highest occupied molecular orbital (HOMO) or lowest unoccupied molecular orbital (LUMO) of the molecules. The electronic structure, as far as that in the neighborhood of the Fermi level can be calculated reasonably well based upon the extended Hückel approximation. The electron band energy is expressed in the tight-binding band approximation as

$$\epsilon(k) = 2t_a \cos(k_x a) + 2t_b \cos(k_y b) + 2t_c \cos(k_z c)$$

where $\epsilon(k)$ is the electron energy, t_i is transfer energy in the i-th direction, k_i is wave number in the i-th direction, and $a, b,$ and c are lattice constants.

For the TMTSF salt, electron transfer energies t_a, t_b, and t_c along the a- , b- and c-directions are given by 0.25, 0.025 and 0.0015 eV, respectively. Because the transfer energies are so different depending on the axes, in accordance with the crystal structure, the electron energy band is regarded to be quasi one-dimensional in the first approximation. In this case the energy dispersion can be represented by

$$\epsilon = \hbar v_F (|k_x| - k_F) + 2t_b \cos(k_y b) + 2t_c \cos(k_z c)$$

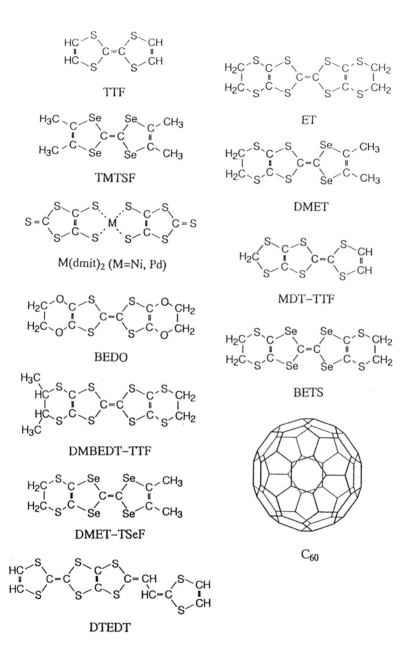

Fig. 1 Chemical structures of TTF and principal molecules consisting charge transfer salts exhibiting superconductivity.

where k_F denotes the Fermi wave number. The transfer energy in the transverse direction, e.g., t_b plays essential roles in the phase transition.

The typical patterns of the Fermi surface cross-sections for TMTSF and BEDT-TTF salts are illustrated schematically in Fig. 3. Validity of the derived Fermi surfaces has been tested by means of the magneto oscillatory effect, such as de Haas-van Alphen effect, Shubnikov-de Haas effect and magneto-geometric resonance effects. The study on the Fermi surface in the normal state has been intensively carried out in these years. The results indicate that the geometrical shapes predicted by the calculation are rather satisfactorily. The readers are recommended to refer to a recent monograph (Wosnitza, 1996), for example..

The restricted dimensionality of the TMTSF and ET salts has been revealed through the investigations of anisotropy in resistivity, optical spectrum, magnetic susceptibility, etc. in the normal state. The Fermi surface study through observation of magneto geometric phenomena, such as angle-dependent magnetoresistance oscillation have demonstrated the restricted dimensionality, as well. However we should remind the results revealed by photoemission study carried out on β-$(ET)_2I_3$, κ-$(ET)_2Cu[N(CN)_2]Br$ and κ-$(ET)_2Cu(NCS)_2$. In these cases a metallic edge typical for metals is not found. The electron correlation is considered to be the likely cause. The importance of the electron-electron interaction is found also in the T^2 dependence of the resistivity in the metallic region for ET salts.

The superconducting properties also reflect the restricted dimensionality. In superconducting TMTSF salts, the Ginzgurg-Landau coherence length $\xi(0)$ shows a pronounced anisotropy ($\xi_a(0) = 70$ nm, $\xi_b(0) = 34$ nm, $\xi_c(0) = 2$nm, for $(TMTSF)_2ClO_4$) In the salt of κ-$(ET)_2Cu[N(CN)_2]Br$ with the highest T_c as organic superconductor, $\xi_a(0) \simeq \xi_b(0) = 2.9$ nm and $\xi_c(0) = 0.58$ nm. In this case the resistive transition under magnetic field near T_c becomes very broad due to the effect of thermal fluctuation. In this case it is noteworthy that the Gintzburg-Landau coherence length in the perpendicular direction to the conducting layer is much shorter than the interlayer spacing of 1.48 nm. The situation is more enhanced for κ-$(ET)_2Cu(NCS)_2$ with $\xi_c = 0.31$ nm, which is much shorter than the interlayer spacing of 1.52 nm. These results indicate that the superconductors can be modeled by Josephson coupled superconducting sheets.

PHASE DIAGRAM

TMTSF Salts

In quasi one-dimensional electron system, the metallic phase becomes unstable and changes to be insulating, with associated modulations having the wave number $2k_F$ (Peierls transition). For $(TMTSF)_2X$ salts with quasi one-dimensional electronic structure as evidenced by the relative ratio of the transfer energies in different crystalline direction, the metallic phase is not stable due Fermi surface nesting. The insulating phase in this case is of SDW ordering due to electron-electron interaction. Application of hydrostatic pressure increase the transverse transfer energy resulting in the violation of the nesting condition. This suppresses the metal-nonmetal transition. The transition temperature T_{NM} decreases with pressure and ultimately tends to zero. Then a transition to a superconducting phase appears at T_c.

The variety of molecular salts can be expanded by simply substituting chemical groups and atoms with similar species. In $(TMTSF)_2X$ salts this has been done by varying the counter anion X^-. The TMTSF molecule by itself can be modified by replacing the Se atom with S atom to produce TMTTF. This molecule can form $(TMTTF)_2X$

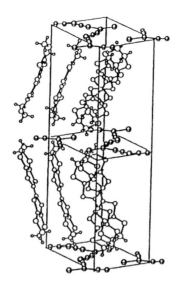

Fig. 2 Crystal structure of κ-(ET)$_2$Cu[N(CN)$_2$]Cl. Two layers formed by ET molecules separated by anion layers act as conductors.

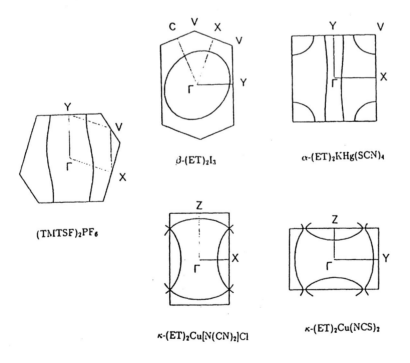

Fig. 3 Fermi surfaces for typical organic superconductors.

143

which is isomorphous to (TMTSF)$_2$X. The difference is found in these two salt types with respect to the degree of dimerization and one-dimensionality. In a one-dimensional spin system with antiferromagnetic interaction, the electrons tend to make singlet pairs by alternating the spacing between them. As a result, the spin-Peierls state becomes more stable than the antiferromagnetic state and, with increase in dimensionality, the SDW phase emerges. A generalized phase diagram of spin-Peierls, SDW, and superconductivity has been proposed by combining (TMTSF)$_2$X and (TMTTF)$_2$X salts against intermolecular spacing or pressure (Jérome, 1991).

In (TMTSF)$_2$X the superconducting phase lies next to SDW phase, which can be suppressed by pressure. In the boundary region, an intrusion of the superconducting phase into the SDW phase is observed in (TMTSF)$_2$AsF$_6$ and the superconducting phase is reentrant. The reentrant phase attracted the interests in relation to the coexistence of magnetism and superconductivity. However, the phase is now considered to be purely superconducting without coexisting SDW. The superconductivity intrudes into the SDW phase because its free energy is lower than that of the SDW phase. Therefore the phase boundary between the SDW and superconducting phase is of first-order nature.

The electronic structure of (TMTSF)$_2$X is remarkably influenced by a magnetic field. That is, the SDW phase suppressed with increasing transverse transfer energy, is revived by strong magnetic field, resulting in field-induced SDW phase, which exhibit successive phase with the field intensity. The extensive studies as functions of magnetic field as well as pressure have provided P-H-T phase diagrams for (TMTSF)$_2$ClO$_4$ and (TMTSF)$_2$PF$_6$ (Kang et al, 1993).

ET Salts

Most of the superconducting ET salts are metallic at ambient pressure, but the normal state stays in the vicinity of nonmetallic phase, presumably of Mott-type insulator due to electron correlation. κ-(ET)$_2$Cu[N(CN)$_2$]Cl whose pressure-temperature phase diagram is given in Fig. 3 provides a typical example (Ito et al., 1996). At zero external pressure κ-(ET)$_2$Cu[N(CN)$_2$]Cl exhibits nonmetallic phase with subphases with

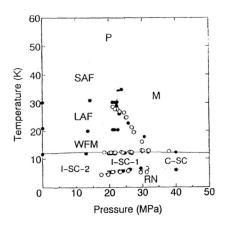

Fig. 4 Pressure phase diagram of κ-(ET)$_2$Cu[N(CN)$_2$]Cl. M denotes metallic phase. Nonmetallic phase is divided into paramagnetic (P), short-range ordered antiferromagnetic (SAF), long-range ordered antiferromagnetic (LAF), weak ferromagnetic (WFM) phases, respectively. I-SC-1 and I-SC-2 denote incomplete superconductivity phases with and without a resistivity decrease due to superconductivity along interlayer direction. C-SC denotes complete SC phases and RN denotes the recurrent nonmetallic phase due to reentrant superconductivity transition.

respect to magnetic properties. The origin and interplay of various electronic states appearing with respect to the arrangement of ET molecules as well as intermolecular distances have been explained in terms of the effect of the on-site Coulomb interaction by taking into account of the transfer integrals between ET molecules. It has been shown by solving a Hubbard-type Hamiltonian within the Hartree-Fock self-consistent technique, that the on-site Coulomb energy to bring about an SDW phase. It is interesting that the relationship among various types of the salts with different types of ET molecule arrangements has been explained on this basis (Kino and Fukuyama, 1995).

For κ-$(ET)_2Cu[N(CN)_2]Cl$ the magnetic studies by means of NMR, ESR and static susceptibility measurements have revealed that the superconducting phase is adjacent to antiferromagnetic phase, which is canted resulting in weak ferromagnetism (Welp et al.1992). It becomes superconducting under an applied pressure of 0.3 kbar but turns to be resistive again below 6 - 7 K (Sushko et al.). The nature of the reentrant resistive state is not yet understood but considered to be likely of that above Tc, implying the two phases are competing. It is interesting that the superconducting phase is subdivided into with and without zero resistance. The incomplete superconducting phase characterized by nonzero resistance is ascribed to superconductivity localization, specific to two-dimensional system: the boundary between two phases is associated with the change in the sheet resistance crossing the universal value of $h/4e^2$.

α-$(ET)_2KHg(SCN)_4$ and α-$(ET)_2RbHg(SCN)_4$ are also intriguing. They are highly two-dimensional, judging from the resistivity anisotropy. Their sister salt α-$(ET)_2NH_4Hg(SCN)_4$ exhibits superconductivity below 1.8 K but the former two salts show only incomplete superconductivity. The incompleteness is ascribed to the of antiferromagnetic-like phase appearing below $\sim$8 K, although it is not clear yet whether it is due to either the magnetism or large sheet resistance. In accordance to the two-dimensionalty, the resistive change of α-$(ET)_2NH_4Hg(SCN)_4$ exhibits Kosterlitz-Thouless transition-like behavior (Taniguchi et al., 1996).

Salts with Magnetic Anion

The magnetic properties dominate the superconductivity phase diagrams as well as intermolecular spacing. To see the roles of the magnetism synthesis to produce molecule-based salts with magnetic counter molecules have been carried out. $(ET)_4Fe(C_2O_4)_3 H_2O PhCN$ is such kind. Among the salts, a pair of λ-$(BETS)_2GaCl_4$ and λ-$(BETS)_2FeCl_4$ formed an interesting combination. That is, λ-$(BETS)_2GaCl_4$ exhibits superconductivity below 8 K. In contrast, λ-$(BETS)_2FeCl_4$ showing similar temperature dependence in the resistivity undergoes insulating transition at the same temperature. ESR measurement demonstrated an antiferromagnetic interaction between Fe^{3+} ions and the metal-nonmetal transition is accompanied by a transition of the magnetic state with respect to anion (Kobayashi et al., 1993). It is also interesting that the sharp metal-nonmetal transition is suppressed by application of a magnetic field exceeding 100 kOe. This is not an orbital but spin-dependent effect. It is believed that a ferromagnetic ordering of Fe^{3+} ions is induced to suppress the antiferromagnetic modulation.

The authors are grateful to G. Saito for providing data base for Tables 1-4.

REFERENCES

Goze, F. et al., Europhys. Lett. 28: 427 (1994).
Ishiguro, T., Yamaji, K. and Saito, G., Organic Superconductors, 2nd ed. (Springer, Heidelberg, 1997).

Ito, H. et al., J. Phys. Soc. Jpn. 65: 2987 (1996).
Jérome, D. and Schurtz, H.J., Adv. Phys. 31: 299 (1982).
Jérome, D., Science 252: 1509 (1991).
Kang, W. et al., Phys. Rev. Lett. 70: 3091 (1993).
Kino, H and Fukuyama, H., J. Phys. Soc. Jpn. 65: 2158 (1996).
Kobayashi, H. et al., Chem. Lett. 1993: 1559 (1993).
Lang, M., Superconductivity Review 2: 1 (1996).
Sushko, Yu.V. et al., Solid Stae Commun. 87: 997 (1993).
Taniguchi, H. et al., Phys. Rev. B 53: R8879 (1996).
Welp et al., Phys. Rev. Lett. 69 840 (1992).
Wosnitza, J., Fermi Surfaces of Low-Dimensional Organic Metals and Superconductors (Springer, Berlin, 1996).

PAIRING AND ITS MECHANISM IN ORGANIC SUPERCONDUCTORS

Takehiko Ishiguro

Department of Physics,
Kyoto University,
Kyoto 606-01, Japan

INTRODUCTION

The theory of superconductivity presented by Bardeen, Cooper and Schrieffer in 1957 stimulated an idea to develop superconductors with electron pairing via exciton mechanism in organic compound (Little, 1964). In that the intra-molecular or inter-molecular electronic excitations are assumed to mediate the attractive electron-electron interaction, which was expected to bring about high critical temperature T_c of superconductivity. The idea was extended to a thin film superconductor (Ginzburg, 1965). These models triggered arguments on the possibility of attaining a high T_c in one- and two-dimensional superconductors, and also many attempts to produce the excitonic superconductor although no success has been reported.

The superconductivity in organic materials was first found in $(TMTSF)_2PF_6$ under pressure (Jérome et al., 1980). At ambient pressure the salt exhibits spin density wave (SDW) phase. With pressure the SDW phase is suppressed and superconductivity appears. As a result, the SDW and superconductivity phases exist in close neighborhood. The phenomena suggested that the mechanism is exotic and may be different from that of the BCS model. In these 17 years, more than 100 organic compounds were found to show superconductivity. They can be categorized by principal molecules acting as π electron conductors. Most of the superconductors are of quasi one- or two-dimensional structure. The highest T_c thus achieved is 13.1 K in κ-$(ET)_2Cu[N(CN)_2]Cl$ with deuterated ET under pressure. The fullerene superconductors first found in 1991 (Haddon et al.,1991) are of three-dimensional structure and not classified as organic compound. However they are regarded as molecular-based superconductors including the traditional organic superconductors. In spite of the difference in the dimensionality, there are lots of similarities in the electronic structure, designated by dominant roles of π electrons, and in the properties, such as the intermolecular spacing dependence of T_c. Therefore it is interesting to see the features of superconductivity of the fullerene compounds in relation to the traditional organic salts.

We describe the experimental studies on the types of superconductivity, such as s-, p- and d-waves as well as the singlet and triplet pairing. Possible violation of the Clogston limit is examined in relation to the triplet superconductivity. Then, the sensitivity to nonmagnetic defects is pointed out as another exotic feature. The effect of isotope substitution is discussed with an interests on the roles of electron-phonon interaction. The proposed mechanisms of the organic superconductors are briefly reviewed.

POSSIBILITY OF ANISOTROPIC PAIRING

In the BCS model, the pairing wave function is s-wave like, with singlet spin state. When a finite attractive interaction is induced in a system with large on-site Coulomb energy, pair functions with non-zero angular momentum become better to produce a superconducting state. If the angular momentum of such a pair function is $1\hbar$ or $2\hbar$, this state is called a p-like or d-like pairing. In the p-pair function, the spin state must be even and triplet. For the d-pair function, the spin state becomes singlet with respect to the spin part. The superconductivity with non-zero angular momentum has an anisotropic superconductivity energy gap and exhibits energy zero part on the Fermi surface for certain wave vector. This can be observed in the excitation related to the density of states. For the case with an isotropic energy gap, the phenomenon exhibits exponential temperature dependence below T_c. For the electromagnetic response, e.g., the spin-lattice relaxation related to NMR, the coherence peak, called Hebel-Slichter peak, appears for the s-wave state. Thus the nuclear relaxation rate has been used as a means to test the anisotropy of the pairing wave function.

Spin-Lattice Relaxation

The proton spin-lattice relaxation rate $(1/T_1)$ of $(TMTSF)_2ClO_4$ was found to show rapid decrease just below T_c (Takigawa et al., 1987). This contrasts remarkably with typical BCS superconductors, where $1/T_1$ increase below T_c reaching a maximum at $T \simeq 0.9T_c$. The feature is ascribed to the vanishing of the anisotropic order parameter along certain lines on the Fermi surface. The feature has been explained by assuming large on-site Coulomb energy, although there exist attractive interactions between neighboring sites as well as on the same site. The origin of the attractive force could be antiferromagnetic fluctuations (Hasegawa and Fukuyama, 1987). With such spin fluctuations, the possibility of an anisotropic singlet state with lines of zero on the Fermi surface for the superconducting order parameter, as well as a triplet state, arises.

Studies on the superconducting state of κ-$(ET)_2Cu[N(CN)_2]Br$ by ^{13}C NMR have brought about the similar result. ^{13}C substituted for ^{12}C in the central part of ET molecule was adapted to probe the superconducting properties. When a magnetic field is applied parallel to the superconducting sheets, the relaxation by superconducting excitation within the sheet is expected. In this case the Hebel-Slichter peak was not observed. The temperature dependence of $1/T_1$ was proportional to T^3 (Mayaffre et al., 1995; DeSoto et al., 1995; Kanoda et al., 1996). This rules out the possibility of the BCS s-wave state accompanying an isotropic energy gap but seems to support a pairing state with possible nodes in the gap function. (We should note that the Hebel-Slichter peak can be suppressed in the case of strong coupling.) On the other hand, the spin susceptibility probed by the Knight shift decreases more quickly than the Korringa relation, which is extended from the normal state, and tends towards zero at low temperature. This implies that the spin pairing is singlet. In the case of a triplet state there should be no difference between the normal and superconducting spin susceptibility. Consequently, the pairing is considered to be of d-like symmetry.

It is noteworthy that the ^{13}C NMR relaxation rate in κ-$(ET)_2Cu[N(CN)_2]Br$ in the perpendicular field orientation is dominated by normal state electrons in the vortex cores. The ^{13}C NMR studies also showed that the normal-state behavior is not that of a simple metal but suggested that antiferromagnetic fluctuation and spin-gap behavior may be present (Mayaffre et al., 1994). This is supported by ESR study yielding an anomalous temperature dependence, exhibiting a decrease below 50 K associated with a linewidth decrease (Kataev et al., 1992) and by observation of magnetic viscosity (Tanatar et al., 1997).

On the other hand, measurements of proton $1/T_1$ in κ-$(ET)_2Cu(NCS)_2$ exhibits an enormous increase below T_c. It forms a sharp peak around 4 K which then drops rapidly at lower temperatures. The peak value of $1/T_1$ was 30 times as large as that at T_c in a 3 kOe field. The enhancement tends to be depressed with magnetic field. The similar peak was observed in β-$(ET)_2I_3$ (Creuzet et al., 1986) and also in κ-$(ET)_2Cu[N(CN)_2]Br$ (Takahashi et al., 1987) when proton is adopted as a probe. To interpret this behavior, we note that the proton exists at the outer end of an ET molecule. As an origin of the anomalous peak the effect of the field fluctuations generated by the vortex motion is considered. The relaxation comes from the Brownian motion of highly diffusive defects in a two-dimensional vortex system. The peak temperature corresponds to the melting of the vortex solid (Mayaffre et al., 1996).

It is interesting that the proton NMR relaxation rate measured in κ-$(MDT-TTF)_2AuI_2$ showed a clear Hebel-Slichter coherence peak just below T_c. The maximum value of $1/T_1$ appeared at $\sim 0.8T_c$ and was 1.4 times as large as the Korringa value just above T_c (Kobayashi et al., 1995). The result indicates that this material is a conventional superconductor with an isotropic energy gap predicted by the BCS theory. The non-single exponential decay of $1/T_1$ below 3 K is interpreted in terms of extra local fields by trapped vortices.

With regard to the fullerene superconductors, ^{13}C NMR investigations in K_3C_{60} and Rb_3C_{60} exhibited the absence or near-absence of Hebel-Slichter peak in the spin-lattice relaxation (Tycko et al., 1992). In a later work, however, the peak was found in Rb_2CsC_{60} at low magnetic field (15 kOe), but to be suppressed at higher field (90 kOe) (Stenger et al., 1995). The suppression of the peak by magnetic field is open for study. It is interesting that the Hebel-Slichter peak was found also by muon-spin relaxation experiment in Rb_3C_{60} (Kiefl et al., 1993). The spin exchange scattering of endohedral muonium (μ^+e^-) with thermal excitation provides the coherence peak just below T_c in $1/T_1$ and can be fitted to the conventional Hebel-Slichter theory for spin relaxation in a superconductor with a broadened BCS density of states.

Magnetic Penetration Depth

The superconductivity symmetry can be studied also by the measurement of the magnetic field penetration depth, which displays an exponential decay below T_c when an isotropic energy gap exists as in a singlet superconductor. Experimentally the penetration depth has been evaluated through the measurements of muon spin relaxation, magnetization, ac susceptibility and electromagnetic response. However the there have been a controversy around the experimental results. Some asserted an s-symmetry (Harshman et al., 1991; Lang et al., 1992; Dressel et al., 1995), while others believed in an unconventional one (Kanoda et al., 1990; Le et al., 1991).

POSSIBLE VIOLATION OF CLOGSTON LIMIT

The singlet superconductivity is broken by a magnetic field which separates the paired electrons. When the contribution of the orbital magnetism is suppressed, the limiting field H_l is given by

$$H_l = \Delta/\sqrt{2}\mu_B$$

where Δ is the gap energy, and μ_B is Bohr magneton, provided that the spin-orbit coupling is negligible (Clogston, 1962). On the other hand, the triplet superconductors does not meet the limitation. In the one- or two-dimensional superconductors, the orbital part can be suppressed by applying the field exactly parallel to the low-dimensional axis, when the transverse transfer energy is negligibly small. In this case provided that the spins of the pairing electrons are antiparallel, the limiting field is given by the magnetic energy ascribable to the sin susceptibility.

In organic superconductors, the upper critical field H_{c2} is anisotropic, due to the low-dimensionality. The H_{c2} becomes smallest when the field lies in the normal direction to the conducting plane due to the large orbital magnetism. On the other hand H_{c2} becomes highest in the field parallel to the conducting plane under which the orbital magnetism becomes negligible when the transfer energy between layers is negligible. In this case when the spins of the paired electrons are parallel, the spin states are not affected by the magnetic field and the superconductivity retains up to higher field.

There have been some reports informing the possibility of the violation of the Clogston limiting value in β-(ET)$_2$I$_3$ (Laukhin et al., 1987) and κ-(ET)$_2$Hg$_{2.98}$Br$_8$ (Lyubobskaya et al., 1990). Recently the anisotropy of upper critical field was measured in pressurized (TMTSF)$_2$PF$_6$ under precisely oriented magnetic field (Lee et al., 1997). When the field is set exactly parallel to the b direction, which lies on the conducting plane but perpendicular to the most conducting direction, H_{c2} exceeds the paramagnetic value by more than $\sim$200 %. This suggests the possibility of triplet pairing and reentrant superconductivity in very high field (Lebed, 1986). Otherwise the singlet superconductivity exists in the Fulde-Ferrell-Larkin-Ovchinikov (FFLO) state, where superconductivity exists inhomogeneously with spatial oscillation of the gap function. It has been pointed out that the FFLO state is enhanced in two-dimensional Fermi surface and the Fermi surface nesting is advantageous to the FFLO state as well as nesting instabilities (Shimahara, 1994).

EFFECT OF ISOTOPE SUBSTITUTION

In order to test the mechanism related to the electron-phonon interaction via molecular vibration, the experimental study on the effect of isotope substitution to T_c has been one of central issue of the superconductivity study.

For ET molecule, either H or C atoms have been replaced with D (deuterium) or ^{13}C, respectively. For κ-(ET)$_2$Cu[N(CN)$_2$]Br, T_c is reduced by the replacement of H with D, in accord with ordinary BCS mechanism. On the other hand for κ-(ET)$_2$Cu(NCS)$_2$, the substitution of H with D enhanced T_c, in contradiction to the expectation based on BCS mechanism (Oshima et al., 1988; Ito et al., 1992). When the ^{12}C at the central C=C part is replaced with ^{13}C in β-(ET)$_2$I$_3$, $\delta T_c/T_c$ = -2.5 % was obtained, which is very large compared to the decrease of C=C stretching mode and ascribed to inelastic scattering which remains still appreciable at low temperature, as evidenced by the continued decrease in resistivity on cooling (Auban-Senzier et al., 1993). The later intensive study revealed that the ^{13}C substituted samples of κ-(ET)$_2$Cu(NCS)$_2$ and κ-(ET)$_2$Cu[N(CN)$_2$]Br do not give measurable shifs of T_c within an uncertainty range of ±0.1 K (Kini et al., 1993). Further no systematic decrease in T_c was observed in β-(ET)$_2$I$_3$ (Carlson et al., 1992). On the other hand by replacing all the eight S atoms, mostly ^{32}S, of ET with ^{34}S, statistical decrease of T_c by $\delta T_c/T_c$ = -0.7 % was observed (Carlson et al., 1993). Further by substituting all S by ^{34}S and peripheral C by ^{13}C of κ-(ET)$_2$Cu(NCS)$_2$, δT_c = - 0.12±0.05 K and α = 0.26±0.11 was reduced (Kini et al., 1996).

For the fullerene superconductors, the effects of isotope substitution have been studied also extensively. For Rb$_3$C$_{60}$, with ~75 % substitution of ^{13}C for ^{12}C a decrease of T_c by ~0.65 K was obtained. For the phonon-mediated pairing, T_c is proportional to M^α, where M is the ionic mass. When we put in the mass of C$_{60}$ neglecting Rb ion, the observed T_c shift of 0.65 K leads to α=0.37. This was interpreted as an evidence for phonon-mediated coupling, via intraball vibration (Ramirez et al., 1992). On the other hand, a similar experiment carried out varying the content of ^{13}C gave a remarkably large value of α=~1.4 (Ebbesen et al., 1992). The cause of the difference is not clear.

In order to find the effect of partial substitution within the molecule, the T_c's for atomically substituted Rb$_3$[^{13}C$_x$ ^{12}C$_{(1-x)}$]$_{60}$ and molecularly substituted Rb$_3$ [^{13}C$_{60x}$ ^{12}C$_{60(1-x)}$] were studied. In the former, T_c interpolates the T_c's of Rb^{13}C$_{60}$ and Rb^{12}C$_{60}$. However, in the latter anomalously large negative shift of T_c is observed. which is even larger than that for complete substitution (Chen and Lieber, 1993). The results were interpreted in terms of polaron pairing rather than the simple BCS pairing or bipolarons (Takada, 1996).

Thus the experimental results on the effect of isotope substitution are not simple. Some aspects suggest that electron-molecular-vibrational coupling (Yamaji, 1987) is effective to T_c. The somewhat complicated dependence can be explained in terms of the contributions of intermolecular vibrations having different energies and coupling modes, although the result is not conclusive. However, some assert that the electron-molecular-vibrational coupling is hampered by the appreciable on-site Coulomb repulsion but intermolecular vibrations are more important .

SENSITIVITY TO DISORDER

It is known that singlet superconductivity is rather insensitive to nonmagnetic defects. However, when weak nonmagnetic disorder was introduced into (TMTSF)$_2$ClO$_4$ by alloying it with (TMTSF)$_2$ReO$_4$ or (TMTTF)$_2$ClO$_4$, the superconductivity is sharply suppressed with increasing alloying ratio and no superconductivity was observed for (TMTSF)$_2$(ClO$_4$)$_{1-x}$ (ReO$_4$)$_x$ for x larger than 0.3 (Tomić et al., 1983). It should be noted that by rapid cooling of (TMTSF)$_2$ClO$_4$ suppresses superconductivity due to weak random potential caused by the random orientation of ClO$_4$ anions. These facts indicate conclusively that the superconductivity in (TMTSF)$_2$X is very sensitive to nonmagnetic defects.

The notable effect of disorder is seen also in the alloy of β-(ET)$_2$I$_{3(1-x)}$(IBr$_2$)$_x$. Both of β-(ET)$_2$I$_3$ and β-(ET)$_2$IBr$_2$ are superconducting but the alloy with $x = 0.25$ or 0.5 is not superconducting (Tokumoto et al., 1987). This can be ascribed to the effect of disorder induced by alloying. Furthermore β-(ET)$_2$I$_2$Br does not show superconductivity although it can be superconducting comparing to the sister salts of β-(ET)$_2$I$_3$ and β-(ET)$_2$IBr$_2$. This is ascribed to the noncetrosymmetric structure of anion as I-I-Br. The suppression of superconductivity in the alloyed ET salts can be understood in terms of the effect of the nonmagnetic disorder as considered in TMTSF salts. However the sensitivity to disorder has also been explained in terms of the enhancement of the Coulomb repulsion due to randomness (Hasegawa and Fukuyama, 1986).

We should note, however, that the resistive measurement of κ-$(ET)_2Cu[N(CN)_2]$-$Cl_{(1-x)}Br_x$ revealed T_c values of 11.3, 11.5 and 10.6 K for $x = 0.5$, 0.75 and 0.85, respectively, at ambient pressure (Bondarenko et al., 1994). A structural study of the salt with $x = 0.5$ showed that disorder is not detected in the anion layer, implying that ordered alternating of $Cu[N(CN)_2]Cl$ and $Cu[N(CN)_2]Br$ may be formed (Kushch et al., 1993).

MECHANISM

Understanding of the mechanism is a matter of great concern for not only physicists but also for synthetic chemists. The excitonic superconductivity mechanism which aroused interest for the organic superconductors has not yet been found experimentally. On the emergence of the superconductivity in $(TMTSF)_2X$, possible presence of high T_c superconductors under fluctuating state due to quasi one-dimensionality was claimed (Jerome and Schultz, 1982). Reported observations on the large energy gap via tunneling junction and far-infrared spectroscopy data were the motivation of the argument. The experiments followed later, however, had not reproduced the anomalous experimental data, but gave the results which is not inconsistent with the BCS model. However, the recent experimental data on the spin-lattice relaxation and the Clogston limit imply the importance of unconventional mechanisms in the organic and molecular-based materials.

We briefly touch on two typical mechanisms, the mediation via antiferromagnetic fluctuation and the electron-molecular vibrational coupling. To know more details of the description and the other mechanisms, such as contribution of intermolecular vibrations, g-ology model, excitonic model, bipolaron model, two-band mechanisms, the readers are to refer to a monograph (Ishiguro et al., 1997).

Effect of Spin Fluctuation

In some of the organic superconductors, typically in TMTSF salts, the SDW phase neighbors on the superconducting phase. Based on this fact the role of attractive interaction between the electrons in adjacent columns via spin fluctuation was pointed by Emery (1986). Since the SDW is induced by Coulomb repulsion and Fermi surface nesting, this mechanism can work for restricted dimensional system. When the $\mathbf{k} - \mathbf{k}'$ becomes equal to the nesting vector, the Coulomb interaction $V(\mathbf{k}, \mathbf{k}')$ working to scatter $(\mathbf{k},\uparrow)$ electron to $(-\mathbf{k}, \downarrow)$ electron diverges due to the electron-electron interaction. Based on this, one can get k-dependent gap parameter whose sign changes with sites on the Fermi surface, resulting in d-like symmetry (Scalapino et al., 1987). However, the numerical evaluation based upon two-dimensional Hubbard model could not derive superconducting phase (Hirsch et al., 1988). The applicability of this kind of mechanism to real organic compound has not yet been established, but some treatment to a quasi one-dimensional system modelling $(TMTSF)_2X$ was given (Shimahara, 1989).

Two-dimensional models of superconductivity of Coulombic origin have been extended in the context of high-T_c cuprate superconductors and heavy fermion superconductors in these years. Serious argument with respect to them is expected to be performed for organic superconductors by taking into account of their structure and property.

Roles of Intramolecular Vibration

All of the molecular-based superconductors, except alkali-metal-doped C_{60}, consist of TTF-type molecules, such as TMTSF, ET, DMET and so on. $M(dmit)_2$ (M = Ni, Pt) molecule is isolobal in the sense that the central metallic atom M corresponds to the central C=C of TTF. The carriers in the superconductors consisting of TTF-derivative exist in the band formed with HOMO. The energy of the HOMO changes proportionally to the change in the normal coordinates of totally symmetric intramolecular modes. This works in the similar way to the electron-phonon interaction via deformation potential. Particularly, in the TTF-derivatives the coupling constants related to the stretching modes of the central C=C and C-S bonds can be appreciably large (Lipari et al., 1977; Bozio et al., 1977). The T_c was calculated after the BCS theory with the attractive force induced by the intramolecular-vibrational coupling (Yamaji, 1987). The model can provide a reasonable T_c and explains some aspects of the superconducting properties. It is asserted that there are possibility of the contribution of this type of mechanism, even if the mechanism is not dominant. The type of superconductivity is s-wave like. With regard to the fullerene superconductors, the LUMO levels are coupled strongly with the intramolecular vibrations (Schlueter et al., 1992; Prassides et al., 1992).

The study on the mechanism is still controversial and we do not yet have any firm conclusion. In addition recent arguments on the strong correlation going on the arguments on the metallic states may mean the significant role to the superconductivity.

REFERENCE

Auban-Senzier, P. et al., 1993, *J. Phys. I (France)* 3: 871.
Béal-Mond, M.T. et al., 1986, *Phys. Rev.* B 34: 7716.
Bondarenko, V.A. et al., 1993, *Physica C* 235-240: 2467.
Bourbonnais, C., Caron, 1988, *Europhys. Lett.* 5: 209.
Bozio, R. et al., 1977, *Chem. Phys. Lett.* 52: 503.
Carlson, K.D. et al., 1992, *Inorg. Chem.* 114: 10069.
Carlson, K.D. et al., 1993, *Physica C* 215:195.
Chen, C.-C. and Lieber, C.M., 1993 *Science* 259: 655.
Clogston, A.M., 1962, *Phys. Rev. Lett.* 9: 266.
Creuzet, F. et al., 1986, *Physica B* 143: 363.
De Soto, S.M. et al., 1995, *Phys. Rev.* B 52:10364.
Dressel, M. et al., 1995, *Phys. Rev.* B 50: 13603.
Ebbesen, T.W et al., 1992, *Nature* 355: 620.
Emery, V., 1986, *Synth. Metals* 13: 71.
Ginzburg, V., 1965, *Sov. Phys. JETP* 2: 1594.
Haddon, R.C. et al., 1991, *Nature* 350: 320.
Harshman, D.R. et al., 1991, *Phys. Rev. Lett.* 64: 655.
Hasegawa, Y and Fukuyama, H., 1986, *J. Phys. Soc. Jpn.* 55: 3717.
Hasegawa, Y and Fukuyama, H., 1987, *J. Phys. Soc. Jpn.* 56: 877.
Hebbard, A.F. et al., 1991, *Phys. Rev.* B 44: 9753.
Hirsch, J.E. et al., 1988, *Physica C* 153: 549.
Ishiguro, T., Yamaji, K. and Saito, G., 1997, *Organic Superconductors 2nd Ed.*, Springer, Heidelberg.
Ito, H. et al., 1992, *Jpn. J. Appl. Phys. Series* 7: 419.
Jérome, D and Schulz, H.J., 1982, *Adv. Phys.* 31: 299.
Kanoda, K. et al., 1996, *Phys. Rev.* B 54: 74.
Kanoda, K. et al., 1990, *Phys. Rev. Lett.* 65: 1271.
Kataev, V. et al., 1992, *Solid State Commun.* 83: 435.
Kiefl, R.F. et al., 1993, *Phys. Rev. Lett.* 70: 3987.
Kini, A.M. et al., 1995, *Physica C* 204: 399.
Kini, A.M. et al., 1996, *Physica C* 24: 81.

Kobayashi, Y. et al., 1995, *Synth. Metals* 70: 871.
Kozlov, M.E., 1989, *Spectrochimica Acta A* 45: 437.
Kushch, N. et al., 1993, *Synth. Metals* 53: 155.
Lang, M. et al., 1992, *Phys. Rev. Lett.* 69: 1443.
Laukhin, V.N. et al., 1987, *JETP Lett.* 45: 399.
Lebed, A.G., 1986, *JETP Lett.* 44: 114.
Lee, I.J. et al., 1997, *Phys. Rev. Lett.* 78: 3555.
Le, L.P. et al., 1992, *Phys. Rev. Lett.* 68: 1923.
Lipari, N.O. et al., 1977, *Int'l J. Quant. Chem. Symp.* 11: 583.
Little, W., 1964, *Phys. Rev.* 134: A1416.
Lyubovskaya, R.N. et al., 1990, *JETP Lett.* 51: 361.
Mayaffre, H. et al., 1994, *Europhys. Lett.* 28: 205.
Mayaffre, H. et al., 1995, *Phys. Rev. Lett.* 75: 4122.
Mayaffre, H. et al., 1996, *Phys. Rev. Lett.* 76: 4951.
Oshima, K. et al., 1988, *Synth. Met.* 27 A473.
Prassides, K. et al., 1992, *Europhys. Lett.* 19: 629.
Ramirez, A.P. et al., 1992, *Phys. Rev. Lett.* 68:1058.
Scalapino, E. et al., 1987, *Phys. Rev.* 35: 6694.
Schlueter, M. et al., 1992, *Phys. Rev. Lett.* 68: 526.
Shimahara, H., 1994, *Phys. Rev. B* 50: 12760.
Shimahara, H., 1989, *J. Phys. Soc. Jpn.* 58: 1735.
Stenger, V.A. et al., 1995, *Phys. Rev. Lett.* 74: 1649.
Takahashi, T. et al., 1988, *Physica C* 153-155: 487.
Takigawa, M. et al., 1987, *J. Phys. Soc. Jpn.* 56: 873.
Takada, T., 1996, *J. Phys. Soc. Jpn.* 65: 3134.
Tokumoto, M. et al., 1986, *Japan. J. Appl. Phys. 26 Suppl.* 26-3: 1977
Tomić, S. et al., 1983, *J. Phys. 44:* C3-1057.
Tycko, R. et al., 1992, *Phys. Rev. Lett.* 68: 1912.
Yamaji, K., 1987, *Solid State Commun.* 61: 413.

SUPERCONDUCTIVITY IN DOPED C_{60} COMPOUNDS

Olle Gunnarsson,[1] Erik Koch,[2] and Richard M. Martin[2]

[1]Max-Planck-Institut für Festkörperforschung, Stuttgart, Germany
[2]Department of Physics, University of Illinois, Urbana, Illinois 61801

ABSTRACT

It is discussed why A_3C_{60} (A =K, Rb) are metals and not Mott-Hubbard insulators, in spite of the strong Coulomb repulsion U. It is shown that the orbital degeneracy N of the t_{1u} band increases the critical value of U for a Mott-Hubbard transition by about a factor of $\sqrt{N}$. Theoretical and experimental estimates of the electron-phonon coupling λ are discussed. In particular, it is shown how photoemission for free C_{60}^- molecules can be used to estimate λ. Finally we discuss the Coulomb pseudopotential μ^*, describing the effects of the Coulomb repulsion. In particular the retardation effects are considered. It is argued that μ^* is relatively large, but that the results are still consistent with the electron-phonon interaction driving the superconductivity.

INTRODUCTION

The fullerene molecules (C_{60}, C_{70},...) have attracted much interest since their discovery by Kroto et al.[1]. This interest increased dramatically when Krätschmer et al.[2] discovered how to produce C_{60} in large quantities, which made it possible to perform traditional solid state experiments. Very soon Haddon et al.[3] found that intercalation of alkali metal atoms in solid C_{60} leads to metallic behavior. Shortly afterwards it was found that some of these alkali-doped C_{60} compounds are superconducting with a transition temperature T_c which is only surpassed by the cuprates[4, 5, 6]. Thus T_c is 33 K for $RbCs_2C_{60}$[6] and for Cs_3C_{60} under pressure $T_c = 40$ has been reported[7]. The great interest in the superconductivity of the alkali-doped C_{60} compounds is in particular due to these systems being a completely new class of superconductors, the large value of T_c and the question whether or not such a large value of T_c can be caused by the coupling to phonons alone.

C_{60} is the most symmetric molecule, having the largest point group (icosahedral) with 120 symmetry operations of the known molecules (see Fig. 1). It has the shape of a soccer ball. The 60 carbon atoms are all equivalent and form 12 pentagons and 20 hexagons.

The C_{60} molecules condense to a solid of weakly bound molecules. While the shortest separation between two atoms on the same molecule is about 1.4 Å, the shortest

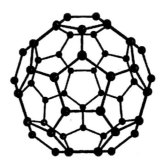

Figure 1: The C_{60} molecule.

separation between two atoms on different molecules is about 3.1 Å. The fullerenes are therefore molecular solids, where many of the molecular properties essentially survive in the solid. The discrete levels of a free C_{60} molecule are only weakly broadened in the solid, leading to a set of essentially nonoverlapping bands with a width of about 1/2 eV as is illustrated in Fig. 2. The system therefore has two very different energy scales, the intramolecular ($E_I \sim 30$ eV) and the intermolecular ($W \sim 1/2$ eV) energy scales. Undoped C_{60} is a band insulator, in which the h_u band is full and the t_{1u} band is empty. When solid C_{60} is doped by alkali atoms, the alkali atoms donate about one electron each to the t_{1u} band[8, 9, 10, 11]. Since the t_{1u} band can take six electrons, it is half-full for A_3C_{60} (A=K, Rb), which is a metal. A_4C_{60} is, however, an insulator[12, 13] although the t_{1u} band is only partly filled and should be a metal according to band theory[14]. A_6C_{60} is a band insulator.

In A_3C_{60} (A=K, Rb) at low temperatures, the C_{60} molecules take one of two likely orientations in an essentially random way[15]. This orientational disorder has a substantial effect on some of the electronic properties[16].

C_{60} has intramolecular vibrations (phonons) with energies up to $\omega_{ph} \sim 0.2$ eV. Only phonons with A_g or H_g symmetry couple to the t_{1u} electrons. It was very early proposed that these phonons drive the superconductivity[17, 18, 19]. Estimates of the electron-phonon interaction $\lambda \sim \frac{1}{2} - 1$ fall in the right range to explain the experimental values of T_c. It has therefore become widely, but not universally, accepted that the intramolecular H_g phonons drive the superconductivity. Other phonon modes, such as librations, C_{60}-C_{60} and alkali-C_{60} vibrations are believed to play a small role. It has, however, also been argued that an electronic mechanism may drive the superconductivity[20, 21, 22].

The effective Coulomb interaction between two electrons on a C_{60} molecule in a solid is about 1-1$\frac{1}{2}$ eV[23, 24, 25, 26]. The long-ranged Coulomb interaction in the A_3C_{60} compounds leads to a charge carrier plasmon, due to the oscillations of the t_{1u} electrons. This plasmon has the energy $\omega_{pl} \sim 1/2$ eV and an intermediate coupling constant $(g/\omega_{pl})^2 \sim 1$[27, 28]. This plasmon has unusual dispersion[29] and broadening[30] behaviour.

The attractive interaction between two electrons induced by the electron-phonon interaction is small ($\sim 1/10$ eV) compared with the Coulomb repulsion between two electrons on the same C_{60} molecule ($U \sim 1 - 1\frac{1}{2}$ eV). For conventional superconductors, however, it is believed that the effects of the Coulomb interaction are drastically reduced by retardation effects, due to the very different energy scales for the phonons and electrons. The dimensionless Coulomb pseudopotential μ^*, describing the effects of the Coulomb repulsion, is therefore believed to be drastically renormalized for such systems. For doped C_{60} solids it has been controversial whether the intramolecular

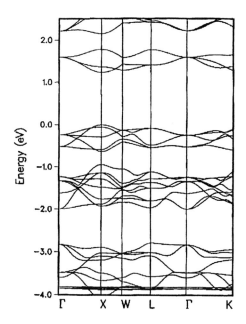

Figure 2: Some of the subbands around the Fermi energy for solid C_{60} in the $Fm\bar{3}$ structure. The bands at about -0.5 eV are the h_u bands, which are occupied in solid C_{60}, and the bands at about 1.5 eV are the t_{1u} bands, which become populated in A_nC_{60}. The bands around -1.5 eV results from the overlapping h_g and g_g bands. (From Erwin[14]).

(E_I) or intermolecular (W)[31] energy scale is relevant for the retardation effects. In the latter case one may expect the retardation effects to be small, since $\omega_{ph} \sim W$. It has therefore been asserted that the phonons alone cannot explain the superconductivity in A_3C_{60}[31]. Later work has provided support for the intermolecular energy scale being the relevant one, but nevertheless found that the electron-phonon mechanism may be sufficient if the Coulomb interaction is screened as efficiently as predicted by RPA[32].

In the A_3C_{60} (A=K, Rb) compounds many energy scales are similar; $\omega_{ph} \lesssim \omega_{pl} \sim W \lesssim U$. The electron-phonon interaction leads to a Jahn-Teller effect and an energy lowering $E_{JT} \sim 0.3$ eV[33] for a C_{60}^{3-} ion which is comparable to the band width W. The multiplet effects may also be of a comparable magnitude.

It can be estimated that $U/W \sim 1.5 - 2.5$[34, 35]. This suggests very strong correlation effects. Actually, it has been suggested that stoichiometric A_3C_{60} is a Mott-Hubbard insulator, and that the experimental samples are metallic only because of small deviations from stoichiometry[23]. This would be a situation similar to the one in the High T_c compounds and it would probably have important implications for the explanation of the properties of these systems. It was later found that the orbital degeneracy of the t_{1u} band increases the the value of U/W where the Mott-Hubbard transition takes place[34], so that one expects even stoichiometric A_3C_{60} to be on the metallic side of a metal-insulator transition.

It is interesting to observe the large role the Coulomb interaction has played in the discussion of A_3C_{60}. Thus the discussion spans the whole range from the issue if it would make stoichiometric A_3C_{60} an insulator, to the suggestion that it actually causes the superconductivity. An intermediate issue is if retardation effects can reduce the effective Coulomb interaction so much that a conventional electron-phonon mechanism can explain the superconductivity.

The similar magnitude of ω_{ph} and W raises doubt about the strength of the retardation effects, as discussed above. It furthermore implies that Migdal's theorem is not valid. The similarity between E_{JT} and W suggests a competition between the Jahn-Teller effect and the kinetic energy. Similarly there may be a competition between the multiplet effects and the kinetic energy. In both cases this competition should be influenced by the Coulomb repulsion, which tends to reduce the importance of hopping. Finally there is a competition between the Jahn-Teller effect, which favors a low-spin state, and the multiplet effects, which according to Hund's rules favor a high-spin state.

These systems are therefore likely to have very interesting physics, and the various competing effects have only been treated theoretically in a rather incomplete way. The superconductivity in the fullerides has recently been reviewed[35].

MOTT-HUBBARD

As mentioned above, U/W can be estimated to be in the range 1.5-2.5[34], where U is the Coulomb interaction between two electrons on the same molecule and W is the t_{1u} band width. On dimensional grounds we expect a metal-insulator transition for $U/W \sim 1$. More detailed considerations suggest a critical ratio of the order of 1.5[36]. These considerations have been made for a system without orbital degeneracy. Below we give qualitative arguments that the orbital degeneracy N should increase the critical ratio by about a factor $\sqrt{N}$ and perform Quantum Monte Carlo calculations to support this.

To discuss the Mott-Hubbard transition, we first consider a model with M sites, the orbital degeneracy N, and hopping integrals $t_{im,jm'} = t\delta_{mm'}$ between orbitals with the same quantum numbers m on neighboring sites i and j. For half-filling the energy gap is then

$$E_g = A - I = [E(NM+1) - E(NM)] - [E(NM) - E(NM-1)], \quad (1)$$

where A is the affinity energy, I is the ionization energy and $E(L)$ is the ground-state energy for L electrons. For $U \gg t$ we have

$$E(NM) = \frac{1}{2}N(N-1)MU + O(\frac{t^2}{U}), \quad (2)$$

since electron hopping costs an extra energy U. For instance, this energy is obtained for a Neel state $|0\rangle$, with the moment N. In the system with $NM+1$ electrons, the extra occupancy can hop without an energy cost U. Thus we introduce the states

$$|1\rangle = \psi_{11\downarrow}^\dagger |0\rangle$$
$$|i\rangle = \frac{1}{\sqrt{N}}\psi_{11\downarrow}^\dagger \sum_m \psi_{im\uparrow}^\dagger \psi_{1m\uparrow}|0\rangle, \quad (3)$$

where $\psi_{jm\uparrow}^\dagger$ creates a spin up electron on site j in orbital m. The sites i are nearest neighbors of site 1, and we have assumed the sites i to have spin down electrons in the state $|0\rangle$. In the state $|i\rangle$ the extra occupancy has hopped from site 1 to a neighboring site i. The matrix element for this process is

$$\langle i|H|1\rangle = \sqrt{N}t, \quad (4)$$

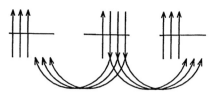

Figure 3: Illustration of how the extra occupancy can hop in N different ways to the neighboring site for the case $N = 3$.

i.e. a factor $\sqrt{N}$ larger than in the one-particle case[34]. This is due to the fact that any of the N spin up electrons can hop from site 1 to i. This is illustrated in Fig. 3. A similar degeneracy factor has also been found in other contexts[37]. We use the analogy with the one-electron case, although it is not rigorous, since in the many-body case there is a string of reduced spins along the path of the moving additional occupancy. In the one-electron case, the hopping of an electron leads to an energy lowering of the order of $W/2$. The close analogy between the one- and many-particle problems then suggests

$$E(NM+1) \approx E(NM) + NU - \sqrt{N}W/2, \quad (5)$$

with an extra factor $\sqrt{N}$ in front of $W/2$. Using a similar result for $E(NM-1)$ we obtain[34]

$$E_g = U - \sqrt{N}W. \quad (6)$$

For $N = 1$ this leads to $E_g = U - W$. If this result is extrapolated to smaller U, it predicts a Mott-Hubbard transition for $U_c/W \sim 1$, close to previous results[36]. Eq. (6) suggests, however, that for a degenerate system the transition takes place for a larger ratio $\sim \sqrt{N}$. We observe that the states in Eq. (3) may not have been chosen in the optimum way, and in some cases we have explicitly constructed states leading to lower energies for the cases with an extra electron or hole[38]. Since, however, the energy in Eq. (2) for the half-filled system is exact in the large U-limit, the variational principle tells us that such improvements can only reduce the gap further and possibly increase the prefactor $\sqrt{N}$. In some special cases we actually find the prefactor N in Eq. (6) in the large U limit[38].

The arguments presented above for $U_c/W \sim \sqrt{N}$ are not rigorous, not even in the large U-limit, and they furthermore have to be extrapolated to intermediate values of U. We therefore need to test these arguments for the systems we are interested in. For this purpose we have studied a model of A_3C_{60}, which includes the three-fold degenerate, partly occupied t_{1u} level. We include the on-site Coulomb interaction U and the hopping integrals $t_{im,jm'}$ between the molecules. This leads to the Hubbard-like model

$$H = \sum_{i\sigma}\sum_{m=1}^{3} \varepsilon_{t_{1u}} n_{i\sigma m} + \sum_{<ij>\sigma mm'} t_{ijmm'} \psi_{i\sigma m}^{\dagger} \psi_{j\sigma m'} + U \sum_{i} \sum_{\sigma m<\sigma' m'} n_{i\sigma m} n_{i\sigma' m'}, \quad (7)$$

where the sum $<ij>$ is over nearest neighbor sites. We have used a fcc lattice. The hopping integrals $t_{imjm'}$ have been obtained from a tight-binding parametrization[39, 11]. The molecules are allowed to randomly take one of two orientations in accordance with experiment[15], and the hopping integrals are chosen so that this orientational disorder is included[40]. The band width of the infinite system is $W = 0.63$ eV. The model neglects multiplet effects. The inclusion of these effects may favor antiferromagnetism, which could reduce the critical value U_c/W for a Mott-Hubbard transition. On

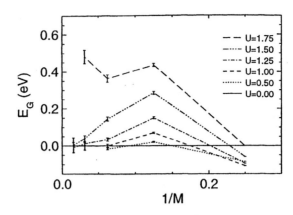

Figure 4: E_G (Eq. (8)) for different U and as a function of $1/M$, where M is the number of molecules. The band width varies between 0.58 eV ($M = 4$) and 0.63 eV ($M = \infty$).

the other hand, the inclusion of the electron-phonon interaction would tend to suppress moment formation, due to the preference for low-spin states when the Jahn-Teller effect is considered.

To study this model, we use a diffusion $T = 0$ Monte Carlo method, which we have developed along the lines of ten Haaf et al.[41]. By introducing a "fixed node" like approximation, the so-called sign problem can be avoided. A trial function is obtained by solving the model in the Hartree-Fock approximation and and by constructing the corresponding Slater determinant. A Gutzwiller Ansatz[42] is made, and the resulting function is used as a trial function in the diffusion Monte Carlo approach.

The energy gap E_g is calculated for a system with M molecules. We then need to extrapolate the result to $M = \infty$ to decide if the the system is a metal or an insulator. To obtain an efficient extrapolation, we correct for finite size effects. Thus we study

$$E_G \equiv E_g - \frac{U}{M} - E_g(U = 0). \tag{8}$$

The Coulomb interaction gives a contribution U/M even for a perfectly itinerant system and $E_g(U = 0)$ is the gap in the one-particle spectrum. Both U/M and $E_g(U = 0)$ go to zero for $M \to \infty$, so the subtracted quantities improve the convergence but do not change the $M \to \infty$ result.

In Fig. 4 we show results for $\tilde{E}_g$ which suggest a Mott-Hubbard transition for U between 1.5 and 1.75 eV[34]. This corresponds to a critical ratio $U_c/W \sim 2.5$. For a fully frustrated system with infinite dimension but without orbital degeneracy $U_c/W = 1.5$ was obtained[36]. Multiplying by the degeneracy factor $\sqrt{3}$ leads to the ratio $\sqrt{3} \times 1.5 = 2.6$ in surprisingly good agreement with our results. It is interesting to note that for U=1.75, which gives an insulator, an antiferromagnetic trial function results in the lowest energy for the larger systems. A stronger degeneracy dependence, $\sim (N + 1)$, was found by Lu[43], who used a Gutzwiller Ansatz together with the Gutzwiller approximation[42].

Since the ratio U/W for A_3C_{60} is believed to be in the range 1.5-2.5, it is very likely that even stoichiometric A_3C_{60} (A= K, Rb) is on the metallic side of a Mott-Hubbard transition. This is also strongly supported by experimental data for T_c as a function of doping[44]. T_c has a pronounced maximum close to filling three and it drops quickly as the filling is increased or decreased[44]. If stoichiometric A_3C_{60} were an insulator, however, one would instead expect T_c to initially increase with increasing deviation in

Table 1: The partial electron-phonon coupling constants $\lambda_\nu/N(0)$ (in eV) according to different theoretical calculations. The energies (in cm^{-1}) of the modes for the undoped system are shown.

Mode	Energy	$\lambda_\nu/N(0)$			
		Varma[17]	Schluter[18]	Antropov[27]	Faulhaber[48]
$H_g(8)$	1575	.011	.009	.022	.009
$H_g(7)$	1428	.034	.013	.020	.015
$H_g(6)$	1250	.000	.003	.008	.002
$H_g(5)$	1099	.006	.001	.003	.002
$H_g(4)$	774	.000	.007	.003	.010
$H_g(3)$	710	.001	.004	.003	.001
$H_g(2)$	437	.001	.007	.006	.010
$H_g(1)$	273	.003	.008	.003	.001
$\sum H_g$		.056	.052	.068	.049

the valence from three, as in the High T_c compounds. These experiments therefore seem to establish that even stoichiometric A_3C_{60} are metals. We note, however, that the doped fullerenes appear to be close to a metal-insulator transition, as expected from the numbers above. Thus $NH_3K_3C_{60}$ at normal pressure and low temperatures is an insulator[45], but becomes a metal under pressure[46]. Similarly, A_4C_{60} are insulators at normal pressure, but Rb_4C_{60} has been shown to be a metal under pressure[47].

ELECTRON-PHONON INTERACTION

The superconductivity in A_3C_{60} is believed to be driven by the electron-phonon interaction. It is therefore of particular interest to obtain the electron-phonon coupling parameters. There are a large number of experimental and theoretical estimates of parameters, but there is still a substantial disagreement between these estimates.

Table 1 shows the results of various theoretical calculations. The calculations of Varma et al.[17] are based on the semi-empirical MNDO method, Schluter et al.[18] used an empirical phonon model together with an *ab initio* (LDA) electronic treatment and Antropov et al.[27] and Faulhaber et al.[48] performed *ab initio* calculations. All these calculations agree that the strongest coupling is to one of the two highest H_g modes. Apart from this, there are, however, substantial variations. Varma et al. find almost no coupling to the low-lying phonons, while the other calculations find a somewhat stronger coupling to these modes. In particular in the calculations of Schluter et al. and Faulhaber et al. the coupling is somewhat more evenly distributed between the modes.

In view of these deviations between the different calculations, it is interesting to consider experimental estimates. There are several estimates based on Raman and neutron scattering against phonons. In the metallic systems, a phonon can decay in an electron-hole pair. This leads to an extra broadening, which can be estimated by comparing the phonon widths in the doped and undoped systems. If Migdal's theorem is valid, the extra broadening can be directly related to $\lambda_\nu N(0)$[49], where $N(0)$ is the density of state at the Fermi energy. Some estimates of this type are collected in Table 2.

An alternative approach is to use photoemission from free negatively charged C_{60}^-

Table 2: The partial electron-phonon coupling constants $\lambda_\nu/N(0)$ (in eV) according to different experimental estimates. The energies (in cm^{-1}) of the modes for the undoped system are also shown.

Mode	Energy		$\lambda_\nu/N(0)$		
		Raman[50]	Neutron[51]	Raman[52]	PES[53]
$H_g(8)$	1575			.002	.023
$H_g(7)$	1428			.003	.017
$H_g(6)$	1250			.001	.005
$H_g(5)$	1099			.001	.012
$H_g(4)$	774		.005	.002	.018
$H_g(3)$	710		.001	.002	.013
$H_g(2)$	437	.022	.023	.014	.040
$H_g(1)$	273		.014	.034	.019

molecules. In this process an electron is emitted from the t_{1u} level. During the emission one or several phonons may be excited. This leads to phonon satellites, and the strength of these satellites are related to the coupling constants. Fig. 5 shows the photoemission spectrum[53]. The coupling to the two lowest, the third and fourth and the two highest H_g phonons show up as three clear structures.

To extract the coupling parameters from the photoemission spectrum, we need to calculate the spectrum for different couplings and vary these couplings until the optimum agreement with experiment is obtained. The corresponding coupling parameters are the estimates derived from photoemission. To perform these calculations, we consider a model with a linear coupling to harmonic phonons

$$H = \varepsilon_0 \sum_{m=1}^{3} \psi_m^\dagger \psi_m + \sum_{\nu=1}^{42} \omega_\nu (b_\nu^\dagger b_\nu + \frac{1}{2}) + \sum_{m=1}^{3} \sum_{n=1}^{3} \sum_{\nu=1}^{42} c_{nm}^\nu \psi_m^\dagger \psi_n (b_\nu + b_\nu^\dagger), \qquad (9)$$

where the first term describes the electron, the second term the phonons and the third term the electron-phonon interaction. $\psi_m^\dagger$ creates an electron in one of the three t_{1u} states, $b_\nu^\dagger$ creates a phonon in one of the 42 phonon modes (eight five-fold degenerate H_g phonons or 2 nondegenerate A_g phonons) and c_{mn}^ν is the coupling constant for the scattering of an electron from state n to state m under the creation or annihilation of a phonon of type ν. The form of the constants c_{nm}^ν for the five degenerate H_g modes is determined by symmetry[54], and there is therefore just one unknown parameter for each of the eight H_g modes and for the two A_g modes describing the absolute strength of the coupling. The relation of these coupling constants to the coupling λ_ν entering in superconductivity is given in the literature[54]. The phonon frequencies were obtained from experiment[55].

The vibration temperature T is of the order 200K, which is substantially lower than the energy of the lowest phonon mode $\sim$ 400 K. We therefore assume that $T = 0$ and consider the ground-state wave function[53],

$$|\Phi> = [\sum_{m=1}^{3} a_m \psi_m^\dagger + \sum_{m=1}^{3} \sum_{\nu=1}^{42} a_{m;\nu} \psi_m^\dagger b_\nu^\dagger + \sum_{m=1}^{3} \sum_{\mu=1}^{42} \sum_{\nu=1}^{\mu} a_{m;\mu,\nu} \psi_m^\dagger b_\mu^\dagger b_\nu^\dagger + ...]|vac>, \qquad (10)$$

where the first term describes a state with no phonon, the second term a state with one phonon and so on. We have considered states with up to five phonons. The Hamiltonian

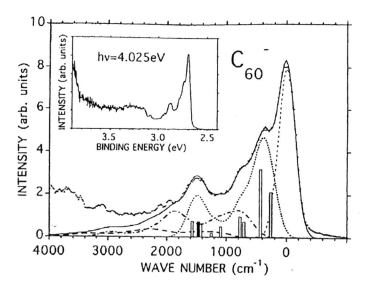

Figure 5: The experimental (dots) and theoretical (full line) photoemission spectrum of C_{60}^-. The theoretical no loss (dashed), single loss (dotted) and double loss (dashed-dotted) curves are also shown. The contributions of the different modes to the single loss curve are given by bars (H_g: open, A_g: solid). The inset shows the experimental spectrum over a larger energy range.

matrix corresponding to these basis states is calculated. The lowest eigenvalue and the corresponding eigenvector are calculated using Lanczos method.

We use the sudden approximation[56], where the emitted electron is assumed not to interact with the system left behind. It is further assumed that the energy dependence of the dipole matrix element to the final state can be neglected[56]. For the low photon energies considered here, the accuracy of these standard approximations is not clear. However, spectra measured with lower energy resolution at different photon energies ranging from 3.5 to 6.4 eV show the same general shape of the photoemission features and therefore support the assumptions above[53].

For the final state of the neutral C_{60} molecule, there is no coupling between states with different number of phonons in the present model, since the electron has been emitted, and the eigenstates are therefore trivial. The photoemission spectrum is then expressed in terms of the coefficients in Eq. (2). A Gaussian broadening with the width (FWHM) 41 meV is introduced to take into account the experimental resolution.

The spectrum in Fig. 5 does not allow us to uniquely determine all the coupling parameters. We have therefore fixed the couplings for the A_g modes to their calculated values[27]. The determination of the remaining eight coupling parameters for the H_g modes is then unique within the present framework. The results are shown in Table 2.

The different experimental estimates agree that there is a large coupling to at least on of the two low-lying H_g modes. This is in strong contrast to most theoretical calculations, which find the strongest coupling to one of the two highest modes, and a rather weak coupling to the lowest modes. This is illustrated in Table 1. The methodology used in the estimates based on photoemission on the one hand and Raman and neutron scattering on the other hand are very different. Since these two types of estimates nevertheless agree that the coupling to the lowest modes is strong, we believe this to be correct. It is then an interesting question why the theoretical calculations

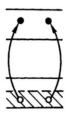

Figure 6: Schematic picture of two subbands and the virtual Coulomb scattering of two electrons up into the higher subband.

find this coupling to be weak. It is less clear how strong the couplings to the upper modes are. The theoretical calculations and the estimates based on photoemission find these couplings to be strong, while the Raman scattering estimates find weak couplings.

COULOMB PSEUDOPOTENTIAL

These estimates of the electron-phonon interaction lead to a phonon induced attractive interaction of the order of 1/10 eV. At the same time there is a repulsive interaction U of the order 1.5 eV[23]. It is usually argued in the context of superconductivity that the effects of the Coulomb interaction are strongly reduced by retardation effects. The energy scales for electrons and phonons are usually very different. An electron can therefore move through the solid and excite phonons to lower its energy. A second electron can follow substantially later, thereby feeling little of the repulsion from the first electron, but still take advantage of the attraction from the nuclei, which may not have had time to return to their original positions. For the A_3C_{60} compounds the relative size of the electronic and phononic energy scales is less clear. An important question in this context is whether the electronic energy scale is determined by the total width E_I of the 2s and 2p states, or just by the width of the t_{1u} band W. In the latter case the electrons and the phonons would have a comparable energy scale and the retardation effects may not be important[31].

A dimensionless quantity

$$\mu = UN(0) \tag{11}$$

is introduced, where U is some typical Coulomb interaction and $N(0)$ is the density of states per spin. Due to retardation and other effects, μ is renormalized to μ^*, the Coulomb pseudopotential.

If multiple scattering of the electrons up into higher energy states is allowed, they can move in a correlated way and thereby reduce the Coulomb interaction but still have an attractive interaction via the phonons. This is shown schematically in Fig. 6 for the scattering up into a higher subband. Mathematically, this is usually expressed by summing ladder diagrams (see Fig. 7). In simple models, e.g., assuming that all Coulomb matrix elements are equal, the ladder diagrams can be summed, giving the effective interaction[57, 58]

$$\mu^* = \frac{\mu}{1 + \mu \log(B/\omega_{ph})}, \tag{12}$$

where B is a typical electron energy (half the band width) and ω_{ph} is a typical phonon energy. If $B/\omega_{ph} >> 1$, μ^* can be strongly reduced relative to μ. In this limit, Eq. (12) simplifies to $\mu^* \approx 1/\log(B/\omega_{ph})$, which may be of the order 0.1-0.2.

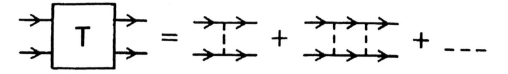

Figure 7: Ladder diagrams describing the repeated scattering of two electrons. A full line represents an electron and a dashed line the (screened) Coulomb interaction.

Table 3: The intraband Coulomb interaction w^{intra} between four Bloch states with equal band indices and the largest interband scattering matrix element w^{inter} ($t_{1u} \to t_{1g}$). We consider both a free molecule (Mol) and a solid (Solid), where in the latter case $q = G/2$ and $q = G/20$, where $G = 2\pi/a(1, 1, 1)$. The unscreened (Unscr) and screened (Screen) interactions in the undoped (Undop) and, for the solid, doped systems are shown.

System		$w^{intra}(q)$		$w^{inter}(q)$	
		G/20	G/2	G/20	G/2
Unscr	Mol	3.81		0.77	
Screen	Mol	3.72		0.14	
Unscr	Solid	176.8	2.03	1.01	0.98
Undop	Solid	78.9	1.68	0.15	0.15
Doped	Solid	0.094	0.065	0.10	0.10

We first discuss the retardation effects on μ^* within the formalism described above, where ladder diagrams in the screened Coulomb interaction are summed. Later we discuss effects beyond this formalism. A crucial quantity is the Coulomb matrix element for scattering two electrons from the t_{1u} band up into higher subbands (see Fig. 6). If these matrix elements are very small, the higher subbands should play a small role in this formalism.

The matrix elements of the screened Coulomb interaction have been studied in the RPA[32]. The system was described in a tight-binding approach with one 2s- and three 2p-orbitals per carbon atom, and the screening was calculated with local-field effects included. A long-range Coulomb interaction was included, which involved an interaction between different atoms both on the same C_{60} molecule and on different molecules. The Coulomb interaction on a carbon atom was set equal to $U_C = 12$ eV. The results are shown in Table 3. The table shows both intraband and interband matrix elements where two electrons are scattered between different bands.

Because of the very efficient metallic screening of the intraband matrix elements, at least within RPA, these matrix elements are smaller than the largest interband matrix elements. These results depend crucially on the fact that these matrix elements have very different character, with the intraband matrix elements involving monopole interactions while the interband scattering matrix elements involving multipole interaction. This behavior is also very different from the electron gas, since there is an abrupt change in character when going from one subband to another, even if they are close in energy, due to the fact that each subband is derived from a different molecular level with a different spatial character. Since Eq. (12) is derived assuming that all matrix elements are equal, it can not be used here. Approximate summing of the ladder diagrams, using the calculated matrix elements for the screened interaction, shows that μ^* is close to zero in this formalism[32].

It is interesting to ask why the intraband matrix elements are so small. In the RPA, the screened interaction can be written as

$$w = (1 - vP)^{-1}v \tag{13}$$

where w is a matrix representing the different screened Coulomb matrix elements, v is the corresponding unscreened matrix elements and P is the polarizability. For the intraband matrix elements, the t_{1u} orbitals dominate the screening, and we can transform to the corresponding basis, neglecting all other orbitals. For small $|q|$, the diagonal elements of P are then related to the density of states $N(0)$ per spin. Since both $N(0)$ and the diagonal elements of v are large, we can assume that the product is much larger than unity. With appropriate assumptions one can then derive [32] that the intraband matrix elements are

$$w^{intra} \approx \frac{1}{2N(0)}, \tag{14}$$

for small $|q|$. To obtain the unrenormalized μ we multiply by $N(0)$ obtaining $\mu \approx 0.5$. Averaging over q gives $\mu \approx 0.4$. We note that w becomes very small in the RPA for large $N(0)$, since in the RPA the cost of screening is purely a kinetic energy cost, and for a large $N(0)$ this cost is very small. For large values of U, we expect RPA to become a poor approximation, and it is therefore an interesting question how accurate RPA is in this case.

The considerations above were all within a framework where ladder diagrams in the screened interaction are summed. This raises the question about the validity of neglecting other diagrams. For instance, it is possible to construct other sets of diagrams, which to a large extent cancel the ladder diagrams. This raises important questions about which diagrams to include.

To circumvent this problem, a two-band model is studied, where some exact results can be obtained[32]. We consider the Hamiltonian

$$H = \sum_i \sum_{n=1}^2 \sum_\sigma \varepsilon_n n_{in\sigma} + t \sum_{\langle ij \rangle} \sum_{n=1}^2 \sum_\sigma \psi_{in\sigma}^\dagger \psi_{jn\sigma}$$
$$+ U_{11} \sum_i \sum_{nn'\sigma\sigma'}{}' n_{in\sigma} n_{in'\sigma'} + U_{12} \sum_i [\psi_{i2\uparrow}^\dagger \psi_{i2\downarrow}^\dagger \psi_{i1\downarrow} \psi_{i1\uparrow} + h.c.], \tag{15}$$

where i is a site index, ε_n the energies of the two levels, t is the hopping integral, U_{11} is the intraband interaction and U_{12} is the interband scattering. The particularly strong interband scattering between the t_{1u} and t_{1g} bands in C_{60} is of a similar type, namely the excitation of two electrons between two bands (see Fig. 6). It is then assumed that

$$\Delta\varepsilon \equiv \varepsilon_2 - \varepsilon_1 \gg U_{12}, t. \tag{16}$$

In this limit, the upper subband can be projected out and a new effective Hamiltonian be obtained, which describes low-energy properties correctly. This process does not generate any new terms for Hamiltonian (15), but it renormalizes the intraband interaction to

$$U_{eff} = U_{11} - \frac{U_{12}^2}{2\Delta\varepsilon}. \tag{17}$$

If it is now assumed that $U_{12}^2/(2\Delta\varepsilon) \ll U_{11}$, the renormalization of the intraband interaction is small. The properties of this effective Hamiltonian then differ very little from the one-band model, where the upper band was completely neglected. Thus the upper band has a small influence on the low energy properties in general and μ^* in particular. For instance, in RPA the screened interaction is $1/[2N(0)]$ in both the one-band and two-band models, if $U_{11}N(0) \gg 1$.

On the other hand, the ladder diagrams in the screened interaction due to scattering into the upper subband can be summed. If it is assumed that

$$\frac{1}{N(0)} \lesssim \frac{U_{12}^2}{2\Delta\varepsilon} \qquad (18)$$

this approach predicts a large renormalization of μ^*. The ladder diagrams subtract a quantity $U_{12}^2/(2\Delta\varepsilon)$, which, although small compared with U_{11}, is large compared with the screened interaction in the limit considered. Thus the summation of ladder diagrams leads to a qualitatively incorrect result in this case. The projection method above and the summation of ladder diagrams describe similar physics. The difference is that in the rigorous projection method, the high energy degrees of freedom were eliminated first, while in the ladder diagram approach the low energy degrees of freedom were treated (approximately) first (by introducing the metallic screening) and the high energy degrees of freedom later (by summing the ladder diagrams). This approach involves uncontrolled approximations.

The model above is too simple to be directly applicable to A_3C_{60}, but it still has important implications. To further support these arguments, we consider a collection of C_{60} molecules, where we have put all hopping and Coulomb matrix elements between different molecules equal to zero. Since there is no hopping between the molecules there are no retardation effects. The reduction of the electron-electron interaction is instead entirely due to intramolecular correlations. These correlations only reduce the Coulomb U for two electrons on the same molecule from about 4 eV (the value of an unscreened Coulomb integral) to the experimental value of about 3 eV. Guided by the results above, we next consider ladder diagrams in the *unscreened* Coulomb interaction. Carrying out this procedure for C_{60} leads to a relatively small correction[32], as it should. Actually, the reduction is too small, since the ladder diagrams leave out important effects. We next include long-ranged Coulomb matrix elements, between different molecules, but still keep the hopping matrix elements equal to zero. This introduces new physics in terms of the electrons on one molecule polarizing the surrounding molecules. Still the reduction of U is moderate from about 3 eV to about 1.5 eV, according to experiment. Such a value of U would still lead to a huge μ of the order of $1.5*7 \sim 10$. Finally we turn on the hopping integrals between the molecules, allowing for retardation effects. We are interested in the interaction between electrons in the t_{1u} band, and in summing ladder diagram we should then consider the corresponding energy arguments. For higher-lying bands, the broadening of a sharp level into a narrow subband then has a very small effect on the energy denominators in the ladder diagrams. Therefore a summation of ladder diagrams in the unscreened interaction would not give a very different result than for the isolated molecules. Since these differences represent the retardation effects, we conclude that the ladder diagrams involving higher subbands should not contribute much to these effects. This does not necessarily mean that these diagrams are negligible, but the contributions in the limit of discrete levels are contributions to the intramolecular correlation. Indeed, we would expect these arguments to hold true for many classes of diagrams. This then reinforces the results of the model calculation that retardation effects due to higher subbands are not very important.

For simplicity one may therefore assume that the renormalization from the higher sub bands due to retardation effects can be neglected completely. If in addition RPA screening is used, one arrives at $\mu^* \sim 0.4$. The approach above has only addressed the renormalization due to the higher subbands but not due to processes within the t_{1u}

band itself. If we assume that the traditional theory is valid for these processes, μ^* is renormalized to 0.3 according to formula (12), if we set $B = 0.25$ eV and $\omega_{ph} = 0.1$ eV. Such a value is then to be used in the McMillan formula, while the value 0.4 applies to the Eliashberg equation including one subband. We notice that only static screening was considered above, although inclusion of dynamic effects may influence the results substantially (Rietschel and Sham, 1983). It should therefore be emphasized that our treatment of the renormalization processes within the t_{1u} band contains a number of uncontrolled approximation, and that this treatment is of a much poorer quality than the treatment of the processes involving higher sub bands.

We have used the Eliashberg theory together with the coupling constants derived from photoemission and $N(0) = 7.2$ states/(eV-spin) derived from NMR. We then find that to obtain the experimental value of T_c for K_3C_{60} we have to empirically increase the value of μ^* to 0.6[53]. We have calculated the isotope effect and found $\alpha = 0.32$ for K_3C_{60} and $\alpha = 0.37$ for Rb_3C_{60}. There are several measurements of the isotope effect for Rb_3C_{60}[59, 60, 61, 62], giving very different results. The probably most reliable measurement, using 99 % substitution, gave $\alpha = 0.30 \pm 0.05$[62]. We have further calculated the reduced gap $2\Delta/T_c$ and found 3.59 for K_3C_{60} and 3.66 for Rb_3C_{60}. Such BCS-like (3.52) values are also found in optical experiments[63] (3.44 and 3.45 for K_3C_{60} and Rb_3C_{60}, respectively) and in muon spin relaxation experiments[64] (3.6 for Rb_3C_{60}), while point contact tunneling experiments give a much larger result[65] (5.3 for Rb_3C_{60}). Recent optical and tunnelling measurements have also given larger values of the gap (4.1)[66].

Although this approach involves substantial uncertainties and approximations, it suggests that the electron-phonon interaction is strong enough to explain the superconductivity, even if retardation effects are much less efficient than normally assumed for conventional superconductors.

CONCLUDING REMARKS

We have shown that the orbital degeneracy N plays an important role, increasing the critical value of the Coulomb interaction U where the Mott-Hubbard metal-insulator takes place. This explains why stoichiometric A_3C_{60} can be metallic in spite of the large ratio U/W in these systems. We have discussed theoretical and experimental estimates of the electron-phonon coupling constants λ_ν. Although most estimates find a total coupling of the order of $\lambda \sim 0.5 - 1$, there are substantial uncertainties in the distribution of the coupling between the different modes. The Coulomb pseudopotential μ^* was discussed, with the emphasis on retardation effects. It was argued that the higher sub bands give a small contribution to these effects. This, however, leaves open questions about Coulomb processes within the t_{1u} band. Although the retardation effects are found to be inefficient at reducing μ^*, the screening, at least within RPA, leads to a large reduction of μ^*. The theory is therefore consistent with the electron-phonon interaction driving the superconductivity.

References

[1] H.W. Kroto, J.R. Heath, S.C. O'Brien, R.F. Curl, and R.E. Smalley, Nature **318**, 162 (1985).

[2] W. Krätschmer, L.D. Lamb, K. Fostiropoulos, and D.R. Huffman, Nature **347**, 354 (1990).

[3] R.C. Haddon, A.F. Hebard, M.J. Rosseinsky, D.W. Murphy, S.J. Duclos, K.B. Lyons, B. Miller, J.M. Rosamilia, R.M. Fleming, A.R. Kortan, S.H. Glarum, A.V. Makhija, A.J. Muller, R.H. Eick, S.M. Zahurak, R. Tycko, G. Dabbagh, and F.A. Thiel, Nature **350**, 320 (1991).

[4] A.F. Hebard, M.J. Rosseinsky, R.C. Haddon, D.W. Murphy, S.H. Glarum, T.T.M. Palstra, A.P. Ramirez, and A.R. Kortan, Nature **350**, 600 (1991).

[5] M.J. Rosseinsky, A.P. Ramirez, S.H. Glarum, D.W. Murphy, R.C., Haddon, A.F. Hebard, T.T.M. Palstra, A.R. Kortan, S.M. Zahurak, and A.V. Makhija, Phys. Rev. Lett. **66**, 2830 (1991).

[6] K. Tanigaki, T.W. Ebbesen, S. Saito, J. Mizuki, J.S. Tsai, Y. Kubo, and S. Kuroshima, Nature **352**, 222 (1991).

[7] T.T.M. Palstra, O. Zhou, Y. Iwasa, P.E. Sulewski, R.M. Fleming, and B.R. Zegarski, Solid State Commun. **93**, 327 (1995).

[8] S.C. Erwin and W.E. Pickett, Science **254**, 842 (1991).

[9] S. Saito, and A. Oshiyama, Phys. Rev. **44**, 11532 (1991).

[10] J.L. Martins, and N. Troullier, Phys. Rev. B **46**, 1766 (1992).

[11] S. Satpathy, V.P. Antropov, O.K. Andersen, O. Jepsen, O. Gunnarsson, and A.I. Liechtenstein, Phys. Rev. B **46**, 1773 (1992).

[12] D.W. Murphy, M.J. Rosseinsky, R.M. Fleming, R. Tycko, A.P. Ramirez, R.C. Haddon, T. Siegrist, G. Dabbagh, J.C. Tully, and R.E. Walstedt, J. Phys. Chem. Solids **53**, 1321 (1992).

[13] R.F. Kiefl, T.L. Duty, J.W. Schneider, A. MacFarlane, K. Chow, J.W. Elzey, P. Mendels, G.D. Morris, J.H. Brewer, E.J. Ansaldo, C Niedermayer, D.R. Noakes, C.E. Stronach, B. Hitti, and J.E. Fischer, Phys. Rev. Lett. **69**, 2005 (1992).

[14] S.C. Erwin, 1993, in *Buckminsterfullerenes*, Eds. W.E. Billups and M.A. Ciufolini (VCH Publishers, New York), p. 217; S.C. Erwin and C. Bruder, Physica B **199-200**, 600 (1994).

[15] P.W. Stephens, L. Mihaly, P.L. Lee, R.L. Whetten, S.-M. Huang, R. Kaner, F. Deiderich, and K. Holczer, Nature **351**, 632 (1991).

[16] M.P. Gelfand and J.P. Lu, Phys. Rev. Lett. **68**, 1050 (1992); Phys. Rev. B **46**, 4367 (1992); Phys. Rev. B **47**, 4149 (1993).

[17] C.M. Varma, J. Zaanen, and K. Raghavachari, Science **254**, 989 (1991).

[18] M. Schluter, M. Lanno, M. Needels, G.A. Baraff, and D. Tomanek, Phys. Rev. Lett. **68**, 526 (1992); J. Chem. Phys. Solids **53**, 1473 (1992).

[19] I.I. Mazin, S.N. Rashkeev, V.P. Antropov, O. Jepsen, A.I. Liechtenstein, and O.K. Andersen, Phys. Rev. B **45**, 5114 (1992).

[20] S. Chakravarty and S. Kivelson, Europhys. Lett. **16**, 751 (1991); S. Chakravarty, M.P. Gelfand and S. Kivelson, Science **254**, 970 (1991).

[21] G. Baskaran and E. Tosatti, Current Science **61**, 33 (1991).

[22] R. Friedberg, T.D. Lee and H.C. Ren, Phys. Rev. B **46**, 14150 (1992).

[23] R.W. Lof, M.A. van Veenendaal, B. Koopmans, H.T. Jonkman, and G.A. Sawatzky, Phys. Rev. Lett. **68**, 3924 (1992).

[24] P.A. Brühwiler, A.J. Maxwell, A. Nilsson, N. Mårtensson, and O. Gunnarsson, Phys. Rev. B **48**, 18296 (1992).

[25] Pederson, M.R., and A.A. Quong, 1992, Phys. Rev. B **46**, 13584.

[26] V.P. Antropov, O. Gunnarsson, and O. Jepsen, 1992, Phys. Rev. B **46**, 13647.

[27] V.P. Antropov, O. Gunnarsson and A.I. Liechtenstein, Phys. Rev. B **48**, 7551 (1993).

[28] M. Knupfer, M. Merkel, M.S. Golden, J. Fink, O. Gunnarsson, V.P. Antropov, Phys. Rev. B **47**, 13944 (1993).

[29] O. Gunnarsson, V. Eyert, M. Knupfer, J. Fink, and J.F. Armbruster, J. Phys.: Cond. Matter **8**, 2557 (1886).

[30] A.I. Liechtenstein, O. Gunnarsson, M. Knupfer, J. Fink, and J.F. Armbruster, J. Phys.: Cond. Matter **8**, 4001 (1886).

[31] P.W. Anderson, preprint (1991).

[32] O. Gunnarsson and G. Zwicknagl, Phys. Rev. Lett. **69**, 957 (1992); O. Gunnarsson, D. Rainer, and G. Zwicknagl, Int. J. Mod. Phys. B **6**, 3993 (1992).

[33] O. Gunnarsson, Phys. Rev. B **51**, 3493 (1995).

[34] O. Gunnarsson, E. Koch, and R.M. Martin, Phys. Rev. B **54**, R11026 (1966).

[35] O. Gunnarssn, Rev. Mod. Phys. **69**, 575 (1997).

[36] See, A. Georges, G. Kotliar, W. Krauth, and M.J. Rozenberg, Rev. Mod. Phys. **68**, 13 (1996).

[37] O. Gunnarsson and K. Schönhammer, Phys. Rev. Lett. **50**, 604 (1983); Phys. Rev. B **28**, 4315 (1983).

[38] O. Gunnarsson, E. Koch, and R.M. Martin, Phys. Rev. B (July 15th, 1997).

[39] O. Gunnarsson, S. Satpathy, O. Jepsen, and O.K. Andersen, Phys. Rev. Lett. **67**, 3002 (1991).

[40] I.I. Mazin, A.I. Liechtenstein, O. Gunnarsson, O.K. Andersen, V.P. Antropov, and S.E. Burkov, Phys. Rev. Lett. **26**, 4142 (1993).

[41] D.F.B. ten Haaf, H.J.M. van Bemmel, J.M.J. van Leeuwen, W. van Saarloos, and D.M. Ceperley, Phys. Rev. B **51**, 353 (1995); H.J.M. van Bemmel, D.F.B. ten Haaf, W. van Saarloos, J.M.J. van Leeuwen, and G. An, Phys. Rev. Lett. **72**, 2442 (1994).

[42] M.C. Gutzwiller, Phys. Rev. **137**, A1726 (1965).

[43] J.P. Lu, Phys. Rev. B **49**, 5687 (1994).

[44] T. Yildirim, L. Barbedette, J.E. Fischer, C.L. Lin, J. Robert, P. Petit, and T.T.M. Palstra, Phys. Rev. Lett. **77**, 167 (1996).

[45] Y. Iwasa et al., Phys. Rev. B **53**, R8836 (1996).

[46] O. Zhou, T.T.M. Palstra, Y. Iwasa, R.M. Flemming, A.F. Hebard, P.E. Sulewski, D.W. Murphy, and B.R. Zegarski, Phys. Rev. B **52**, 483 (1995).

[47] R. Kerkoud et al.R. Kerkoud et al., J. Phys. Chem Solids **57**, 143 (1996).

[48] J.C.R. Faulhaber, D.Y.K. Ko, P.R. Briddon, Phys. Rev. B **48**, 661 (1993).

[49] P.B. Allen, Phys. Rev. B **6**, 2577 (1972). Observe a misprint, which was corrected in Allen, P.B., Solid State Commun. **14**, 937 (1974).

[50] M.G. Mitch, S.J. Chase, and J.S. Lannin, Phys. Rev Lett. **68**, 833 (1992); Phys. Rev. B **46**, 3696 (1992).

[51] K. Prassides, C. Christides, M.J. Rosseinsky, J. Tomkinson, D.W. Murphy, and R.C. Haddon, Europhys. Lett. **19**, 629 (1994).

[52] J. Winter and H. Kuzmany, Phys. Rev. B **53**, 655 (1996).

[53] O. Gunnarsson, H. Handschuh, P.S. Bechthold, B. Kessler, G. Ganteför and W. Eberhardt, Phys. Rev. Lett. **74**, 1875 (1995).

[54] M. Lannoo, G.A. Baraff, M. Schluter, and D. Tomanek, Phys. Rev. B **44**, 12106 (1991).

[55] D.S. Bethune, G. Meijer, W.C. Tang, H.J. Rosen, W.G. Golden, H. Seki, C.A. Brown, and M.S. de Vries, Chem. Phys. Lett. **179**, 181 (1991).

[56] L. Hedin and S. Lundqvist, in *Solid State Physics*, edited by H. Ehrenreich, D. Turnball, and F. Seitz (Academic, New York, 1969), Vol. 23, p. 1.

[57] N.N. Bogoliubov, Nuovo Cimento **7**, 794 (1958); P. Morel and P.W. Anderson, Phys. Rev. **125**, 1263 (1962); J.R. Schrieffer, *Theory of superconductivity* (Benjamin, New York, 1964).

[58] V.L. Ginzburg and D.A. Kirzhnits, *High-Temperature Superconductivity*, Consultants Bureau, New York, 1982.

[59] A.P. Ramirez, A.R. Kortan, M.J. Rosseinsky, S.J. Doclos, A.M. Mujsce, R.C. Haddon, D.W. Murphy, A.V. Makhija, S.M. Zahurak, and K.B. Lyons, Phys. Rev. Lett. **68**, 1058 (1992).

[60] T.W. Ebbesen, J.S. Tsai, K. Tanigaki, J. Tabuchi, Y. Shimakawa, Y. Kubo, I. Hirosawa, and J. Mizuki, Nature **355**, 620 (1992).

[61] A.A. Zakhidov, K. Imaeda, D.M. Petty, K. Yakushi, H. Inokuchi, K. Kikuchi, I. Ikemoto, S. Suzuki, and Y. Achiba, Phys. Lett. A **164**, 355 (1992).

[62] C.C. Chen and C.M. Lieber, Science **259**, 655 (1993).

[63] L. Degiorgi, G. Briceno, M.S. Fuhrer, A. Zettl, and P. Wachter, Nature **369**, 541 (1994).

[64] R.F. Kiefl, W.A. MacFarlane, K.H. Chow, S. Dunsiger, T.L. Duty, T.M.S. Johnston, J.W. Schneider, J. Sonier, L. Brard, R.M. Strongin, J.E. Fischer, and A.B. Smith III, Phys. Rev. Lett. **70**, 3987 (1993).

[65] Z. Zhang, C.-C. Chen, S.P. Kelty, H. Dai, and C.M. Lieber, Nature **353**, 333 (1991).

[66] D. Koller, M.C. Martin, L. Mihaly, G. Mihaly, G. Oszlanyi, G. Baumgartner, and L. Forro, Phys. Rev. Lett. **77**, 4082 (1996).

ON THE PAIR CORRELATIONS BETWEEN ELECTROSOLITONS

L. S. Brizhik, A. A. Eremko

Bogolyubov Institute for Theoretical Physics
252143 Kyiv, Ukraine

INTRODUCTION

The concept of 'molecular soliton' [1] is intensively used to explain various phenomena in biological physics and condensed matter physics [2-4]. In this respect there arises the question of soliton interactions in low-dimensional molecular systems. It was shown in [5,6] that in the adiabatic approximation at moderate enough values of electron-phonon interaction constant two electrosolitons with opposite spins due to their interaction with the local deformation of the chain bind in a localized bound singlet spin state called 'bisoliton'. The binding energy of such state depends both on the exchange and electron-phonon interaction constants, as well as on the velocity of the bisoliton. As it was shown in [7], in the strong nonadiabatic limit, two extra electrons with opposite spins form a bound delocalized state. The interaction of two electrosolitons with parallel spins was investigated in the zero adiabatic approximation in [8] where it was shown that the triplet bound state is unstable with respect to the decay into two solitons. Here we investigate the interaction of two extra electrons with parallel spins and the interaction of two singlet bisolitons, with account of the nonadiabaticity of the system and show that the latter effect results in the 'direct' interaction between the solitons which stabilizes the two-soliton triplet state.

1 General description of the model

Let us consider a chain of the length $\mathcal{L} = Na$ (a is the lattice constant, and we assume $N \gg 1$) with few extra electrons in the conductive band. Neglecting the Coulomb repulsion between electrons, such chain can be described by the Fröhlich Hamiltonian

$$H = \sum_{\sigma,k} E(k) A^+_{\sigma,k} A_{\sigma,k} + \frac{1}{\sqrt{N}} \sum_{\sigma,k,q} \chi(q) A^+_{\sigma,k} A_{\sigma,k-q} (b_q + b^+_{-q}) + \sum_q \hbar\omega_q b^+_q b_q. \quad (1.1)$$

Here $A^+_{\sigma,k}$, $A_{\sigma,k}$ are creation and annihilation operators of the electron with spin projection σ and wave number k in the conductive band with the dispersion law $E(k)$, b^+_q, b_q are creation and annihilation operators of phonons with wave number q and frequency ω_q, the value $\chi(q)$ is determined by a short-range interaction of electrons with atom displacements (in a simple chain such displacements correspond to acoustic phonons). In a long-wave approximation, counting the energy from the electron band bottom, one can write

$$E(k) = \frac{\hbar^2 k^2}{2m}, \quad \omega_q = V_a |q|, \quad \chi(q) = i\chi qa \sqrt{\frac{\hbar}{2M\omega_q}} \quad (1.2)$$

where m is the effective mass of an electron in the conductive band, $V_a = a\sqrt{w/M}$ is sound velocity in a chain, w is the elasticity coefficient, and χ is the electron-phonon interaction constant.

Consider the Schrödinger equation for the state vector of the system described by the Hamiltonian (1)

$$i\hbar \frac{\partial}{\partial t} |\Psi(t)> = H|\Psi(t)>. \quad (1.3)$$

The operator of the total momentum of the system

$$P = \sum_{\sigma,k} \hbar k A^+_{\sigma,k} A_{\sigma,k} + \sum_q \hbar q b^+_q b_q \quad (1.4)$$

which is connected with the translation operator $T = exp(iPx/\hbar)$, commutes with the Hamiltonian (1). The energy of a stationary state depends on the eigenvalue p of the operator P, and every value of p corresponds to the movement with group velocity $V = dE(p)/dp$. We can pass to the coordinate system moving with this velocity

$$|\Psi(t)> = \exp(-iPVt/\hbar) |\tilde{\Psi}(t)>, \quad (1.5)$$

and rewrite the Schrödinger equation for stationary states in the moving reference frame

$$(H - VP) |\tilde{\Psi}> = E |\tilde{\Psi}>. \quad (1.6)$$

Let us look for the state vector in the form

$$|\tilde{\Psi}> = |\tilde{\Psi}_{ph}> |\tilde{\Psi}_e> = U(b_q)|\tilde{\Psi}_0>, \quad |\tilde{\Psi}_{ph}> = U|0_{ph}>, \quad |\tilde{\Psi}_0> = |\Psi_e> |0_{ph}> \quad (1.7)$$

which corresponds to the multiplicative adiabatic approximation.

The unitary operator, incoming in (1.7)

$$U = exp\left[\frac{1}{\sqrt{N}} \sum_q (\beta_q b^+_q - \beta^*_q b_q)\right] \quad (1.8)$$

describes the ground state renormalization caused by the electron-phonon interaction. Substituting (1.7) into (1.6) we find

$$\tilde{H}|\tilde{\Psi}_0> = E|\tilde{\Psi}_0> \qquad (1.9)$$

where

$$\tilde{H} = U^+(H - Vp)U$$

$$= W + \sum_{k,\sigma}[[E(k) - \hbar V k]A^+_{\sigma,k}A_{\sigma,k} + \frac{1}{N}\sum_q \chi(q)(\beta_q + \beta^+_{-q})A^+_{\sigma,k}A_{\sigma,k-q}] - \sum_q \hbar(\omega_q - Vq)b^+_q b_q$$

$$+ \frac{1}{\sqrt{N}}\sum_{k,q,\sigma}\chi(q)A^+_{\sigma,k}A_{\sigma,k-q}(b_q + b^+_{-q}) + \frac{1}{\sqrt{N}}\sum_q \hbar(\omega_q - Vq)(\beta_q b^+_q + \beta^*_q b_q), \qquad (1.10)$$

and

$$W = \frac{1}{N}\sum_q \hbar(\omega_q - Vq)|\beta_q|^2. \qquad (1.11)$$

With the help of the unitary transformation

$$A_{\sigma,k} = \sum_\lambda \Psi_\lambda(k)a_{\sigma,\lambda} \qquad (1.12)$$

we can introduce new Fermi operators. The coefficients $\Psi_\lambda(k)$ can be chosen from the condition that the part of the operator (1.10) which is quadratic with respect to Fermi operators, is diagonal:

$$[E(k) - \hbar V k]\Psi_\lambda(k) + \frac{1}{N}\sum_q \chi(q)(\beta_q + \beta^*_{-q})\Psi_\lambda(k-q) = E_\lambda \Psi_\lambda(k). \qquad (1.13)$$

This transforms the Hamiltonian (1.10) into the form $\tilde{H} = H_0 + H_1$, where H_0 is the operator which describes the adiabatic states of the electron-phonon system,

$$H_0 = W + \sum_{\lambda,\sigma}\left[E_\lambda + \frac{1}{\sqrt{N}}\sum_q f_{\lambda\lambda}(q)(b_q + b^+_{-q})\right]a^+_{\sigma,\lambda}a_{\sigma,\lambda}$$

$$+ \frac{1}{\sqrt{N}}\sum_q \hbar(\omega_q - Vq)(\beta_q b^+_q + \beta^*_q b_q) + \sum_q \hbar(\omega_q - Vq)b^+_q b_q, \qquad (1.14)$$

and the operator H_1 is the nonadiabaticity operator,

$$H_1 = \frac{1}{\sqrt{N}}\sum_{\lambda\neq\lambda',q}f_{\lambda\lambda'}(q)a^+_{\sigma,\lambda}a_{\sigma,\lambda'}(b_q + b^+_{-q}). \qquad (1.15)$$

Here

$$f_{\lambda\lambda'}(q) = \chi(q)\sum_k \Psi^*_\lambda(k)\Psi_{\lambda'}(k-q). \qquad (1.16)$$

2 Adiabatic approximation

In the zero adiabatic approximation neglecting the term H_1 of the Hamiltonian, Eq.(1.9) reads as

$$H_0|\tilde{\Psi}_0> = E|\tilde{\Psi}_0>. \qquad (2.1)$$

The Hamiltonian (1.1) conserves the number of electrons, and the eigenstate of Eq. (2.1) with one extra electron has the form of the Davydov's soliton [9]. In [10] it was shown that the nonadiabaticity is not important when the inequality $Jg^2 > \hbar\omega(q = ga)$ takes place, which is equivalent to the following inequality

$$\chi^2 > w\hbar V_a/a. \tag{2.2}$$

Here

$$J = \frac{\hbar^2}{2ma^2}, \quad g = \frac{\chi^2}{2Jw}. \tag{2.3}$$

We shall assume the condition (2.2) being fulfilled and consider zero adiabatic approximation (2.1).

2.1 Two-particle states

In the presence of two extra electrons in a chain, the eigenstate of the Schrödinger equation (2.1) can be written in the general form as

$$|\Psi_0> = a^+_{\alpha\sigma}a^+_{\alpha'\sigma'}|0>. \tag{2.4}$$

For the sake of simplicity we shall consider the states with zero total momentum. Substituting (2.4) into Eq.(2.1), we get

$$H_0|\Psi_0> = \left\{W + E_1 + E_2 + \frac{1}{\sqrt{N}}\sum_q [[\hbar\omega_q\beta_q^* + f_{11}(q) + f_{22}(q)]b_q + h.c.]\right\}|\Psi_0>, \tag{2.5}$$

from where we conclude that the state (2.4) is the eigenstate with the energy

$$E = W + E_1 + E_2, \tag{2.6}$$

provided the equality fulfills

$$\beta_q = \beta_{12}(q) = -\frac{\chi^*(q)}{\hbar\omega_q}\left[\sum_k \Psi_1^*(k)\Psi_1(k-q) + \sum_k \Psi_2^*(k)\Psi_2(k-q)\right]. \tag{2.7}$$

We shall look for the solution of Eq.(1.13) with account of (2.7) at small values of k, and use the long-wave approximation (1.2) when the transformation is valid

$$\Psi_\lambda(x) = \frac{1}{\sqrt{L}}\sum_k \Psi_\lambda(k)\exp(ikxa), \tag{2.8}$$

Now Eq. (1.13) written in terms of the continuous variable x, takes the form

$$-\frac{\hbar^2}{2ma^2}\frac{d^2\Psi_\lambda}{dx^2} + U(x)\Psi_\lambda = E_\lambda\Psi_\lambda, \tag{2.9}$$

which coincides with the Schrödinger equation for an electron in the potential $U(x)$ created by the lattice deformation. Accounting for (2.8) and (2.7), we find

$$U(x) = \frac{1}{N}\sum_q \chi(q)[\beta_{12}(q) + \beta_{12}^*(-q)]\exp(-iqxa) = -\frac{\chi^2}{w}(|\Psi_1|^2 + |\Psi_2|^2). \tag{2.10}$$

Thus, the self-consistent states of electrons and lattice deformation obey the stationary two-component nonlinear Schrödinger equations [11,12]

$$\frac{d^2\Psi_1}{dx^2} = \mu_1^2\Psi_1 - 2g(\Psi_1^2 + \Psi_2^2)\Psi_1,$$

$$\frac{d^2\Psi_2}{dx^2} = \mu_2^2 \Psi_2 - 2g(\Psi_1^2 + \Psi_2^2)\Psi_2. \tag{2.11}$$

Here we have taken into account that the bound states of electrons in the deformation potential U(x) correspond to real wave functions, and introduced the notations

$$E_i = -\frac{\hbar^2 \mu_i^2}{2ma^2}, \quad i = 1, 2. \tag{2.12}$$

The parameter g is determined in Eq.(2.3).

The system of Eqs. (2.11) admits the solution $\Psi_1 = \Psi_2$ which corresponds to the lowest energy

$$E_{bs} = -\frac{2}{3}Jg^2, \quad \mu = g \tag{2.13}$$

and describes the bisoliton [5,6]. In the physical system considered here, this solution has the sense only in the case of opposite spins of electrons in the initial state, i.e., when the inequality $\sigma \neq \sigma'$ takes place in (2.4).

The system (2.11) can be solved using the inverse scattering transformation [11] for the two-soliton state, as it was done in [6], or, since it belongs to the Liouville-type class [13], using the method of quadraturs, as it was done in [8]. This solution reads as follows

$$\Psi_1 = \sqrt{\frac{\mu_1^2 - \mu_2^2}{g}} \frac{\mu_1 \cosh \phi_2}{S}, \quad \Psi_2 = \sqrt{\frac{\mu_1^2 - \mu_2^2}{g}} \frac{\mu_2 \sinh \phi_1}{S}, \tag{2.14}$$

where

$$S = \mu_1 \cosh \phi_1 \cosh \phi_2 - \mu_2 \sinh \phi_1 \sinh \phi_2, \quad \phi_1 = \mu_1(x+\lambda), \quad \phi_2 = \mu_2(x-\lambda). \tag{2.15}$$

and λ is the constant of integration. Introducing the operator of the number of electrons on the n-th site of the chain

$$N_{n,\sigma} = A_{n,\sigma}^+ A_{n,\sigma}$$

we can calculate the electron density distribution

$$\rho = \sum_\sigma <\Psi|A_{n,\sigma}^+ A_{n,\sigma}|\Psi> = \Psi_1^2(na) + \Psi_2^2(na)$$

$$= \frac{\mu_1^2 - \mu_2^2}{g} \frac{\mu_1^2 \cosh^2 \phi_1 + \mu_2^2 \sinh^2 \phi_2}{S^2}. \tag{2.16}$$

The energy of the corresponding state, according to (2.6), equals

$$E = -\frac{\hbar^2 \mu_1^2}{2ma^2} - \frac{\hbar^2 \mu_2^2}{2ma^2} + \frac{\hbar^2 g}{2ma^2} \int_{-L/2}^{L/2} (\Psi_1^2 + \Psi_2^2)^2 \, dx$$

$$= -J(\mu_1^2 + \mu_2^2) + \frac{4J}{3g}(\mu_1^3 + \mu_2^3). \tag{2.17}$$

The parameters μ_1 and μ_2 which determine the energies of bound states, should be found from the normalization conditions

$$\int_{-L/2}^{L/2} \Psi_i^2 \, dx = 1,$$

and, hence, depend on the length of the chain. At long enough chain when $L \to \infty$ the normalization conditions result in $\mu_1, \mu_2 \to \mu = g/2$. Let $\mu_1 = \mu + \Delta$, $\mu_2 = \mu - \Delta$. Then at infinitely small Δ we get

$$\Psi_1 = \frac{\mu}{\sqrt{2g \cosh(2\mu\lambda)}} \left[\frac{e^{-\mu\lambda}}{\cosh[\mu(x-R)]} + \frac{e^{\mu\lambda}}{\cosh[\mu(x+R)]} \right],$$

$$\Psi_2 = \frac{\mu}{\sqrt{2g \cosh(2\mu\lambda)}} \left[\frac{e^{\mu\lambda}}{\cosh[\mu(x-R)]} - \frac{e^{-\mu\lambda}}{\cosh[\mu(x+R)]} \right] \quad (2.18)$$

from where it follows

$$\rho = \Psi_1^2 + \Psi_2^2 = \frac{\mu^2}{g} \left[\cosh^{-2}[\mu(x-R)] + \cosh^{-2}[\mu(x+R)] \right], \quad (2.19)$$

where

$$R = \frac{1}{2\mu} \ln \frac{2\mu \cosh(2\mu\lambda)}{\Delta} \to \infty \quad \text{at} \quad \Delta \to 0. \quad (2.20)$$

The integration constant λ as it follows from Eq.(2.25), does not income into ρ but it determines the electron state. At $\lambda = 0$ the electron state is absolutely collectivized and both electrons participate in the formation of the collective state with equal probabilities, while at $\lambda \neq 0$ the symmetry is broken and at large enough value of λ functions Ψ_1 and Ψ_2 describe individual solitons respectively. The value of μ determined from the normalization condition, equals $\mu = g/2$, and two electrons have the energy

$$E = -\frac{1}{6} Jg^2 + 2Jg^2 \cosh^2(g\lambda) \exp(-2gR) \quad (2.21)$$

from where it follows that there is the repulsion between solitons, caused by the exponentially decreasing force

$$F = \frac{\partial E}{\partial R} \propto \exp(-2gR). \quad (2.22)$$

This is confirmed by the numeric calculations of wave functions (2.14), from which it follows that in the long enough chain the two solitons are repelled from one another so that the distance between them exceeds their size. The integration constant λ characterizes the level of the hybridization of electrons in the bound state. At $|\lambda| \geq 5$ the electrons with parallel spins form the unbound state, and their wavefunctions can be described as independent free solitons.

2.2 Four-particle states

In the case when there are four extra electrons in a chain the eigenstate can be chosen in the form

$$|\Psi_4\rangle = A^+_{1,\uparrow} A^+_{1,\downarrow} A^+_{2,\uparrow} A^+_{2,\downarrow} |0\rangle \quad (2.23)$$

and calculations similar to given in Section 2.1, show that the equality takes place

$$H_0 |\Psi_4\rangle = \left[W + 2E_1 + 2E_2 + \frac{1}{\sqrt{N}} \sum_q [\hbar\omega_q \beta_q^* + 2(f_{11}(q) + f_{22}(q))b_q + h.c.] \right] |\Psi_4\rangle \quad (2.24)$$

where
$$\hbar\omega_q \beta_q = -2\left[f_{11}(q) + f_{22}(q)\right]. \tag{2.25}$$

This leads to the results similar to the one obtained above with the substitution $g \to G = 2g$. The energy corresponding to this state, is

$$E = 2E_{bs} + 8Jg^2 \cosh^2(2g\lambda)\exp(-4gR) \tag{2.26}$$

where E_{bs} and R are determined in (2.13) and (2.20), respectively.

This means that in the zero adiabatic approximation there is the exponentially decreasing with distance repulsion between two bisolitons, due to which the bisolitons are separated by the distance as maximum as possible for the given chain length.

3 The account of nonadiabatic terms

Below we shall show that the account of the nonadiabatic term H_1 (1.15) of the Hamiltonian results in additional direct interaction between the two solitons via the phonon field which stabilizes the two-soliton solution. We shall consider only the triplet state of two extra electrons since this result can be easily generalized for the case of few bisolitons, as it has been shown above. We shall consider the nonadiabatic term of the Hamiltonian H_1 determined in Eq. (1.15) as the perturbation. One can write the first order energy correction in the form

$$\Delta E = <\Psi_0 \mid H_1 \frac{1}{E - H_0} H_1 \mid \Psi_0 >, \tag{3.1}$$

where the operator notation is used

$$\frac{1}{E - H_0} = \lim_{\epsilon \to 0} \Re \frac{1}{i} \int_0^\infty \exp\left\{i(E - H_0 + i\epsilon)\tau\right\} d\tau. \tag{3.2}$$

Here the wavefunction of the triplet state and the corresponding energy in the zero adiabatic approximation are determined in Eqs. (2.4), (2.6), respectively. Notice, that the sum in (1.15) should be taken over all states, including the bound states $\lambda = 1,2$, and continuous spectrum characterized by the wave number κ.

The expression (3.1) can be simplified with respect to the electron operators and transformed to the following one

$$\Delta E = \lim_{\epsilon \to 0} \Re \frac{1}{i} \int_0^\infty d\tau \, M \, \exp\left\{i(E + i\epsilon)\tau\right\}, \tag{3.3}$$

where

$$M = \sum_j \frac{1}{N} \sum_{\kappa,q,q'} f_{j,\kappa}(q') f_{j,\kappa}^*(q) M_{ph}(q,q') \exp\left\{-i(W + E_{i \neq j} + E_\kappa)\tau\right\}, \tag{3.4}$$

$$M_{ph}(q,q') = {}_{ph}<0|b_{q'}e^{-iV_j\tau}b_q^+|0>_{ph}, \tag{3.5}$$

Here V_j is the part of the Hamiltonian (1.10) which contains the phonon operators only, namely,

$$V_j = \sum_q \left[\hbar\omega_q b_q^+ b_q - \frac{1}{\sqrt{N}}[\alpha_j^*(q)b_q^+ + \alpha_j(q)b_q]\right], \tag{3.6}$$

where
$$\alpha_j(q) = f_{j,j}(q) - f_{\kappa,\kappa}(q). \tag{3.7}$$

Substituting Eq. (3.6) into (3.5), we get

$$M_{ph}(q,q') = e^{iW_j\tau + g(\tau)}\left[e^{-i\hbar\omega_{q'}\tau}\delta_{qq'} + \frac{\alpha_j^*(q')}{N\hbar\omega_{q'}}\left(1 - e^{-i\hbar\omega_{q'}\tau}\right)\frac{\alpha_j(q)}{\hbar\omega_q}\left(1 - e^{-i\hbar\omega_q\tau}\right)\right] \tag{3.8}$$

where

$$W_j = \frac{1}{N}\sum_q \frac{|\alpha_j(q)|^2}{\hbar\omega_q}, \quad g(\tau) = \frac{1}{N}\sum_q \frac{|\alpha_j(q)|^2}{(\hbar\omega_q)^2}\left(e^{-i\hbar\omega_q\tau} - 1\right). \tag{3.9}$$

Now we can rewrite the energy correction in the following form:

$$\Delta E = \frac{1}{N}\sum_{j,\kappa}\left\{\sum_q |f_{j\kappa}(q)|^2 \Phi_j(\Delta_{j\kappa} + \hbar\omega_q) + \sum_{q'}\frac{f_{j\kappa}(q')\alpha_j^*(q')}{\hbar\omega_{q'}}\sum_q \frac{f_{j\kappa}^*(q)\alpha_j(q)}{N\hbar\omega_q}I_j\right\} \tag{3.10}$$

where the notations are used for the real parts of integrals

$$\Phi_j(z) = \Re\frac{1}{i}\int_0^\infty e^{-iz\tau} e^{g(\tau)} d\tau, \quad z = E_\kappa - E_j - W_j$$

$$I_j = \Re\frac{1}{i}\int_0^\infty d\tau \exp\left[-i(E_\kappa - E_j - W_j - i\epsilon)\tau + g(\tau)\right]\left(1 - e^{-i\hbar\omega_{q'}\tau}\right)\left(1 - e^{-i\hbar\omega_q\tau}\right) \tag{3.11}$$

To calculate the functions (3.8)-(3.9), we need to know the coefficients (3.7) determined via the functions of complete spectrum of the two-component NLSE (2.11) by the relations (1.16), (2.8). The functions corresponding to bound states are given by expressions (2.18)-(2.19). The functions of continuous spectrum, according to (2.16), can be found as the solutions of the PDE

$$\frac{d^2\psi_\kappa}{dx^2} = -E_\kappa\psi_\kappa - 2\frac{d^2}{dx^2}(\ln S)\psi_\kappa, \tag{3.12}$$

where the potential function S is given by expression (2.15), and, hence, read as

$$\psi_\kappa = be^{i\kappa x}\left(i\kappa + \frac{1}{2}\frac{d}{dx}\ln U\right) \tag{3.13}$$

where

$$U = \frac{d^2}{dx^2}\ln S = g(\psi_1^2 + \psi_2^2). \tag{3.14}$$

The integration constant b can be found from the normalization condition of functions (3.13):

$$b = \frac{1}{\sqrt{L(\kappa^2 + \epsilon_0)}}, \quad \epsilon_0 = \frac{\mu_1^2 + \mu_2^2}{2} \tag{3.15}$$

Substituting now functions (2.18), (3.13) into (1.15) and using the transformation (2.8), we find the complete set of coefficients (1.16):

$$f_{j,\kappa}(q) = -\frac{2iq\pi\chi(q)}{\cosh\frac{k\pi}{2\mu}\sqrt{2gL(\kappa^2 + \epsilon_0)}\cosh(2\mu\lambda)} \times \begin{cases} \cosh(\mu\lambda - ikR), & j = 1 \\ \sinh(\mu\lambda + ikR), & j = 2 \end{cases} \tag{3.16}$$

$$f_{j,j}(q) = \frac{\pi q \chi(q) \cosh(2\mu\lambda \pm iqR)}{g \cosh(2\mu\lambda) \sinh\frac{q\pi}{2\mu}}, \quad j = 1,2 \qquad (3.17)$$

where $k = \kappa + q$. These, together with (3.7), allow us to calculate g-factor (3.9):

$$g(\tau) \approx -iW_j\tau - A_j^2\tau^2, \quad A_j^2 = \frac{1}{N}\sum_q |f_{jj}(q)|^2. \qquad (3.18)$$

Substituting this result into the function in (3.11), we find that it is determined by the hypergeometric function $\Phi(a,b;z)$:

$$\Phi_j(z) = -\frac{z+W_j}{2A^2} e^{-\frac{(z+W_j)^2}{4A^2}} \Phi\left(\frac{1}{2};\frac{3}{2};\frac{(z+W_j)^2}{4A^2}\right). \qquad (3.19)$$

Here

$$A^2 \equiv \frac{1}{N}\sum_q |f_{j,j}(q)|^2 \approx \frac{16\hbar\chi^2 a}{\pi M V_a g^2} \mu^4 \frac{\cosh 4\mu\lambda}{\cosh^2 2\mu\lambda}\left(1 + \frac{\sin 2\beta}{4\beta \cosh 4\mu\lambda}\right), \qquad (3.20)$$

$$\beta = \frac{2\mu R}{\pi}, \qquad (3.21)$$

where R is determined in Eq.(2.20).

Since the phonon part of energy W_j (see Eq. (3.9)) is small as compared with the eigenenergies of the localized states in the zero adiabatic approximation, we have

$$W_j \ll E_j, \quad z \approx E_\kappa - E_j, \qquad (3.22)$$

and, hence,

$$\frac{(z+W_j)^2}{4A^2} \approx \frac{z^2}{4A^2} = \frac{g\pi^2}{2^7\gamma} \gg 1, \qquad (3.23)$$

where γ is the nonadiabaticity parameter:

$$\gamma = \frac{\hbar V_a \pi}{Ja}. \qquad (3.24)$$

Taking into account inequality (3.23), we find from (3.19)

$$\Phi_j(z) \approx -\frac{1}{z+W_j} \qquad (3.25)$$

which, together with the definition (3.11), transforms the expression for energy correction (3.10) into the following one:

$$\Delta E = \sum_{j,\kappa}\left\{-\frac{\sum_q |f_{j\kappa}(q)|^2}{N(z+W_j)} + \frac{\sum_q \hbar\omega_q |f_{j\kappa}(q)|^2}{N(z+W_j)^2} - 2\frac{|\sum_q f_{j\kappa}(q)\alpha_j(q)|^2}{N^2(z+W_j)^3}\right\}, \qquad (3.26)$$

and, further,

$$\Delta E \approx -\frac{2m}{N\hbar^2} \sum_{j,\kappa,q} \frac{|f_{j\kappa}(q)|^2}{\mu_j^2 + \kappa^2}. \qquad (3.27)$$

Using now in the latter expression the explicit form of the coefficients (3.16) and changing the sum by integral over wavenumbers, we finally get:

$$\Delta E = \Delta E_1 + \Delta E_2, \qquad (3.28)$$

181

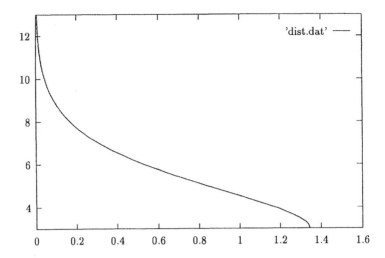

Figure 1: The dependence of the equilibrium distance between the solitons c.c.c. R (in units 2g) on the parameter B connected with the nonadiabaticity of the system by Eq.(4.3)

where

$$\Delta E_1 = -\frac{\alpha}{2J\pi^2} \int_{-\infty}^{\infty} \frac{d\kappa}{(\kappa^2 + \mu^2)^2} \int_{-\infty}^{\infty} \frac{|q|^3 dq}{\cosh^2 \frac{\pi(\kappa+q)}{2\mu}} \approx -J\frac{\gamma g^3}{2^3 \pi^6}, \quad (3.29)$$

$$\Delta E_2 = -\frac{\alpha}{2J\pi^2} \int_{-\infty}^{\infty} \frac{d\kappa}{(\kappa^2 + \mu^2)^2} \int_{-\infty}^{\infty} \frac{|q|^3 \cos[2(\kappa+q)R]}{\cosh^2 \frac{\pi(\kappa+q)}{2\mu}} \cosh(2\mu\lambda) \, dq$$

$$\approx -\frac{3Jg\gamma}{(2\pi)^3 \cosh(\lambda g)} \frac{1}{R^2} \left(1 - \frac{1}{3}\cos\frac{2gR}{\pi}\right). \quad (3.30)$$

4 Conclusion

Present considerations confirm that the lowest state of two extra electrons in a chain correspond to the singlet bisoliton. In a triplet state there is the repulsion between two electrons as well as between two bisolitons. This situation is opposite to what happens with two small bipolarons which, as it is shown in [12,13], are attracted to one another and form a bipolaron drop.

The repulsion between two (bi)solitons is partly compensated by the interaction between electrosolitons via the phonon field. The latter is described by the nonadiabatic terms of the Hamiltonian. The corresponding lowering of energy of the system consists of two parts, one of which, given by expression (3.29), is constant and small (proportional to the nonadiabaticity of the system), and another one, (3.30), depends on the distance between the solitons c.m.c., R, calculated in the zero adiabatic approximation.

Taking into account the results obtained in sections 2 and 3, we can write the total energy of a chain with two extra electrons in it within the first order perturbation accuracy (see Eqs.(2.26), (3.28)-(3.30)):

$$E = -\frac{1}{6}Jg^2\left(1+\frac{3\gamma}{8\pi^6 g}\right) + 2Jg^2\frac{e^{-2gR}}{\cosh^2(\lambda g)} - \frac{3Jg\gamma}{(2\pi)^3 R^2 \cosh(\lambda g)}\left(1-\frac{1}{3}\cos\frac{2gR}{\pi}\right). \tag{4.1}$$

As usually, the minimization of the expression (4.1) with respect to R, which is equivalent to the solution of the equation

$$x^3 e^{-x} = B, \tag{4.2}$$

where

$$B = \frac{6g\gamma \cosh(\lambda g)}{(2\pi)^3}, \quad x = 2gR, \quad R = \frac{1}{g}\ln\frac{g\cosh(2\mu\lambda)}{\sqrt{\mu_1^2 - \mu_2^2}}, \tag{4.3}$$

determines the equilibrium distance between the solitons c.m.c. The numeric solution to the latter transcendent equation is plotted on Fig.1, from where we conclude that the more is the nonadiabaticity of the system (remaining, of course, small, $B < B_{cr}$), the less is the repulsion between the two electrosolitons in the triplet state, or, respectively, between two bisolitons, and, hence, the less is the distance between the (bi)solitons c.m.c.

As a result, in the presence of N_e extra electrons in chain, the periodic lattice of bisolitons is formed in it with the distribution period $l = N/N_{bs}$, $N_{bs} = N_e/2$ [14,15] that corresponds to the many-electron solution of the Peierls-Fröhlich problem at zero temperature [15].

References

1. A.S. Davydov and N.I. Kislukha, Solitary excitations in one-dimensional molecular chains, Phys. Stat Sol.(b) 59:465 (1973).

2. A.S. Davydov. *Solitons in Molecular Systems*, Reidel, Dordrecht (1985).

3. P.L. Christiansen and A.C. Scott, eds.*Davydov's Soliton Revisited*, Plenum, New York (1990).

4. A.C. Scott, Davydov's soliton, Phys. Rep. 217:1 (1992).

5. L.S. Brizhik and A.S. Davydov, The electrosoliton pairing in soft molecular chains, Fiz. Nizk. Temp. 10:748 (1984).

6. L.S. Brizhik, Bisolitons in one-dimensional molecular systems, Fiz. Nizk. Temp. 14:437 (1986).

7. L.S. Brizhik and A.A. Eremko, Bound states of electrons in one-dimensional chain, Phys. Stat. Sol. (b) 182:89 (1994).

8. L.S. Brizhik and A.A. Eremko, Electron autolocalized states in molecular chains, Physica D 81:295 (1995).

9. A.S. Davydov and A.A. Eremko, Radiative lifetime of solitons in molecular chains, Ukr. Fiz. Zh. 22:881 (1977).

10. J.P. Cottingham and J.W. Schweitzer, Calculation of the lifetime of a Davydov soliton at finite temperature, Phys. Rev. Lett. 62:1792 (1989).

11. S.V. Manakov, On the theory of two-dimensional stationary self-focusing of the electromagnetic waves, Zh.Eksper.Teor.Fiz. 65:735 (1976).

12. M.V. Tratnik and J.E. Sipe, Bound solitary waves in a birefrigent optical fiber, Phys.Rev.A 38:2011 (1988).

13. E.T. Whittaker, *A Treatise on the Analytical Dynamics of Particles and Rigid Bodies* Cambridge, at the University Press (1927).

14. L.S. Brizhik L.S. and A.S. Davydov, Nonlinear theory of the conductivity in quasi-one-dimensional molecular crystals, Fiz. Nizk. Temp. 10:358 (1984).

15. A.A. Eremko, Peierls-Fröhlich problem in continuum approximation, Phys. Rev. B 46:3721 (1992).

III. SUPERFLUID He^3.

THE POMERANCHUK EFFECT*

Robert Richardson, Dept. of Physics, Cornell University, Ithaca, NY 14853

INTRODUCTION

A central part of the story of the discovery of superfluid ^{3}He is the cooling technique used for the experiments, the Pomeranchuk Effect. Although it is not an especially useful technique for obtaining low temperatures today, it contains my favorite example of the use of the Clausius-Clapeyron equation. The cooling technique is fun to describe in undergraduate physics classes on thermodynamics.

In 1950, I. Pomeranchuk, a well known particle theorist, suggested that melting ^{3}He could be cooled by squeezing it [1]. At the time of his suggestion ^{3}He was quite rare and had not yet even been liquefied. He observed that at low enough temperatures the thermal phenomena in condensed ^{3}He would be dominated by spin properties instead of phonon properties. The liquid of ^{3}He would obey Fermi statistics with an entropy proportional to the temperature, much like the free electrons in a good metal. On the other hand, the entropy of solid ^{3}He would be that of the disordered collection of weakly interacting spin 1/2 nuclei. At temperatures greater than the [then] expected nuclear magnetic ordering temperatures less than 1 µK, the entropy per mole of solid ^{3}He would be $S = R \log_e 2$, independent of temperature until the high temperature phonon modes of the solid become important. (The Debye temperature of solid ^{3}He is approximately 30 K.)

The idea of the method is represented in Figure 1. The entropy of solid ^{3}He exceeds that of liquid 3He at temperatures less than 0.3K. If the mixture is compressed without heat input it will cool as liquid is converted into solid.

DISCUSSION OF POMERANCHUK'S PROPOSAL

Fifteen years passed before anyone took up the suggested cooling technique [2]. There were several reasons. The most important was the availability of ^{3}He. It comes from tritium decay. Tritium was being produced for the most deadly part of the weapons industry. By 1965 copious quantities had been made. Low temperature physicists took advantage of the waste product of the arms race, the ^{3}He extracted from the gases prepared for hydrogen bombs.

The second reason for the late date of attempts at the method was the skepticism of experimentalists about practical considerations. The entropies of the liquid and solid phases are illustrated in Figure 2. Liquid ^{3}He is rather accurately described by Landau's Fermi Liquid Theory[3,4]. At low temperatures[5,6] the entropy per mole of the liquid at melting pressure is approximately given by $S \approx 3RT$. The entropies of the liquid and solid phases

*Nobel lecture (1996)

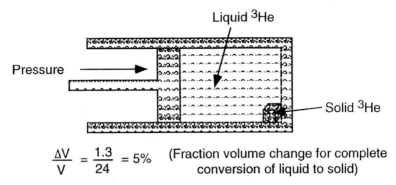

$$\frac{\Delta V}{V} = \frac{1.3}{24} = 5\%$$ (Fraction volume change for complete conversion of liquid to solid)

Figure 1. Pomeranchuk's suggestion for cooling a melting mixture of ^{3}He. The solid phase has a higher entropy than the liquid at low temperatures. As the liquid-solid mixture is compressed, heat is removed from the liquid phase as solid crystallites form. The fractional change of volume required to completely convert liquid into solid is approximately 5%. Unlike melting water, the solid phase forms at the *hottest* part of the container.

are equal at 0.32 K. At lower temperatures Pomeranchuk's suggestion for cooling will work.

The adiabatic cooling path is indicated with the arrow **A** marked on the vertical axis. In the example, liquid compressed at an initial temperature of 0.1 K, with an entropy of 0.2R, will form a liquid-solid mixture which eventually cools to very low temperatures. The maximum

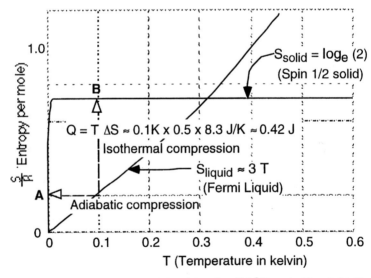

Figure 2. The entropies of liquid and solid ^{3}He. At T < 0.32 K, liquid ^{3}He has a lower entropy than the solid phase. The figure show an example at T = 0.1 K. The latent heat associated with converting 1 mole of ^{3}He liquid into solid is 0.42 joules, a substantial amount of heat removal at these low temperatures.

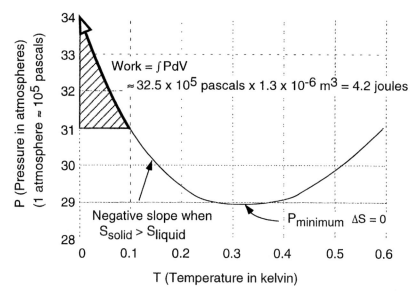

Figure 3. The melting pressure of ^{3}He. The figure can be constructed from Figure 2 through the Clausius-Clapeyron equation [see text]. The molar volume of liquid ^{3}He exceeds that of the solid by 1.3 cm^3 per mole. Thus, the slope of the melting curve is negative at temperatures less than 0.32 K. The work of compression in forming the solid is approximately 4.2 joules, an order of magnitude large than the heat which might be extracted. .

amount of heat which can be removed is the latent heat of conversion of the liquid to solid and is indicated by the isothermal path, labelled **B**. The latent heat per mole at T = 0.1 K is 0.42 joules.

The cooling effectiveness of the method must be compared with the possible heat losses in the process. At this point the natural skepticism of experimentalists arises. The amount of work involved during the compression is large. The melting pressure versus temperature of ^{3}He is illustrated in Figure 3. The melting curve may be calculated from the liquid and solid entropies using the Clausius-Clapeyron equation,

$$\left.\frac{dP}{dT}\right)_{melting} = \frac{S_{liquid} - S_{solid}}{V_{liquid} - V_{solid}}, \qquad [1]$$

where, $S(T)_{liquid}$ and $S(T)_{solid}$ are the molar entropies at melting. V_{liquid} and V_{solid} are the molar volumes of the two phases at melting. The difference, $V_{liquid} - V_{solid}$ is nearly independent of temperature and has the value 1.3 cm^3 per mole. We will return to Equation [1] later when I describe our experiments designed to measure the entropy of solid ^{3}He.

The work performed in converting the liquid to solid, starting a 0.1 K can be obtaining from the integral, $\int P dV$ along the melting curve. It's value is approximately 4.2 Joules. An order of magnitude more work must be done than will be extracted during the cooling process. The ratio, W/Q, of work to heat extracted is near a minimum at the temperature illustrated. When the process is performed at lower starting temperatures, as was the usual practice, W/Q becomes larger than 100. The challenge of the experimental design, thus apparently became the avoidance of frictional heat losses during the compression process.

COMPRESSIONAL COOLING IN PRACTICE

In the Spring of 1966, David Lee invited me to join him at Cornell to begin experiments on the cooling of solid ^{3}He using the compressional cooling technique. The goal was to reach the temperature of the nuclear magnetic ordering transition in solid ^{3}He. My Ph. D. thesis at Duke University with Horst Meyer[7] had been concerned with nmr measurements of the size of the exchange interaction in solid ^{3}He. We knew from these measurements that the magnetic phase transition in solid ^{3}He at the melting pressure should occur at temperatures closer to 1 mK than 1 µK. Despite my certainty that Pomeranchuk's technique for cooling was probably doomed to failure, I was anxious to join Dave in searching for the transition. As a back-up we would attempt to cool solid 3He with magnetic cooling schemes.[2] The latter had not yet been used successfully to obtain such low temperatures in liquid or solid helium but we began a parallel effort to use nuclear demagnetization[8].

By the time I arrived at Cornell in October 1966, the method had been successfully demonstrated by Anufriev in Moscow.[9] A cross section of Anufriev's apparatus is represented in Figure 4. The volume available for liquid ^{3}He is exaggerated in the figure. The ^{3}He space contained 30 cm^2 of metal foil to act as heat exchangers to cool the outside wall of the chamber. The outer chamber was first filled with liquid ^{3}He at the melting pressure at temperatures greater

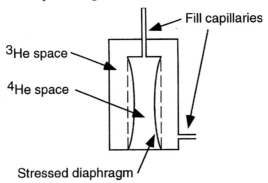

Figure 4. Cross sectional representation of the compressional cooling cell used by Anufriev. Both inner and outer chambers have a rectangular shape. Thin stainless diaphragms of the inner cell were displaced inward with the application of ^{3}He at the melting pressure. The walls of the stressed diaphragm were then forced outward by filling the inner chamber with pressurized ^{4}He.

than 0.32 K.. As the cell was cooled using the demagnetization of a paramagnetic salt, a block formed in the fill capillary trapping a fixed quantity of ^{3}He in the cell. At low temperatures carefully cooled ^{4}He was admitted within the inner chamber to act as a hydraulic fluid. Pressurized liquid ^{4}He forced the diaphragm walls outward to decrease the volume available for the melting ^{3}He. The cell cooled to temperatures less than 15 mK, the minimum temperature which could be measured with the thermometers in contact with the exterior of the cell.

It is interesting to notice the use of liquid ^{4}He as the hydraulic fluid for forcing changes in the ^{3}He cell dimensions. At temperatures less than 0.3 K, the normal fluid fraction of ^{4}He liquid is very small. The heat capacity and thermal conductivity of liquid ^{4}He are negligible. Anufriev's pioneering experiment tested the method, showed that it would work, and demonstrated the use of liquid ^{4}He as a hydraulic fluid at low temperatures. The fears of excessive frictional heating associated with the movement of a metal diaphragm were unfounded. We now know that practically all metals have a very high quality factor at temperatures less than 4 K. In the subsequent years, *every* apparatus built upon the principle of forcing a metal diaphragm or bellows to move produced successful cooling. There is an important caveat in using liquid ^{4}He. The melting pressure of ^{4}He is 4 atmospheres less than that of ^{3}He. Spring tension or pressure amplifiers must be used in the experimental design so that ^{4}He does not solidify during the compression process.

POMERANCHUK CELLS AT CORNELL

The Cooling Cell of Jim Sites

Our first venture with Pomeranchuk cooling came with the thesis project of Jim Sites. The goal of his experiment was to measure the magnetic susceptibility of melting solid ^{3}He at temperatures near or below the nuclear magnetic phase transition. We had long discussions about the design of an apparatus. I have often regretted that we did not follow one of Dave Lee's original suggestions, the use of a weight to compress a bellows filled with melting ^{3}He. A heavy mass would be suspended by a wire and slowly lowered on the bellows in a controlled manner. The mass would have to be a metal of high density with negligible magnetic properties. Gold seemed to be the only really suitable metal. Imagine the profit if we had purchased 5 kilograms of gold at the 1967 price, $30 per troy ounce!

Before we completed our first experiments at Cornell the Wheatley group in La Jolla reported highly successful compressional cooling of ^{3}He down to temperatures less than 2 mK. [10, 11] Their design employed tube of ^{3}He with an ellipsoidal cross section. The tube was surrounded by pressurized liquid ^{4}He. The entire cooling assembly was placed within the mixing chamber of a dilution refrigerator[2] and pre-cooled, over a matter of days, to 24 mK.

The design first used at Cornell was one suggested by our colleague John Reppy. It is illustrated in Figure 5. A simplified schematic of the apparatus is shown in Figure 6. The cell contained two concentric Beryllium-Copper bellows and three helium chambers.[12, 13] The ^{3}He sample was contained in the innermost chamber, Chamber I, and ^{4}He in the outer two chambers. Initially all three chambers were pressurized to approximately the melting curves of their contents, and the sample cell was precooled with a dilution refrigerator to 25 mK. Compression of the ^{3}He was achieved by releasing the pressure on the ^{4}He in Chamber II. We had thought that there might be some advantage in removing fluid from the cell. The position of the bellows assembly was monitored by measuring the change in

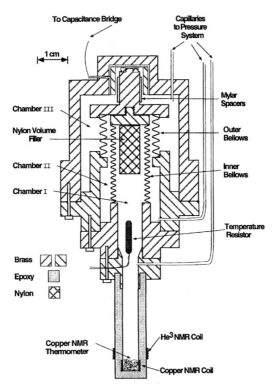

Figure 5. The Pomeranchuk cooling cell of Jim Sites.

capacitance of a short rod attached to the 'Top Plate' of the assembly. The minimum temperature recorded by the copper NMR thermometer was 7 mK. Extrapolating from the size of the magnetic susceptibility of the solid 3He, the average temperature of the solid was sometimes as low as 2 mK. Like the Wheatley group[10], our attention was focused solely on the solid ^{3}He. The liquid component was viewed merely as the cooling agent!

Our first Pomeranchuk cell had several significant disadvantages. The most important was that it tended to warm up quite rapidly after only several hours and achieved far less volume change than the maximum 5% required for conversion of all of the liquid into solid. Apparently, solid helium was being trapped within the convolutions of the bellows to become crushed as the bellows contracted. With regard to our desire to measure the temperature dependence of the magnetic susceptibility of the solid the cell had two further design flaws. The place where solid grew in the cell was unpredictable and the time constant for the copper nmr thermometer was very long. Fortunately, some of the solid ^{3}He nucleated on the "rat's nest" of copper wires at the bottom of the NMR tail section. The thermal equilibrium time for nuclear magnetization in metals, T_1, is inversely proportional to temperature. In copper the product $T_1 \circ T$ is approximately 1 sec-K. With only 10 minutes available at the bottom temperatures near 2 mK, the copper thermometer never caught up with the temperature changes.

A final conceptual mistake in our first compressional cooling cell is that there was no provision for the direct measurement of the ^{3}He pressure. We could monitor the volume change by keeping track of the bellows displacement. But the valuable thermodynamic information available with a knowledge of the melting pressure was not available in this set of experiments.

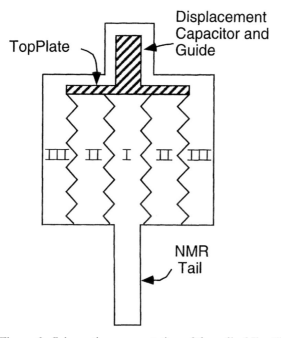

Figure 6. Schematic representation of the cell of Jim Sites.

The Corruccini-Osheroff Cooling Cell

Our next cooling attempts were made with a cell which was designed to investigate some unusual spin diffusion phenomena predicted by Leggett and Rice.[14] The cell is illustrated in Figure 7. We used the cold liquid ^{3}He in a Pomeranchuk cell to cool a separate chamber of either liquid ^{3}He or dilute liquid mixtures of ^{3}He in ^{4}He.[15,16] Doug worked on the development of the cooling technique while Linton Corruccini worked on the design of the chamber for NMR measurements. Following our experience with Jim Site's cell, Osheroff decided to try compressing the ^{3}He with a bellows which expands. The idea was to avoid crushing solid ^{3}He within contracting bellows. The ^{3}He compression cell was filled with liquid at the melting pressure and ^{4}He pressure was applied to the upper chamber. The differences in the melting pressures of the two isotopes were taken into account by having the diameter of the ^{4}He bellows larger than that for the ^{3}He by a ratio of 3.5:1. The ^{3}He could be completely solidified without raising the ^{4}He pressure above 10 bar.

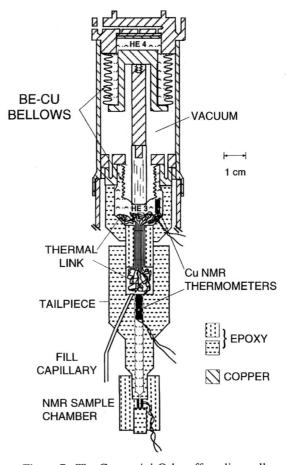

Figure 7. The Corruccini-Osheroff cooling cell.

A bundle of small copper wires made a thermal connection between the ^{3}He compression region and the NMR sample region. The temperatures of the compression region and sample region were measured with copper NMR signals.

The apparatus worked very well indeed, as did the Leggett-Rice theory. It was an easy experiment. Before compression, the cell was first cooled to 25 mK with a dilution refrigerator. We succeeded in cooling the sample region to temperatures as cold as 4 mK. We typically spent 8 to 12 hours in compressing the liquid ^{3}He in order to maintain thermal equilibrium and maximize the cooling efficiency between the cooling cell and sample cell. The chamber would remain at the lowest temperature for up to 4 hours, then slowly warm up to 10 mK over a period of 5 hours. Further warming was achieved by partial decompression of the cooling cell. The minimum temperature recorded by the copper NMR thermometer in the cooling cell was 3 mK. The cell was a precursor for the one used in the discovery of superfluid ^{3}He. Since it was not intended for studies of the ^{3}He in the there was minimal instrumentation in the compression cell.

Osheroff's Compression Cell

The compressional cooling cell used by Osheroff[17] during the course of the experiments on melting ^{3}He was a variation of the one we used in Corruccini's spin diffusion measurements. The epoxy bottom of the cell could be readily replaced and the course of a six months no less than five different epoxy tail sections were used. The cell illustrated in Figure 8 is the version published with the results purporting to measure the phase transition in solid ^{3}He. It had two very important design changes from the previous cells. The first was that we changed the metal NMR thermometer from copper to platinum. The thermal equilibrium time for ^{195}Pt nuclei is a factor of 30 shorter than that of copper. It is far less susceptible to small eddy current heating effects than copper. It eventually read temperatures well under 2 mK.

The second change was probably more important. It was the inclusion of a gauge to measure the pressure of the melting ^{3}He. A thin metal diaphragm on the bottom of the cell deflected as the pressure changed. The amount of the deflection was measured capacitively. One plate was attached to the center of the diaphragm while the other was fixed to a mounting arrangement on the epoxy tail section. The design is one which had been invented by Straty and Adams[18] and became widely used at Cornell and elsewhere. The melting pressure is a unique function of temperature. At higher temperatures the vapor pressure of ^{3}He and ^{4}He gases in equilibrium with liquid helium are routinely used to calibrate other thermometers. Adams had previously suggested that the ^{3}He melting pressure be used as a temperature standard.[19] During the interval in late November 1971 in which Doug Osheroff was 'practicing' the use of the apparatus the pressure measurements gave information about changes in temperature of the apparatus.

Dave Lee and Doug Osheroff have described many of the details about those early measurements in their Nobel Prize lectures. The well known 'pressure versus time' curve is reproduced in Figure 9. The experiment was conducted with a constant rate of compression of the ^{3}He bellows, that is with a constant cooling rate. The pressure scale can be interpreted as a measurement of temperature change. The temperature scale on the right was our best guess at the thermodynamic temperature and was based upon measurements of the magnetic susceptibility of the ^{195}Pt magnetic susceptibility. The pressure measurements are relative to the maximum melting pressure of ^{3}He.

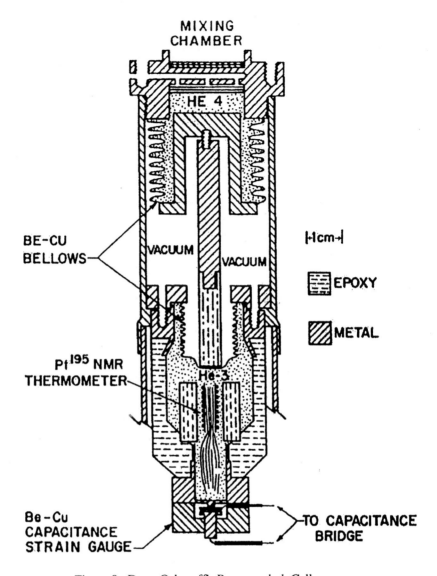

Figure 8. Doug Osheroff's Pomeranchuk Cell

The points labelled A and A' are transitions of liquid 3He from the normal liquid phase to the superfluid A phase and then back again. The cooling or warming rate changes at these points because of the change in heat capacity of the liquid ^{3}He. $dT = (1/C) \cdot dQ$, where T is the temperature, C is the heat capacity, and Q is the heat input. For a constant rate of heating dQ/dt, the rate of temperature change becomes $(dT/dt) = (1/C) \cdot (dQ/dt)$. A sudden increase in heat capacity will cause the rate of cooling at A to decrease. At the time, we mistakenly identified the heat capacity change with the long sought nuclear magnetic transition in the solid phase. Points B and B' are related to another thermal event. At point B there must be an evolution of latent heat because there is a sudden but small decrease in the temperature in the cell. The pressure at which the B type event took place varied, generally depending upon the cooling rate. We attributed this, correctly to a super cooling. Point B' is the equilibrium transition, we now know, from the superfluid B phase to the superfluid A phase. The temperature change pauses briefly as the B phase absorbs extra heat to pass through a first order phase change (like the melting of ice, or the freezing of liquid ^{3}He). Points C and D in Figure 9 correspond to the maximum melting pressure achieved and the time at which a slow decompression was begun.

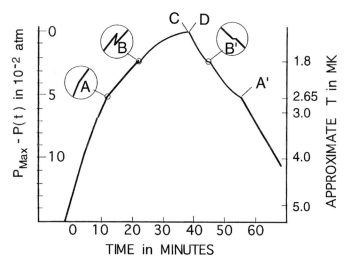

Figure 9. Measurement of the change in ^{3}He melting pressure with time with a constant cooling rate in a Osheroff's Pomeranchuk cell.

This measurement in Figure 9 contained an embarrassing amount of contradictory detail. Since nothing was previously known about the nuclear ordering process we supposed that points B and B' marked the transition to a second magnetic phase. Still, the total pressure change between point A and the maximum melting pressure was surprisingly large. In SI units the pressure difference is 0.00527 MPa. In this connection we made the following observation, "In order to obtain sufficient pressure change from 2.7 mK to 0 mK through integration of the Clausius-Clapeyron equation,

$$\Delta P = \int \frac{(S_{solid} - S_{liquid})dT}{(V_{solid} - V_{liquid})} \approx \int \frac{S_{solid}}{\Delta V}dT, \qquad [2]$$

to agree with the value presented above, one is forced to hold the solid entropy nearly constant over a broad temperature region below the 2.7 mK transition temperature. This possible behavior of the solid entropy is, in fact, also suggested by the nearly constant slope of P(t) between A and B in Figure [9.] We know of no physical system which furnishes a precedent for the entropy behavior we postulate here."[20] Using our 'approximate' temperature scale would require the solid entropy to remain at the value R•log$_e$2 over the temperature interval between 2.7 mK and 1.5 mK!

Our misgivings about the interpretation of the data were well founded. In our subsequent paper about the NMR properties of the ^{3}He in the compression cell we finally got it right.[21] The change in heat capacity signaled at point A corresponded with a change in the properties of liquid ^{3}He and point B (or B') marked the phase boundary to a liquid phase with even different behavior.

Less than a year after the report of the A and B transitions in ^{3}He, the group in Helsinki[22] used Pomeranchuk cooling to study the pressurized liquid phase in the cell to show that the viscosity of liquid ^{3}He decreases by a factor of 1000 in the new phases. The viscosity measurements were made with a vibrating wire immersed within the liquid.

Finally, the Real Phase Transition in Solid ^{3}He

After Anufriev, Jim Sites, and the Wheatley group had shown us that the compressional cooling method was an effective way to cool melting helium, we decided to begin a second set of

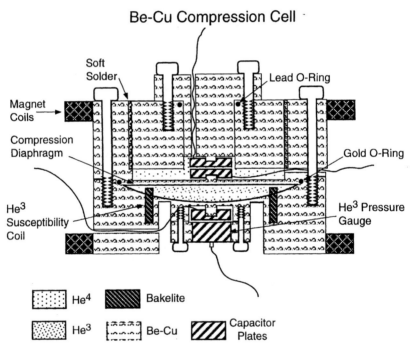

Figure 10. The Pomeranchuk Cell of Bill Halperin

measurements using a cell designed primarily for the optimization of studies of solid ^{3}He.[23] The compression cell designed by Bill Halperin is shown in Figure 10. The design thought was that by having a lens shaped compression region there would be a minimum of heating related to crushing solid ^{3}He during compression. The cell contained both a pressure gauge for the ^{3}He and a method for measuring the absolute volume changes, a second set of capacitor plates attached to the moving diaphragm. We also had a provision for measuring the changes in the magnetization of the ^{3}He. The basic operation of the cell was the same as the others I have described. ^{3}He at the melting pressure was trapped in the lower region while the upper was filled with liquid ^{4}He. The assembly was precooled to 25 mK and compression was achieved by forcing liquid ^{4}He in the upper region.

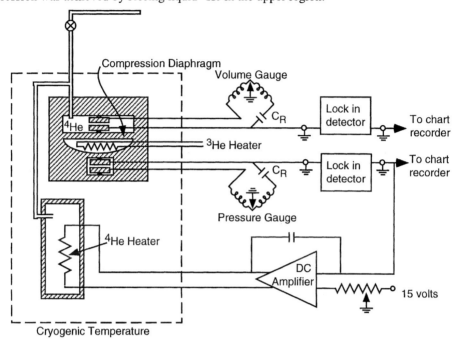

Figure 11. Servo Loop for Controlling Compression Rate

In the sequence of measurements with this cell, we measured the entropy of solid ^{3}He down to temperatures below the phase transition. In addition we were able to measure the heat capacity of liquid 3He through the superfluid transition, the latent heat of the transition between the A and B phases, and to determine a 'first principles' temperature scale.[23-25] Through experience, we grew to understand the different time constants for thermal equilibrium of the liquid and solid phases. A separate container for liquid ^{4}He was located in the cryostat. Heat applied to the ^{4}He in that vessel would rapidly change the ^{3}He volume in the compression cell.

Heat pulses and 'cool pulses' could be applied by means of short bursts of decompression or compression of the diaphragm. Heating was also accomplished by passing a calibrated current through a heater wire in the ^{3}He compression cell. The melting pressure (and hence temperature) could be maintained at a constant value by passing the error signal from the pressure gauge back through the DC amplifier to the ^{4}He heater. An example of such a measurement is shown in Figure 12. A signal from the pressure bridge has been sent to the ^{4}He heater to maintain a constant ^{3}He pressure. At t = 0 a short heat pulse of 32.26 ergs was applied to the ^{3}He heater. In response, the servo control system briefly accelerated the rate of ^{3}He compression. The volume change associated with the heat pulse was 2.56×10^{-4} cm³ as additional liquid was suddenly converted into solid ³He. The output of the pressure bridge is also shown the pressure has been converted to temperature units. During the measurement the maximum temperature excursion of the cell was less than 5 μK.

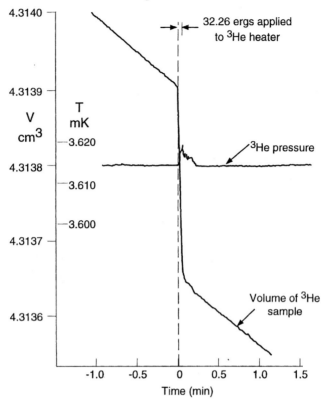

Figure 12. Measurement of volume change following a heat pulse with the compression cell constrained at constant pressure.

The Clausius-Clapeyron equation can be invoked, once again, with the data obtained in Figure 12. If we multiply both sides of Equation [1] by the temperature, T, we obtain, $T \cdot \left.\frac{dP}{dT}\right)_{melting} = \frac{T \cdot \Delta S}{\Delta V} = \frac{\Delta Q}{\Delta V}$. The quantity on the right is the ratio of the heat input to the volume change. The measurement was repeated from temperatures near 25 mK down to the temperature of the maximum melting pressure to generate a table of values of P versus T(dP/dT).

With regard to the elusive phase transition of solid ³He, measurements like that illustrated in Figure 12 and other non-equilibrium measurements with pulsed volume changes were used to generate the data shown in Figure 13.[24] At a pressure near the maximum melting pressure, the value of T(dP/dT) decreases rapidly, corresponding to an entropy decrease of more than half of the spin entropy. The experiment was the first quantitative identification of the point of magnetic order.

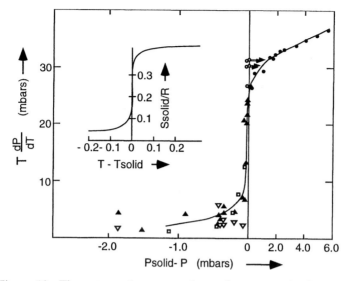

Figure 13. The entropy decrease at the nuclear magnetic phase transition of solid ^{3}He. The entropy versus temperature data was calculated from the measurements of T(dT/dP) versus pressure.

The entropy versus temperature curve shown in the inset curve in Figure 13 was obtained through integration of the P versus T(dP/dT) data. A relative temperature scale is given by $\dfrac{T}{T_{solid}} = \exp \int_{P_{solid}}^{P} \left[T\dfrac{dP}{dT} \right]^{-1} dP'$. A single fixed point in temperature is sufficient to generate the complete low temperature melting curve from the data. The fixed point we used was that of the high temperature solid entropy, $S_{solid} \to R \cdot \log_e 2$. [The Debye temperature for solid helium is of order 30 K so that the phonon contribution is negligible at 30 mK.]

The Pomeranchuk Effect and Melting Curve in 1996

The Pomeranchuk effect is no longer the preferred cooling technique for studying either liquid or solid ^{3}He. It is too restrictive because it permits measurements of ^{3}He only at the melting pressure. Moreover, an even more important reason is that the method of nuclear demagnetization is far more efficient and permits cooling of ^{3}He to temperatures in the range of 10 μK.[26]

The ^{3}He melting plays the same role in thermometry as it did in the days of our discovery of the superfluid transitions. There are four easily measured and reproduced fixed points on the melting curve: the minimum in the melting pressure; the superfluid A transition; the superfluid B transition; and the magnetic ordering transition of solid ^{3}He. Melting curve thermometers[27] have become the temperature standard for very low temperature work. In the days since Halperin's integration of T(dP/dT) along the melting curve there have been many independent measurements of the melting curve fixed points.[6, 28] The most recent[28] give: $T_{minimum}$= 0.31517 mK; T_A=2.41 mK; T_B=1.87 mK; and T_{solid}= 0.88 mK. The associated pressures are also know with great precision so that the strain gauge can also be calibrated with the fixed points. It is amusing to realize that many modern low temperature physicists routinely reproduce the data of Figure 9 as a standard temperature calibration of their apparatus.

References

1. I. Pomeranchuk, *Zh. Eksp. i Teor. Fiz.* (USSR) **20**, 919 (1950).

2. There are several elegant and systemattic discussions of the compressionai cooling technique. Some especially useful discussions about compressional cooling and crygenic methods in general are contained in the textbooks by Betts and Lounasmaa. Betts, D. S., *Refrigeration and Thermometry Below 1K*, Sussex University Press, London (1974), and O. V. Lounasma, *Experimental Principles and Methods Below 1K*, Academic Press, NY (1974).

3. L. D. Landau, *Zh. Eksp. i Teor. Fiz.* (USSR) **30**, 1058 (1956).

4. My first encounter with a discussion of the ^{3}He melting curve came through **Exercise 9** of Pippard's excellent undergraduate thermodynamics book, *The Elements of Classical Thermodynamics*, A. B. Pippard, Cambridge University Press, London, 161 (1957). There are only 14 problem in the 'Exercise' section for the entire book!

5. A good source of data about liquid ^{3}He remains the compilation of Wheatley in the review: J. C. Wheatley, "Experimental Properties of Superfluid ^{3}He", *Review of Modern Physics*, **47**, 415 (1975).

6. D. S. Greywall, *Phys. Rev.* **B33**, 7520 (1986).

7. R. C. Richardson, E. R. Hunt, and H. Meyer, *Physical Review* **138**, A1326 (1965).

8. N. Kurti, *Cryogenics* **1**, 2 (1960).

9. Yu. D. Anufriev, *Sov. Phys. J. E. T. P. Letters* **1**, 155 (1965).

10. R. T. Johnson, R. Rosenbaum, O. G. Symko, and J. C. Wheatley, *Phys. Rev. Lett.* **22**, 449 (1969).

11. R. T. Johnson, O. V. Lounasmaa, R. Rosenbaum, O. G. Symko, and J. C. Wheatley, *J. Low Temp. Phys.* **2**, 403 (1970).

12. James R. Sites, "Magnetic Susceptibility of Solid Helium Three Cooled by Adiabatic Compression," Ph. D. Thesis, Unpublished, Cornell University (1969).

13. J. R. Sites, D. D. Osheroff, R. C. Richardson, and D. M. Lee, *Phys. Rev. Lett.* **23**, 836 (1969).

14. A. J. Leggett and M. J. Rice, *Phys. Rev. Lett.* **20,** 586; 21, 506 (1968).

15. L. R. Corruccinni, "Spin Wave Phenomena in Liquid ^{3}He Systems", Ph. D. Thesis, Unpublished, Cornell University (1972).

16. L. R. Corruccini, D. D. Osheroff, D. M. Lee, and R. C. Richardson, *J. Low Temp. Phys.* **8**, 119 (1972).

17. D. D. Osheroff, "Compressional Cooling and Ultralow Temperature Properties of ^{3}He ", Ph. D. Thesis, Unpublished, Cornell University (1972).

18. G. C. Straty and E. D. Adams, *Rev. Sci. Instrum.* **40**, 1393 (1969).

19. R. A. Scribner, M. Panzick, and E. D. Adams, *Phys. Rev. Lett.* **21**, 427 (1968).

20. D. D. Osheroff, R. C. Richardson, and D. M. Lee, *Phys. Rev. Lett.* **14**, 885 (1972).

21. D. D. Osheroff, W. J. Gully, R. C. Richardson, and D. M. Lee, *Phys. Rev. Lett.* **14**, 920 (1972).

22. Yu. D. Anufriev, T. A. Alvesalo, H. K. Collan, N. T. Opheim, P. Wennerström, *Phys. Lett.* **43A**, 175 (1973); T. A. Alvesalo, Yu. D. Anufriev, H. K. Collan, O.V. Lounasmaa, P. Wennerström, *Phys. Rev. Lett.* **30**, 962 (1973).

23. W. P. Halperin, *"Melting Properties of 3He: Specific Heat, Entropy, Latent Heat, and Temperature "*, Ph. D. Thesis, Unpublished, Cornell University (1975).

24. W. P. Halperin, C. N. Archie, F. B. Rasmussen, R. A. Buhrman, and R. C. Richardson, *Phys. Rev. Lett.* **32**, 927 (1974).

25. W. P. Halperin, C. N. Archie, F. B. Rasmussen, and R. C. Richardson, *Phys. Rev. Lett.* **34**, 718 (1975).

26. An excellent review of nuclear magnetic cooling is given in the text by F. Pobell, Matter and Methods at Low Temperatures, Springer-Verlag, New York (1992).

27. J. S. Souris and T. T. Tommila, *Experimental Techniques in Condensed Matter Physics at Low Temperatures*, [R. C. Richardson and E. N. Smith, editors; Addison-Wesley, New York], 245 (1988).

28. G. Schuster, A. Hoffmann, and D. Hechtfischer, *Czech. J. Phys.* **46-S1**, 481 (1996).

PAIR CORRELATIONS IN SUPERFLUID HELIUM 3

Dieter Vollhardt

Theoretische Physik III
Elektronische Korrelationen und Magnetismus
Universität Augsburg
86135 Augsburg, Germany

ABSTRACT

In 1996 Lee, Osheroff and Richardson received the Nobel Prize for their 1971 discovery of superfluid helium 3 – a discovery which opened the door to the most fascinating system known in condensed matter physics. The superfluid phases of helium 3, originating from pair condensation of helium 3 atoms, turned out to be the ideal test-system for many fundamental concepts of modern physics, such as macroscopic quantum phenomena, (gauge-)symmetries and their spontaneous breakdown, topological defects, etc. Thereby they enriched condensed matter physics enormously and contributed significantly to our understanding of various other physical systems, from heavy fermion and high-T_c superconductors all the way to neutron stars and the early universe. A pedagogical introduction is presented.

1 THE HELIUM LIQUIDS

There are two stable isotopes of the chemical element helium: helium 3 and helium 4, conventionally denoted by ^{3}He and ^{4}He, respectively. From a microscopic point of view, helium atoms are structureless, spherical particles interacting via a two-body potential that is well understood. The attractive part of the potential, arising from weak van der Waals-type dipole (and higher multipole) forces, causes helium gas to condense into a liquid state at temperatures of 3.2 K and 4.2 K for ^{3}He and ^{4}He, respectively, at normal pressure. The pressure versus temperature phase diagrams of ^{3}He and ^{4}He are shown in Figs. 1.1 and 1.2. When the temperature is decreased even further one finds that the helium liquids, unlike all other liquids, do not solidify unless a pressure of around 30 bar is applied. This is the first remarkable indication of macroscopic quantum effects in these systems. The origin of this unusual behaviour lies in the quantum-mechanical uncertainty principle, which requires that a quantum particle can never be completely at rest at given position, but rather performs a zero-point motion about the average position. The smaller the mass of the particle and the weaker the

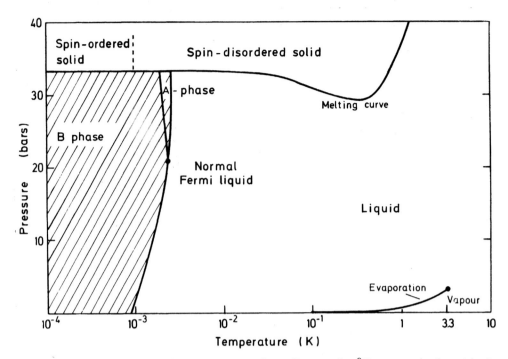

Figure 1.1 Pressure versus temperature phase diagram for ^{3}He; note the logarithmic temperature scale.

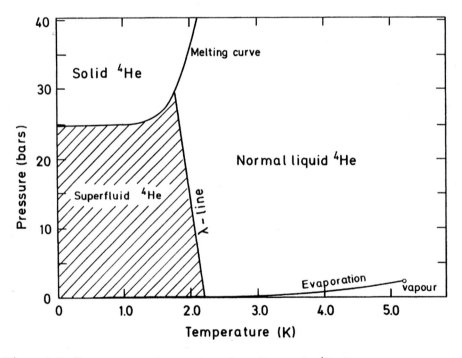

Figure 1.2 Pressure versus temperature phase diagram for ^{4}He; linear temperature scale.

binding force, the stronger these oscillations are. In most solids the zero-point motion is confined to a small volume of only a fraction of the lattice-cell volume. In the case of helium, however, two features combine to prevent the formation of a crystalline solid with a rigid lattice structure: (i) the strong zero-point motion arising from the small atomic mass (helium is the second-lightest element in the periodic table); and (ii) the weakness of the attractive interaction due to the high symmetry of these simple atoms. It is this very property of helium – of staying liquid – that makes it such a valuable system for observing quantum behavior on a macroscopic scale. Quantum effects are also responsible for the strikingly different behaviors of ^{4}He and ^{3}He at even lower temperatures. Whereas ^{4}He undergoes a second-order phase transition into a state later shown to be superfluid, i.e. where the liquid is capable of flowing through narrow capillaries or tiny pores without friction, no such transition is observed in liquid ^{3}He in the same temperature range (see Figs. 1.1 and 1.2). The properties of liquid ^{3}He below 1 K are nevertheless found to be increasingly different from those of a classical liquid. It is only at a temperature roughly one thousandth of the transition temperature of ^{4}He that ^{3}He also becomes superfluid, and in fact forms *several* superfluid phases, each of which has a much richer structure[1] than that of superfluid ^{4}He.

The striking difference in the behaviors of ^{3}He and ^{4}He at low temperatures is a consequence of the laws of quantum theory as applied to systems of identical particles, i.e. the laws of quantum statistics. The ^{4}He atom, being composed of an even number of electrons and nucleons, has spin zero and consequently obeys Bose-Einstein statistics. In contrast, the ^{3}He nucleus consists of *three* nucleons, whose spins add up to give a total nuclear spin of $I = \frac{1}{2}$, making the total spin of the entire ^{3}He atom $\frac{1}{2}$ as well. Consequently liquid ^{3}He obeys Fermi-Dirac statistics. So it is the tiny nuclear spin, buried deep inside the helium atom, that is responsible for all the differences of the macroscopic properties of the two isotopes.

Since in a Bose system single-particle states may be multiply occupied, at low temperatures this system has a tendency to condense into the lowest-energy single-particle state (Bose-Einstein condensation). It is believed that the superfluid transition in ^{4}He is a manifestation for Bose-Einstein condensation. The all-important qualitative feature of the Bose condensate is its phase rigidity, i.e. the fact that it is energetically favourable for the particles to condense into a single-particle state of fixed quantum-mechanical phase, such that the global gauge symmetry spontaneously broken. As a consequence, macroscopic flow of the condensate is (meta-)stable, giving rise to the phenomenon of superfluidity.

In a Fermi system, on the other hand, the Pauli exclusion principle allows only single occupation of fermion states. The ground state of the Fermi gas is therefore the one in which all single-particle states are filled up to a limiting energy, the Fermi energy E_F. As predicted by Landau (1956, 1957, 1958) and later verified experimentally (for a review see Wheatley (1966)), the properties of ^{3}He well below its Fermi temperature $T_F = E_F/k_B \approx 1K$ are similar to those of a degenerate Fermi gas. In particular the formation of a phase-rigid condensate is not possible in this framework. Until the mid-1950s a superfluid phase of liquid ^{3}He was therefore believed to be ruled out. On the other hand, it is most remarkable that the property of superfluidity (London, 1950, 1954) was indeed first discovered experimentally in a *Fermi* system, namely that of the "liquid" of conduction electrons in a superconducting metal (Kamerlingh Onnes, 1911). The superfluidity of ^{4}He was only found more than 25 years later.

---

[1] A comprehensive treatment of superfluidity in ^{3}He with a very extensive reference list can be found in the book by Vollhardt and Wölfle (1990).

2 PAIR CONDENSATION IN A FERMI LIQUID

The key to the theory of superconductivity (Bardeen, Cooper and Schrieffer (BCS) 1957) turned out to be the formation of "Cooper pairs", i.e. pairs of electrons with opposite momentum **k** and spin projection σ: (**k** ↑, −**k** ↓). In the case of conventional superconductors the Cooper pairs are structureless objects, i.e. the two partners form a spin-singlet state in a relative s-wave orbital state. These Cooper pairs have total spin zero and may therefore be looked upon in a way as composite bosons, which all have the same pair wave function and are all in the same quantum-mechanical state. In this picture the transition to the superconducting state corresponds to the Bose-condensation of Cooper pairs, the condensate being characterized by macroscopic quantum coherence. The concept of Bose-Einstein condensation is appealing since key features of superconductivity like the Meissner effect, flux quantization and superfluid mass currents in conventional superconductors are naturally implied. Nevertheless, since the theory of conventional superconductivity is firmly based on BCS theory, the concept of a Bose-Einstein condensation of Cooper pairs traditionally did not receive much attention (or was even considered to be downright wrong). Within the context of superfluid ^{3}He, this notion was taken up again by Leggett (1980a,b), who argued that tightly bound Bose-Einstein-condensed molecules on the one hand and Cooper pairs on the other may be viewed as extreme limits of the same phenomenon. This approach, which was quite provocative at the time, is now well accepted (Zwerger, 1992; Nozières, 1995; Randeria, 1995). However, the original idea that at T_c Cooper pairs form and automatically Bose-condense has been revised since then. Apparently Cooper pair formation is not a separate phase transition but is rather a matter of thermal equilibrium: for any finite coupling there exists a finite density of pairs even above T_c, although in conventional superconductors – and even in high T_c materials – their density is negligibly small. At weak coupling (BCS limit) the condensation temperature and the (not well-defined) temperature of pair formation practically coincide; they become different only at very strong coupling (Bose limit). Similar ideas are also implicit in several theoretical approaches to high-T_c superconductivity.

While in free space an attractive force has to be sufficiently strong to bind two electrons, inside the metal the presence of the filled Fermi sea of conduction electrons blocks the decay of a Cooper pair, so that an *arbitrarily* small attractive interaction leads to the formation of stable Coopers pairs. The attractive interaction between the electrons of a Cooper pair in a conventional superconducting metal is due to the exchange of virtual phonons (electron-phonon interaction). If the phonon-mediated interaction is strong enough to overcome the repulsive Coulomb interactions between the two electrons then a transition into a superconducting state may occur. On the other hand, any other mechanism leading to attraction between electrons at the Fermi surface is equally well suited for producing superconductivity.

Given the success of the BCS theory in the case of superconductivity, it was natural to ask whether a similar mechanism might also work for liquid ^{3}He. Since there is no underlying crystal lattice in the liquid that could mediate the attractive force, the attraction must clearly be an intrinsic property of the one-component ^{3}He liquid itself. The main feature of the interatomic ^{3}He potential is the strong repulsive component at short distances, and the weak van der Waals attraction at medium and long distances. It soon became clear that, in order to avoid the hard repulsive core and thus make optimal use of the attractive part of the potential, the ^{3}He atoms would have to form Cooper-pairs in a state of *nonzero* relative angular momentum l. In this case the

Cooper-pair wave function vanishes at zero relative distance, thus cutting out the most strongly repulsive part of the potential. In a complementary classical picture one might imagine the partners of a Cooper pair revolving about their centre of gravity, thus being kept away from each other by the centrifugal force.

When the superfluid phases of ^{3}He were finally discovered in 1971 at temperatures of about 2.6 mK and 1.8 mK respectively (Osheroff, Richardson and Lee, 1972a), in an experiment actually designed to observe a magnetic phase transition in solid ^{3}He, the results came as a great surprise.

3 PROPERTIES OF SUPERFLUID ^{3}He

Soon after the discovery of the phase transitions by Osheroff, Richardson and Lee (1972a), it was possible to identify altogether *three* distinct stable superfluid phases of bulk ^{3}He ; these are referred to as the A, B and A_1 phases. In zero magnetic field only the A and B phases are stable. In particular, in zero field the A phase only exists within a finite range of temperatures, above a critical pressure of about 21 bar. Hence its region of stability in the pressure-temperature phase diagram has a roughly triangular shape as shown in Fig. 1.1. The B phase, on the other hand, occupies the largest part of this phase diagram and is found to be stable down to the lowest temperatures attained so far. Application of an external magnetic field has a strong influence on this phase diagram. First of all, the A phase is now stabilized down to zero pressure. Secondly, an entirely new phase, the A_1 phase, appears as a narrow wedge between the normal state and the A and B phases.

Owing to the theoretical work on anisotropic superfluidity that had been carried out before the actual discovery of superfluid ^{3}He, progress in understanding the detailed nature of the phases was very rapid. This was clearly also due to the excellent contact between experimentalists and theorists, which greatly helped to develop the right ideas at the right time; for reviews see Leggett (1975), Wheatley (1975), Lee and Richardson (1978). In particular, it fairly soon became possible to identify the A phase and the B phase as realizations of the states studied previously by Anderson and Morel (1960, 1961) and Balian and Werthamer (1963) respectively. Therefore the A phase is described by the so-called "Anderson-Brinkman-Morel" (ABM) state, while the B phase is described by the "Balian-Werthamer" (BW) state. Consequently, "A phase" and "ABM state" are now used as synonyms; the same is true in the case of "B phase" and "BW state". (The fact that the ABM state describes the A phase and the BW state the B phase is a very fortunate coincidence – if it was the other way around, it would be quite confusing!).

Although the three superfluid phases all have very different properties, they have one important thing in common: the Cooper pairs in all three phases are in a state with *parallel* spin (S = 1) and relative orbital angular momentum $l = 1$. This kind of pairing is referred to as "spin-triplet p-wave pairing". In contrast, prior to the discovery of the superfluid phases of ^{3}He, Cooper pairing in superconductors was only known to occur in a state with opposite spins (S = 0) and $l = 1$, i.e. in a "spin-singlet s-wave state". It should be noted that Cooper pairs in a superconductor and in superfluid ^{3}He are therefore very different entities: in the former case pairs are formed by pointlike, structureless electrons and are spherically symmetric, while in the case of ^{3}He Cooper pairs are made of actual atoms (or rather of quasiparticles involving ^{3}He atoms) and have an internal structure themselves.

3.1 The internal structure of Cooper pairs

Quantum-mechanically, a spin-triplet configuration (S = 1) of two particles has three substates with different spin projection S_z. They may be represented as $|\uparrow\uparrow\rangle$ with $S_z = +1$, $2^{-1/2}(|\uparrow\downarrow\rangle + |\downarrow\uparrow\rangle)$ with $S_z = 0$, and $|\downarrow\downarrow\rangle$ with $S_z = -1$. The pair wave function Ψ is in general a linear superposition of all three substates, i.e.

$$\Psi = \psi_{1,+}(\mathbf{k})\,|\uparrow\uparrow\rangle + \psi_{1,0}(\mathbf{k})(|\uparrow\downarrow\rangle + |\downarrow\uparrow\rangle) + \psi_{1,-}(\mathbf{k})\,|\downarrow\downarrow\rangle \qquad (1)$$

where $\psi_{1,+}(\mathbf{k}), \psi_{1,0}(\mathbf{k})$ and $\psi_{1,-}(\mathbf{k})$ are the three complex-valued amplitudes of the respective substates. In the case of a superconductor, where S = 0 and $l = 0$, the pair wave function is much simpler, i.e. it is given by only a single component

$$\Psi_{sc} = \psi_0(|\uparrow\downarrow\rangle - |\downarrow\uparrow\rangle) \qquad (2)$$

with a single amplitude ψ_0.

So far we have only taken into account that, since S = 1, there are three substates for the spin. The same is of course true for the relative orbital angular momentum $l = 1$ of the Cooper pair, which also has three substates $l_z = 0, \pm 1$. This fact is important if we want to investigate the amplitudes $\psi_{1,+}(\mathbf{k})$ etc. further. They still contain the complete information about the space (or momentum) dependence of Ψ. The pair wave function Ψ is therefore characterized by three spin substates and three orbital substates, i.e. by altogether 3 x 3 = 9 substates with respect to the spin and orbital dependence. Each of these nine substates is connected with a complex-valued parameter. Here we see the essential difference betweeen Cooper pairs with $S = l = 0$ (conventional superconductors) and $S = l = 1({}^3He)$: their pair wave functions are very different. In the former case a single complex-valued parameter is sufficient for its specification, in the latter case of superfluid ^{3}He *nine* such parameters are required. This also expresses the fact that a Cooper pair in superfluid ^{3}He has an internal structure, while that for a conventional superconductor does not: because $l = 1$, it is intrinsically *anisotropic*. This anisotropy may conveniently be described by specifying some direction with respect to a quantization axis both for the spin and the orbital component of the wave function.

In order to understand the novel properties of superfluid ^{3}He, it is therefore important to keep in mind that there are two characteristic directions that specify a Cooper pair. Here lies the substantial difference from a superconductor and the origin of the multitude of unusual phenomena occurring in superfluid ^{3}He: the structure of the Cooper pair is characterized by *internal degrees of freedom*. Nevertheless, in both cases the superfluid/superconducting state can be viewed as the condensation of a macroscopic number of these Cooper pairs into the same quantum-mechanical state, similar to a Bose-Einstein condensation, as discussed above.

3.2 Broken symmetry and the order parameter

In the normal liquid state Cooper pairs do not exist. Obviously in the superfluid a new state of order appears, which spontaneously sets in at the critical temperature T_c. This particular transition from the normal fluid to the superfluid, i.e. into the ordered state, is called "continuous", since the condensate – and hence the state of order – builds up continuously. This fact may be expressed quantitatively by introducing an "order-parameter" that is finite for $T < T_c$ and zero for $T \geq T_c$. A well-known example of such a transition is that from a paramagnetic to a ferromagnetic state of a metal when the system is cooled below the Curie temperature. In the paramagnetic regime the spins of the particles are disordered such that the average magnetization $\langle \mathbf{M} \rangle$ of the system is zero. By contrast, in the ferromagnetic phase the spins are more or less aligned and $\langle M \rangle$ is thus finite. In this case the system exhibits long-range order of the spins. The degree of ordering is quantified by $|\langle M \rangle|$, the magnitude of the magnetization. Hence $|\langle M \rangle|$ is called the "order parameter" of the ferromagnetic state. Clearly, the existence of a preferred direction M of the spins implies that the symmetry of the ferromagnet under spin rotations is reduced ("broken") when compared with the paramagnet: the directions of the spins are no longer isotropically distributed, and the system will therefore no longer be invariant under a spin rotation. This phenomenon is called "spontaneously broken symmetry"; it is of fundamental importance in the theory of phase transitions. It describes the property of a macroscopic system (i.e. a system in the thermodynamic limit) that is in a state that does not have the full symmetry of the microscopic dynamics.

The concept of spontaneously broken symmetry also applies to superconductivity and superfluid ^{3}He. In this case the order parameter measures the existence of Cooper pairs and is given by the probability amplitude for a pair to exist at a given temperature. It follows from the discussion of the possible structure of a Cooper pair in superfluid ^{3}He that the associated order parameter will reflect this structure and the allowed internal degrees of freedom. What then are the spontaneously broken symmetries in superfluid ^{3}He?

As already mentioned, the interparticle forces between the ^{3}He atoms are rotationally invariant in spin and orbital space and, of course, conserve particle number. The latter symmetry gives rise to a somewhat abstract symmetry called "gauge symmetry". Nevertheless, gauge symmetry is spontaneously broken in any superfluid or superconductor. In addition, in an odd-parity pairing superfluid, as in the case of ^{3}He, where $l = 1$, the pairs are necessarily in a spin-triplet state, implying that rotational symmetry in spin space is broken, just as in a magnet. At the same time, the anisotropy of the Cooper-pair wave function in orbital space calls for a spontaneous breakdown of orbital rotation symmetry, as in liquid crystals. All three symmetries are therefore simultaneously broken in superfluid ^{3}He. This implies that the A phase, for example, may be considered as a "superfluid nematic liquid crystal with (anti)ferromagnetic character". One might think that a study of the above mentioned broken symmetries could be performed much more easily by investigating them separately, i.e. within the isotropic superfluid, the magnet, the liquid crystal etc. itself. However, the combination of several *simultaneously* broken continuous symmetries is more than just the simple sum of the properties of all these known systems. Some of the symmetries broken in superfluid ^{3}He are "relative" symmetries, such as spin-orbit rotation symmetry or gauge-orbit symmetry (Leggett, 1972,1973b; Liu and Cross, 1978). Because of this, a rigid connection is established between the corresponding degrees of freedom of the condensate, leading to long-range order only in the combined (and not in the individual) degrees of freedom. This particular kind of broken symmetry, for example the so-called "spontaneously broken spin-orbit symmetry", gives rise to very unusual behaviour, as will be discussed later.

It is clear that in principle the internal degrees of freedom of a spin-triplet p-wave state allow for many different Cooper-pair states and hence superfluid phases. (This is again different from ordinary superconductivity with $S = 0, l = 0$ pairing, where only a *single* phase is possible). Of these different states, the one with the lowest energy for given external parameters will be realized. In fact, Balian and Werthamer (1963) showed, that, within a conventional "weak-coupling" approach, of all possible states there is precisely one state (the "BW state") that has the lowest energy at *all* temperatures. This state is the one that describes the B phase of superfluid ^{3}He.

The state originally discussed by these authors is one in which the orbital angular momentum l and spin **S** of a Cooper pair couple to a total angular momentum **J** = l + **S** = 0. This $^3P_0$ state is, however, only a special case of a more general one with the same energy (in the absence of spin-orbit interaction), obtained by an arbitrary rotation of the spin axes relative to the orbital axes of the Cooper-pair wave function. Such a rotation may be described mathematically by specifying a rotation axis $\hat{n}$ and a rotation angle θ. In the BW state all three spin substates in (1) occur with equal measure. This state has a rather surprising property: in spite of the intrinsic anisotropy, the state has an *isotropic* energy gap. (The energy gap is the amount by which the system lowers its energy in the condensation process, i.e. it is the minimum energy required for the excitation of a single particle out of the condensate.) Therefore the BW state resembles ordinary superconductors in several ways. On the other hand, even though the energy gap is isotropic, the BW state is intrinsically anisotropic. This is clearly seen in dynamic experiments in which the Cooper-pair structure is distorted. For this reason the BW state is sometimes referred to as "pseudo-isotropic". Owing to the quantum coherence of the superfluid state, the rotation axis $\hat{n}$ and angle θ characterizing a Cooper pair in the BW state are macroscopically defined degrees of freedom, whose variation is physically measurable.

Since in weak-coupling theory the BW state always has the lowest energy, an explanation of the existence of the A phase of superfluid ^{3}He obviously requires one to go beyond such an approach and to include "strong–coupling effects"(Anderson and Brinkmann, 1973, 1978; for a review of a systematic approach see Serene and Rainer (1983)). In view of the fact that at present microscopic theories are not capable of computing transition temperatures for ^{3}He, it is helpful to single out a particular effect that can explain the stabilization of the A phase over the B phase at least qualitatively. As shown by Anderson and Brinkman (1973), there is such a conceptually simple effect, which is based on a feedback mechanism: the pair correlations in the condensed state change the pairing interaction between the ^{3}He quasiparticles, the modification depending on the actual state itself. As a specific mechanism, these authors considered the role of spin fluctuations and showed that a stabilization of the state first considered by Anderson and Morel (1960, 1961) is indeed possible. This only happens at somewhat elevated pressures, since spin fluctuations become more pronounced only at higher pressures. This "ABM state" (from the initals of the above three authors) does indeed describe the A phase. It has the property that, in contrast with ^{3}He-B, its magnetic susceptibility is essentially the same as that of the normal liquid. This is a clear indication that in this phase the spin substate with $S_z = 0$, which is the only one that can be reduced appreciably by an external magnetic field, is absent. Therefore ^{3}He-A is composed only of $|\uparrow\uparrow\rangle$ and $|\downarrow\downarrow\rangle$ Cooper pairs. This implies that the anisotropy axis of the spin part of the Cooper pair wave function, called $\hat{d}$, has the same fixed direction in every pair. (More precisely, $\hat{d}$ is the direction along which the total spin of the Cooper pair vanishes: $\hat{d} \cdot \mathbf{S} = 0$). Likewise, the direction of the relative orbital angular momentum $\hat{l}$ is the same for all Cooper pairs. Therefore in the A phase the anisotropy axes $\hat{d}$ and $\hat{l}$ of the Cooper-pair wave function are long-range-ordered, i.e. are preferred directions in the whole macroscopic sample. This implies a pronounced anisotropy of this phase in all its properties. In particular, the value of the energy gap now explicitly depends on the direction in k space on the Fermi sphere and takes the form

$$\Delta_{\hat{k}}(T) = \Delta_0(T)[1 - (\hat{k} \cdot \hat{l})^2]^{1/2}. \tag{3}$$

Hence the gap vanishes at two points on the Fermi sphere, namely along $\pm\hat{l}$. Because of the existence of an axis $\hat{l}$, this state is also called the "axial state". The existence of nodes implies that in general quasiparticle excitations may take place at arbitrarily low temperatures. Therefore, in contrast with ^{3}He-B or ordinary superconductors, there is a finite density of states for excitations with energies below the average gap energy, leading for example to a specific heat proportional to T^3 at low temperatures.

The third experimentally observable superfluid phase of ^{3}He, the A_1 phase, is only stable in the presence of an external magnetic field. In this phase Cooper pairs are all in a single spin substate, the $|\uparrow\uparrow\rangle$ state, corresponding to $S_z = +1$; the components with $|\uparrow\downarrow\rangle + |\downarrow\uparrow\rangle$ and $|\downarrow\downarrow\rangle$ states are missing. It is therefore a "magnetic" superfluid, the first ever observed in nature.

3.3 Orientational effects

For a pair-correlated superfluid, the pairing interaction is the most important interaction, since it is responsible for the formation of the condensate itself. Nevertheless, there also exist other, much weaker, interactions, which may not be important for the actual transition to the pair-condensed state, but which do become important if their symmetry differs form the aforementioned. In particular, they may be able to break remaining degeneracies.

The dipole–dipole interaction. The dipole–dipole interaction between the nuclear spins of the ^{3}He atoms leads to a very weak, spatially strongly anisotropic, coupling. The relevant coupling constant $g_D(T)$ is given by

$$g_D(T) \approx \frac{\mu_0^2}{a^3} \left(\frac{\Delta(T)}{E_F} \right)^2 n \qquad (4)$$

Here μ_0 is the nuclear magnetic moment, such that μ_0^2/a^3 is the average dipole energy of two particles at relative distance a (the average atomic distance), while the second factor measures the probability for these two particles to form a Cooper pair and n is the overall particle density. Since μ_0^2/a^3 corresponds to about 10^{-7}K, this energy is extremely small and the resulting interaction of quasiparticles at temperatures of the order of 10^{-3}K might be expected to be completely swamped by thermal fluctuations. This is indeed true in a normal system. However, the dipole-dipole interaction implies a spin-orbit coupling and thereby has a symmetry different from that of the pairing interaction. In the condensate the symmetries with respect to a rotation in spin and orbital space are spontaneously broken, leading to long-range order (for example of $\hat{\mathbf{d}}$ and $\hat{\mathbf{l}}$ in the case of ^{3}He-A). Nevertheless, the pairing interaction does not fix the *relative* orientation of these preferred directions, leaving a continuous degeneracy. As pointed out by Leggett (1973a,b, 1974, 1975), in this situation the tiny dipole interaction is able to lift the degeneracy, namely by choosing that particular relative orientation of the long-range ordered preferred directions for which the dipolar energy is minimal. Thereby this interaction becomes of *macroscopic* importance. One may also view this effect as a *permanent* local magnetic field of about 3 mT at any point in the superfluid (in a liquid!). In ^{3}He-A the dipolar interaction is minimized by a parallel orientation of $\hat{\mathbf{d}}$ and $\hat{\mathbf{l}}$.

Effect of a magnetic field. An external magnetic field acts on the nuclear spins and thereby leads to an orientation of the preferred direction in spin space. In the case of ^{3}He-A the orientation energy is minimal if $\hat{\mathbf{d}}$ is perpendicular to the field $\mathbf{H}$, since (taking into account $\hat{\mathbf{d}} \cdot \mathbf{S} = 0$) this orientation guarantees $\mathbf{S} \| \mathbf{H}$.

Walls. Every experiment is performed in a volume of finite size. Clearly, the walls will have some effect on the liquid inside. In superfluid ^{3}He this effect may readily be understood by using a simple picture. Let us view the Cooper pair as a kind of giant "molecule" of two ^{3}He quasiparticles orbiting around each other. For a pair not to bump into a wall, this rotation will have to take place in a plane parallel to the wall. In the case of ^{3}He-A, where the orbital angular momentum $\hat{l}$ has the same direction in all Cooper pairs (standing perpendicular on the plane of rotation), this means that $\hat{l}$ has to be oriented perpendicular to the wall. So there exists a strict orientation of $\hat{l}$ caused by the walls (Ambegaokar et al., 1974). In the B phase, with its (pseudo) isotropic order parameter, the orientational effect is not as pronounced, but there are qualitatively similar boundary conditions.

3.4 Textures

From the above discussion, it is clear that the preferred directions $\hat{l}$ and $\hat{d}$ in ^{3}He-A are in general subject to different, often competing, orientational effects (for simplicity, we shall limit our description to ^{3}He-A). At the same time, the condensate will oppose any spatial variation of its long-range order. Any "bending" of the order-parameter field will therefore increase the energy, thus giving an internal stiffness or rigidity to the system. While the orientational effects might want $\hat{d}$ and $\hat{l}$ to adjust on the smallest possible lengthscale, the bending energy wants to keep the configuration as uniform as possible. Altogether, the competition between these two opposing effects will lead to a smooth spatial variation of $\hat{d}$ and $\hat{l}$ throughout the sample, called a "texture". This nomenclature is borrowed from the physics of liquid crystals, where similar orientational effects of the preferred directions occur.

The bending energy and all quantitatively important orientational energies are invariant under the replacement $\hat{d} \to -\hat{d}, \hat{l} \to -\hat{l}$. A state where $\hat{d}$ and $\hat{l}$ are parallel therefore has the same energy as one where $\hat{d}$ and $\hat{l}$ are antiparallel. This leads to two different, degenerate, ground states. There is then the possibility that in one part of the sample the system is in one ground state and in the other in a different ground state. Where the two configurations meet they form a planar "defect" in the texture, called a "domain wall" (Maki, 1977). This is in close analogy to the situation in a ferromagnet composed of domains with different orientations of the magnetization. Domain walls are spatially localized and are quite stable against external perturbations. In fact, their stability is guaranteed by the specific nature of the order-parameter structure of ^{3}He-A. Mathematically, this structure may be analysed according to its topological properties; for reviews see Mermin (1979) and Mineev (1980). The stability of a domain wall can then be traced back to the existence of a conserved "topological charge". Using the same mathematical approach, one can show that the order-parameter fields of the superfluid phases of ^{3}He not only allow for planar defects but also for point and line defects, called "monopoles" and "vortices" respectively. Defects can be "nonsingular" or "singular", depending on whether the core of the defect remains superfluid or whether it is forced to become normal liquid. The concept of vortices is of course well known from superfluid ^{4}He. However, since the order-parameter structure of superfluid ^{3}He is so much richer than that of superfluid ^{4}He, there exist a wide variety of different vortices in these phases. Their detailed structure has been the subject of intensive investigation, in particular in the context of experiments on rotating superfluid ^{3}He, where they play a central role (Hakonen and Lounasmaa, 1987; Salomaa and Volovik, 1987).

3.5 Dynamic properties

From the discussion presented so far, we have already seen that the static properties of an anisotropic superfluid are very unusual. Clearly, the dynamic properties can be expected to be at least as new and diverse. Indeed, the fact that in superfluid ^{3}He Cooper pairs have an internal structure can only be investigated in detail by studying the dynamics, i.e. the frequency and momentum dependence, of the condensate. One may roughly distinguish between magnetic and nonmagnetic dynamic properties, depending on whether the magnetization of the system is probed or whether properties such as mass transport or the propagation of sound are studied.

For the investigation of dynamical effects it is instructive to have an idea of the typical frequencies inherent to the superfluid condensate. Both the normal and the superfluid components are essentially characterized by a single timescale each: for the normal component this is the quasiparticle lifetime τ, and for the superfluid component it is $\hbar/\Delta(T)$, where $\Delta(T)$ is the average of the temperature-dependent energy gap. The orders of magnitude of the equivalent frequencies are given by $\tau^{-1} \approx 10$ MHz and $\Delta(T)/\hbar \approx 10^3(1-T/T_c)^{1/2}$ MHz, i.e. usually one has $\tau^{-1} \ll \Delta(T)/\hbar$. For frequencies ω much smaller than either of these characteristic values, the liquid is always in local thermodynamic equilibrium, since the system always has sufficient time to adjust to any change induced on the timescale ω^{-1}. This is called the "hydrodynamic regime", which is important for a couple of reasons: (i) in this regime knowledge of the conserved quantities and of those describing the broken symmetries is sufficient to describe the properties of the system; and (ii) this regime is experimentally well accessible. The multitude of broken symmetries in superfluid ^{3}He consequently leads to very rich hydrodynamics, which describes the various low-frequency collective excitations of the system. Here the word "collective" (as opposed to "single-particle") means that a macroscopic number of particles is involved in a coherent fashion.

Spin dynamics. Investigations of the collective magnetic (i.e. spin-dependent) properties of the superfluid phases of ^{3}He by nuclear magnetic resonance (NMR) were particularly useful in identifying the explicit order-parameter structure of these phases (Lee and Richardson, 1978). In usual NMR experiments the system under investigation is brought into a strong constant external magnetic field $\mathbf{H}_0 = H_0\hat{\mathbf{z}}$, which forces the (nuclear) spin $\mathbf{S}$ to precess about $\mathbf{H}_0$. By applying a weak high-frequency magnetic field $\mathbf{H}_{rf}$ perpendicular to $\mathbf{H}_0$, one is able to induce transitions in S_z, the component along $\mathbf{H}_0$, of magnitude $\pm\hbar$. This effect is observed as an energy absorption from the magnetic field. In the case of noninteracting spins these transitions occur *exactly* at the energy $\gamma\hbar H_0$, i.e. at the Larmor frequency $\omega_L = \gamma H_0$, where γ is the gyromagnetic ratio of the nucleus. How does this change in the presence of interactions? For a spin of magnitude $\frac{1}{2}\hbar$, as in the case of the ^{3}He nucleus, a very general statement is possible (Leggett, 1972, 1973b): as long as the interactions are spin-*conserving*, there is no change at all – the resonance remains at ω_0. On the other hand, for spin-nonconserving interactions, such as the spin-orbit interaction caused by the dipole coupling of the nuclear spins, a frequency shift may indeed occur. However, such a "nonsecular" shift will usually be very small, namely at most of the order of the linewidth. The experimental data obtained by Osheroff et al. (1972b) in connection with their discovery of the superfluid phases therefore came as a great surprise - they found that the resonance, although still very sharp, occurred at frequencies substantially higher than ω_L. The origin of this large shift was especially mysterious, since it obviously corresponded to a *constant* local magnetic field of order 3 mT surrounding the nuclear spins in the liquid.

The solution to this puzzle was found by Leggett (1972, 1973b, 1974, 1975), who showed that the NMR shifts are a consequence of the broken symmetries of the spin-triplet p-wave condensate, which he named "spontaneously broken spin-orbit symmetry". As explained earlier, the meaning of this concept is that the preferred directions in spin and orbital space are long-range-ordered (individually so, or in a combined way) and the tiny dipole interaction may take advantage of this situation by lifting the remaining degeneracy. The macroscopic quantum coherence of the condensate therefore raises the dipole coupling to macroscopic importance. In this way, Leggett (1974) was able to calculate the general NMR response of the spin-triplet p-wave condensate. In particular, in the A phase the transverse NMR frequency ω_t is given by

$$\omega_t^2 = \omega_L^2 + \Omega_A^2(T) \tag{5}$$

where $\Omega_A^2(T)$ is proportional to the dipole coupling constant (see (4)). It should be noted that the field and temperature dependences of ω_t are neatly separated in a "Pythagorean" form: ω_L only depends on H_0 and Ω_A only on T. In fact, Leggett (1974) worked out a complete theory of spin dynamics, whose predictions were experimentally confirmed in every detail. For example, the equation of motion of the total spin $\mathbf{S}$ is given by

$$\dot{\mathbf{S}} = \gamma \mathbf{S} \times \mathbf{H} + \mathbf{R}_D, \tag{6}$$

where $\mathbf{H} = \mathbf{H}_0 + \mathbf{H}_{rf}$ is the total external magnetic field and a dot over a symbol indicates the time derivative. Here $\mathbf{R}_D$ is the anisotropic "dipole torque", which itself depends on the change of the dipole energy under a reorientation of the order parameter. In the normal phase $\mathbf{R}_D$ is always zero. In the superfluid, however, one has $\mathbf{R}_D \neq 0$, except for static situations. If the system is displaced from static equilibrium (for example by applying $\mathbf{H}_{rf}$), $\mathbf{R}_D$ acts as restoring force. For example, in the A phase a periodic oscillation of $\mathbf{S}$ will lead to an oscillation of $\hat{\mathbf{d}}$, the preferred direction in spin space, around the orbital degree of freedom $\hat{\mathbf{l}}$ (which may be assumed to remain fixed because it cannot move very quickly). Equation (6) led Leggett to a spectacular prediction: even if the high-frequency field $\mathbf{H}_{rf}$ is oriented *parallel* to $\mathbf{H}_0$, there is a resonance, i.e. there exists a *longitudinal* spin resonance! Since in this case $(\mathbf{S} \times \mathbf{H})_z = 0$, (6) yields $dS_z/dt = R_{Dz}$. In a normal system there can be no resonance since there is no restoring force: the z-component of the magnetization will simply relax exponentially but will not oscillate. How then can we understand the nature of the longitudinal oscillation in the case of superfluid ^{3}He-A? The A phase only consists of the two spin substates $|\uparrow\uparrow\rangle$ and $|\downarrow\downarrow\rangle$. They may be viewed as essentially independent interpenetrating superfluids, which are only very weakly coupled by the spin-nonconserving dipole coupling. This coupling allows for a transition of $|\uparrow\uparrow\rangle$ pairs into $|\downarrow\downarrow\rangle$ pairs, and vice versa. (The situation is quite similar to a pair of weakly coupled superconductors, where Cooper pairs can tunnel from one superconductor to the other (the "Josephson effect"); the difference here is that the two subsystems fill the same volume, i.e. they are not spatially separated). Applying a high-frequency magnetic field parallel to the static field $\mathbf{H}_0$ leads to oscillatory nonequilibrium between the two spin subsystems, with the dipole interaction acting as a restoring force. The resonant frequency of this longitudinal oscillation occurs at

$$\omega_l = \Omega_A(T), \tag{7}$$

where Ω_A is the frequency that has already appeared in the expression for the transverse frequency (5).

Any texture formed by the order-parameter field changes the dipole torque $\mathbf{R}_D$ in a very specific way. Therefore the measurement of NMR shifts, in combination with the corresponding theory, provides the most versatile, and at the same time sensitive, tool for the investigation of order-parameter textures.

NMR frequencies are generally considerably smaller than the characteristic frequencies τ^{-1} and $\Delta(T)/\hbar$ of the normal and superfluid components. Hence such experiments take place in the hydrodynamic regime. At such low frequencies, i.e. energies, the magnitude of the order parameter $\Delta(T)$ does not change at all - only the orientation of its spin part varies. Hence the *structure* of the order parameter is left intact - the dynamics is due to a "rigid" excitation of the order parameter. At higher frequencies, $\omega \approx \Delta(T)$, this changes dramatically. To understand the consequences of this, it is again helpful to view a Cooper pair as some kind of diatomic molecule. As in the case of a molecule, an energy of the order of the binding energy will lead to internal excitations such as rotational and vibrational states.

Ultrasound excitations. Such a situation occurs in experiments measuring the attenuation of ultrasound at sound frequencies close to $\Delta(T)/\hbar$. Quite unexpectedly, one finds that the sound attenuation of the superfluid has a sharp maximum directly below the transition temperature T_c; this maximum depends strongly on the frequency ω. These and other phenomena are explained by collective excitations of the order-parameter structure of the condensate. They owe their existence to pair correlations in a state with nonzero relative orbital angular momentum, which imply an internal structure of the Cooper pair (Wölfle, 1973, 1978). This structure allows for the excitation of high-frequency ($\omega \approx \Delta(T)/\hbar$) collective oscillations (pair-vibration modes). Besides this, there is also the possibility of a break-up of the Cooper pair. Pair breaking is only possible if the energy $\hbar\omega$ of the sound wave is larger than the minimum energy for breaking a pair, $2\Delta_{\hat{k}}(T)$. Here $\Delta_{\hat{k}}(T)$ is the energy gap, which in general depends on $\hat{k}$, the position on the Fermi sphere. For smaller energies, only vibrations can be excited. A detailed theory of sound absorption, including damping effects etc., has been developed (for details see Vollhardt and Wölfle (1990)) and is in good agreement with experiments (Halperin and Varoquaux, 1990). In particular, the existence of isotropic and anisotropic energy gaps in the B and A phase, respectively, led to an early identification of these phases. Indeed, in the B phase sound attenuation is independent of the direction of the sound entering the probe. By contrast, in the A phase it strongly depends on the relative orientation of the sound wave to the anisotropy axis $\hat{l}$. This orientation dependence is very remarkable: by coupling to the nuclear spins, a weak external magnetic field of the order of 3 mT is able to change the direction of $\hat{l}$ and thereby to modify the sound absorption. It is the coherent ordering of *nuclear* spins that is ultimately responsible for the anisotropy of sound absorption!

4 RELATION TO OTHER FIELDS AND RECENT DEVELOPMENTS

Why spend so much effort on sorting out the strange behavior of states of matter that are not even found in nature, at temperatures well outside the reach of even a well-equipped low-temperature laboratory? Partly, of course, "because it's there", and because - like any other system - superfluid ^{3}He deserves to be studied in its own right. However, what is even more important is that superfluid ^{3}He is a model system that exemplifies many of the concepts of modern theoretical physics and, as such, has given us, and will further provide us, with new insights into the functioning of quantum-mechanical many-body systems close to their ground state.

As discussed in Section 3.2, the key to understanding superfluid ^{3}He is "spontaneously broken symmetry". In this respect there are also fundamental connections with particle physics, deriving from the interpretation of the order-parameter field as a quantum field with a rich group structure. The collective modes of the order-parameter as well as the localized topological defects in a given ground-state configuration are the particles of this quantum field theory. Various anomalies known from particle physics can be identified in the ^{3}He model system, and one may hope that insights gained from the study of superfluid ^{3}He will turn out to be useful in elementary particle theory (Volovik, 1987, 1992).

There are several other physical systems for which the ideas developed in the context of superfluid ^{3}He are relevant. One or them is an anisotropic superfluid system that already exists in nature but is not accessible for laboratory experiments: this is the nuclear matter forming the cores of neutron stars. There the pairing of neutrons has been calculated to be of p-wave symmetry. Because of the strong spin-orbit nuclear force, the total angular momentum of the Cooper pairs is $J = 2$ (Sauls et al., 1982; Pines and Alpar, 1985).

Above all, an anisotropic superconducting state is particularly exciting. There are now strong indications that superconductivity in the so-called "heavy-fermion" systems and in high-T_c cuprates, is, at least in some cases, due to the formation of anisotropic pairs with d-wave symmetry (Cox and Maple, 1995). Many of the concepts and ideas developed for superfluid ^{3}He have been adapted to these systems.

The above discussion shows that superfluid ^{3}He is a field of continuing interest. Indeed, most recently superfluid ^{3}He has been used as a test system for the creation of "cosmic strings" in the early stages of the universe. According to Kibble (1976) and Zurek (1985) the observed inhomogeneity of matter in the universe may be understood as the result of the creation of defects generated by a rapid cooling through second-order phase transitions, which led to the present symmetry-broken state of the universe. In two different experiments with superfluid ^{3}He, performed at the low-temperature laboratories in Grenoble (Bäuerle et al., 1996) and Helsinki (Ruutu et al., 1996), a nuclear reaction in the superfluid, induced by neutron radiation, caused a local heating of the liquid into the normal state. During the subsequent, rapid cooling back into the superfluid state the creation of a vortex tangle was observed. The experimentally determined density of this defect state was found to be consistent with Zurek's estimate and thus gives important support to this cosmological model. Furthermore, a recent experimental verification of momentogenesis in ^{3}He-A by Bevan et al. (1997) was found to support current ideas on cosmological baryogenesis. (Baryogenesis during phase transitions in the early universe is believed to be responsible for the observed excess of matter over antimatter). In view of these exciting new developments it may become possible in the future to model and study cosmological problems in the low-temperature laboratory in much more detail.

Due to the intense experimental and theoretical research since 1971 the superfluid phases of ^{3}He now belong to the best-understood states of matter (Vollhardt and Wölfle, 1990). The unique richness of their structure continues to lead to new aspects whose investigation provides unexpected insights.

ACKNOWLEDGMENT

I am grateful to Peter Wölfle for many useful discussions.

REFERENCES

Ambegaokar, V., de Gennes, P.G., and Rainer, D., 1974, *Phys. Rev.* **A9**, 2676.
Anderson, P.W., and Brinkman, W.F., 1973, *Phys. Rev. Lett.* **30**, 1108.
Anderson, P.W., and Brinkman, W.F., 1978, in *The Physics of Liquid and Solid Helium, Part II*, K.H. Bennemann and J.B. Ketterson, eds., Wiley, New York, p. 177.
Anderson, P.W., and Morel, P., 1960, *Physica* **26**, 671.
Anderson, P.W., and Morel, P., 1961, *Phys. Rev.* **123**, 1911.
Balian, R., and Werthamer, N.R., 1963, *Phys. Rev.* **131**, 1553.
Bardeen, J., Cooper, L.N., and Schrieffer, J.R., 1957, *Phys. Rev.* **108**, 1175.
Bäuerle, C., Bunkov, Yu.M., Fisher, S.N., Godfrin, H., and Pickett, G.R., 1996, *Nature* **382**, 332.
Bevan, T.D.C., Manninen, A.J., Cook, J.B., Hook, J.R., Hall, H.E., Vachaspati, T., and Volovik, G.E., 1997, *Nature* **386**, 689.
Cox, D.L., and Maple, M.B., 1995, *Physics Today*, **48**, No. 2, 32.
Hakonen, P., and Lounasmaa, O.V., 1987, *Phys. Today* **40**, 70.
Halperin, W.P., and Varoquaux, E., 1990, in *Helium Three*, W.P. Halperin and L.P. Pitaevskii, eds., North-Holland, Amsterdam, p. 353.
Kamerlingh Onnes, H., 1911, *Proc. R. Acad. Amsterdam* **11**, 168
Kibble, T.W.B., 1976, *J. Phys.*, **A9**, 1387.
Landau, L.D., 1956, *Zh. Eksp. Teor. Fiz.* **30**, 1058 [Sov. Phys. JETP **3**, 920 (1957)].
Landau, L.D., 1957, *Zh. Eksp. Teor. Fiz.*, **32**, 59 [Sov. Phys. JETP **5**, 101 (1957)].
Landau, L.D., 1958, *Zh. Eksp. Teor. Fiz.*, **35**, 97 [Sov. Phys. JETP **8**, 70 (1959)].
Lee, D.M., and Richardson, R.C., 1978, in *The Physics of Liquid and Solid Helium, Part II*, K.H. Bennemann and J.B. Ketterson, eds., Wiley, New York, p. 287.
Leggett, A.J., 1972, *Phys. Rev. Lett.* **29**, 1227.
Leggett, A.J., 1973a, *Phys. Rev. Lett.* **31**, 352.
Leggett, A.J., 1973b, *J. Phys.* **C6**, 3187.
Leggett, A.J., 1974, *Ann. Phys. (N.Y.)* **85**, 11.
Leggett, A.J., 1975, *Rev. Mod. Phys.* **47**, 331.
Leggett, A.J., 1980a, in *Proceedings of the XVI Karpacz Winter School of Theoretical Physics* (Lecture Notes in Physics, Vol. 115), A. Pekalski and J. Przyslawa, eds., Springer, Berlin, p. 13.
Leggett, A.J., 1980b, *J. Physique* **41**, Colloq. C-7, p. 19.
Liu, M., and Cross, M.C., 1978, *Phys. Rev. Lett.* **41**, 250.
London, F., 1950, *Superfluids, Vol. I*, Wiley, New York.
London, F., 1954, *Superfluids, Vol. II*, Wiley, New York.
Maki, K., 1977, *Physica* **90B**, 84.
Mermin, N.D., 1979, *Rev. Mod. Phys.* **51**, 591.
Mineev, V.P., 1980, in *Soviet Scientific Reviews, Sect. A., Physics Reviews, Vol. 2* Harwood Academic Publishers, Chur, p. 173
Nozières, P., 1995, in *Bose-Einstein Condensation*, A. Griffin et al., eds., Cambridge University Press.
Osheroff, D.D., Richardson, R.C., and Lee, D.M., 1972a, *Phys. Rev. Lett.* **28**, 885.
Osheroff, D.D., Gully, W.J., Richardson, R.C., and Lee, D.M., 1972b, *Phys. Rev. Lett.* **29**, 920.

Pines, D., and Alpar, A., 1985, *Nature* **316**, 27.

Randeria, M., 1995, in *Bose-Einstein Condensation*, A. Griffin et al., eds., Cambridge University Press.

Ruutu, V.M.H., Eltsov, V.B., Gill, A.J., Kibble, T.W.B., Krusius, M., Makhlin, Yu. G., Plaçais, B., Volovik, G.E., and Xu, W., 1996, *Nature* **328**, 334.

Salomaa, M.M., and Volovik, G.E., 1987, *Rev. Mod. Phys.*, **59**, 533.

Sauls, J. A., Stein, D.L., and Serene, J.W., 1982, *Phys. Rev.* **D25**, 976.

Serene, J.W., and Rainer, D., 1983, *Phys. Rep.* **101**, 221.

Vollhardt, D., and Wölfle, P., 1990, *The Superfluid Phases of Helium 3*, Taylor and Francis, London.

Volovik, G.E., 1987, *J. Low Temp. Phys.* **67**, 301.

Volovik, G.E., 1992, *Exotic properties of superfluid ^3He*, World Scientific, Singapore.

Wheatley, J.C., 1966, in *Quantum Fluids, Vol. VI*, D.F. Brewer, ed., North-Holland, Amsterdam, p. 183.

Wheatley, J.C., 1975, *Rev. Mod. Phys.* **47**, 415.

Wölfle, P., 1973, *Phys. Rev. Lett.* **30**, 1169.

Wölfle, P., 1978, in *Progress in Low Temperature Physics, Vol. VIIA*, D.F. Brewer, ed., North-Holland, Amsterdam, p. 191.

Zurek, W.H., 1985, *Nature* **317**, 505.

Zwerger, W., 1992, *Ann. der Physik* **1**, 15.

IV. FINITE SYSTEMS. CLUSTERS.

NON-ADIABATICITY AND PAIRING IN THE FINITE SYSTEMS

Vladimir Kresin

Lawrence Berkeley Laboratory
University of California
Berkeley, CA 94720

INTRODUCTION

This paper is concerned with pair correlation in finite Fermi systems; this phenomenon is different from macroscopic superconductivity. We will focus on the π-electron systems of large aromatic molecules, including fullerenes, and on small metallic clusters. The polyatomic aromatic molecules contain a finite number of delocalized π-electrons. For example, the hexabenzcoronene molecule (fig.1a) has 42 π-electrons, the ovalene has 40 π-electrons. The pairing in molecular systems has been discussed by the author et al. in [1-3]. As for metallic clusters, they represent a bridge between isolated atoms and bulk metals, and the number of delocalized electrons varies as a function of its size [4-6].

The finite size of the system makes the phenomenon similar to that in atomic nuclei [7, 8]. Unlike bulk metals, we are dealing with discrete energy spectra. On the other hand, π-electrons as well as the electrons in metallic clusters have a number of similarities with conduction electrons in metals: Coulomb forces, the presence of light (electrons) and heavy (ions) particles. As a result, one can employ the adiabatic approximation and introduce the electron-vibrational coupling which is similar to electron-phonon interaction in metals.

SUPERCONDUCTIVITY AS A NONADIABATIC PHENOMENON

The pair correlation is given rise to by an effective attractive interaction between electrons. The attraction is caused by the electron-lattice coupling, and its strength must be sufficient to overcome the Coulomb repulsion forces. The latter are actually somewhat weakened by a peculiar logarithmic factor; still, the interelectron attraction mediated by phonon exchange must be sufficiently strong: its energy must be on the order of E_F. One may wonder whether the electron-phonon interaction is capable of producing such a large effect. It turns out that the answer is yes, and the reason has to do with the presence of giant non-adiabaticity [10]. From this point of view the terminology accepted in theory of superconductivity such as "superconductors with weak coupling" is somethat misleading. The small value of the coupling constant which determines the value of Tc does not eliminate the necessity to overcome the Coulomb repulsion.

Consider electrons near the Fermi level performing transitions between energy states in a region $\approx h\Omega$, where Ω is the vibrational frequency (in bulk metals $\Omega \cong \Omega_D$), so that $\Delta\varepsilon \cong \Omega$. Transitions of this kind are non-adiabatic, since the change in the electronic energy is smaller that the characteristic phonon energy. As we shall see, these non-adiabatic transitions are responsible for the pairing electron-electron attraction.

Let us estimate the strength of the phonon-mediated electron-electron interaction. The matrix elements of interest is equal to

$$M_{p_1,p_2;p_3,p_4} = \frac{|H_1|^{p_3,p_2; N_{q\lambda}=1}_{p_1,p_2; N_{q\lambda}=0}|H_1|^{p_3,p_4; N_{q\lambda}=0}_{p_3,p_2; N_{q\lambda}=1}}{\varepsilon_{p_1} - \varepsilon_{p_3} - \Omega_{q\lambda}}$$
$$+ \frac{|H_1|^{p_1,p_4; N_{-q\lambda}=1}_{p_1,p_2; N_{-q\lambda}=0}|H_1|^{p_3,p_4; N_{-q\lambda}=0}_{p_1,p_4; N_{-q\lambda}=1}}{\varepsilon_{p_2} - \varepsilon_{p_4} - \Omega_{q\lambda}}. \qquad (1)$$

here $p_i \equiv (\vec{p}_1, \vec{p}_2)$ and $p = (\vec{p}_3, \vec{p}_4)$ correspond to the initial and final electronic states. Note that the expression contains non-diagonal terms of the non-adiabatic operator H_{na}. These nondiagonal terms are finite. Since we are considering virtual phonon exchange processes, there is no need to conserve energy for the elementary act of electron-phonon scattering.

The main contribution to the interaction comes from the non-adiabatic region. Let us estimate the matrix element in this region. The Hamiltonian, describing the electron-vibrational coupling, can be written as a series in a power of the ionic displacements. Rigorous derivation [11, 12] leads to the expression:

$$H_{na}|_{nv}^{n'v'} = H^{(1)}|_{nv}^{n'v'} + H^{(2)}|_{nv}^{n'v'} + \cdots \qquad (2)$$

where

$$H^{(1)}|_{nv}^{n'v'} = \frac{\partial V}{\partial R_{i0}} \delta R|_{nv}^{n'v'} - \frac{\partial \varepsilon_n}{\partial R_{i0}} \delta R|_{nv}^{n'v'} \qquad (2a)$$

$$H^{(2)}|_{nv}^{n'v'} = \left[\frac{1}{2} \frac{\partial^2 V}{\partial R_i \partial R_{\kappa|0}} \delta R_i \delta R_\kappa - \frac{1}{2} \frac{\partial^2 \varepsilon_n}{\partial R_i \partial R_{\kappa|0}} \delta R_i \delta R_\kappa \right]_{nv}^{n'v'} \qquad (2b)$$

The parameter of the expansion is $\kappa = (a/L)$; here a is a vibrational amplitude, and L is a length of the bond. Note that, despite of being an expansion in a series of a small parameter $\kappa = (a/L)$, the Hamiltonian H_{na} corresponds to large quantity. Indeed, it is easy to see, that H_2 $\hbar\Omega$, that is has an order of the vibrational energy. As for H_1, this term is large than H_2, that is $H_1 \cong H_2 / \kappa = \hbar\Omega(L/a) >> \hbar\Omega$. Therefore, the non-adiabatic term, generally speaking exceeds the vibrational energy. This is a special feature of the non-adiabaticity; a more detailed discussion see in [10-12].

Using Eq. (1), we obtain:

$$M \cong \frac{(\hbar\omega/\kappa)^2}{\hbar\omega} = \frac{\hbar\omega}{(\hbar\omega/\varepsilon_F)} = \varepsilon_F. \qquad (3)$$

Hence the attractive interaction is very strong. The qualitative arguments presented here explain how a strong attraction can arise and overcome the Coulomb repulsion. A detailed quantitative analysis must be based on the methods of many-body theory, since effects of this magnitude cannot be rigorously treated by perturbation theory.

Therefore, a large value of the attractive interaction is due to the large contribution which comes from the nonadiabatic region. The existence of such a region in bulk metals is guaranteed by the continuous nature of the electron and phonon spectra. Situation in clusters and molecules is different. Indeed, the energy spectra in these systems are discrete, and in small particles the spacing between the electronic levels exceeds that for vibrational motion. As a result, there is no non-adiabatic region. A special case which makes, nevertheless, the pairing realistic, corresponds to presence of unfilled set of degenerate levels. Another possibility (see below) corresponds to a non-phonon mechanism of pairing.

DEGENERATE LEVELS; JAHN-TELLER SPLITTING

If we have degenerate states, then it is necessary to take into account the Jahn-Teller (JT) effect. We consider here the situation which corresponds to dynamic JT effect (see, e.g. ref. [13]), that is the case when the JT energy is comparable with the vibrational energy. In this case we are dealing with ion delocalization between two minima. The JT crossing leads to the total wave function being a superposition of two configurations, α and β: α corresponds to left minimum, and the β configuration is near the right minimum. The total Hamiltonian has a form:

$$H = H_e + \hat{T}_R \qquad (4)$$

where

$$H_e = \hat{T}_r + V(\mathbf{r}, \mathbf{R}) \qquad (5)$$

$\hat{T}$ is the kinetic energy operator, **r** and **R** are the electronic and nuclear coordinates, correspondingly; V(**r**, **R**) is the total potential energy.

The total wave function ψ (**r**, **R**) can be written in the form:

$$\psi(\mathbf{r},\mathbf{R}) = C_\alpha \varphi_\alpha(\mathbf{r},\mathbf{R}) \phi_\alpha(\mathbf{R}) + C_\beta \varphi_\beta(\mathbf{r},\mathbf{R}) \phi_\beta(\mathbf{R}) \tag{6}$$

Here φ_i and ϕ_i ($i \equiv \alpha, \beta$) are the electronic and nuclear wavefunctions, corresponding to the crossing α and β terms. It is essential that the wavefunction φ_α coincides with the eigenfunction of the total electronic Hamiltonian $H_e = T_\mathbf{r} + V(\mathbf{r},\mathbf{R})$ in the region α and it is not an eigenfunction of H_e in the region β; analogous behavior is true for φ_β (diabatic representation, see [14]). The electronic transition is accompanied by transfer to another electronic term; this process is similar to Landau-Zener effect (see [15]). It is essential that the JT crossing leads to the total wavefunction (see Eq.(6)) can not be written as a product of electronic and nuclear wavefunctions. Contrary to the usual adiabatic picture (Born-Oppenheimer approximation), the JT crossing leads to impossibility to separate the electronic and nuclear motions; we are dealing with a vibronic state (see, e.g. [14]).

Presence of two minima leads to appearance of symmetric and asymmetric terms. One can evaluate the energy splitting ΔE between these terms. Indeed, substituting (6) into the stationary Schrodinger equation, and performing straightforward calculations, we obtain:

$$C_\alpha \varepsilon_\alpha + C_\beta \varepsilon_{\alpha\beta} = C_\alpha E \tag{7}$$

$$C_\beta \varepsilon_\beta + C_\alpha \varepsilon_{\beta\alpha} = C_\beta E \tag{7'}$$

Here

$$\varepsilon_i = \int d\mathbf{R}\, \phi_i [T_\mathbf{R} + H_{e;ii}(\mathbf{R})]\phi_i \tag{8}$$

$$\varepsilon_{ik} = \int d\mathbf{R}\, \phi_i L(\mathbf{R})\phi_k \tag{9}$$

$$H_{e;ii}(\mathbf{R}) = \int d\mathbf{r}\, \varphi_i^*(\mathbf{r},\mathbf{R}) H_e(\mathbf{r},\mathbf{R}) \varphi_i(\mathbf{r},\mathbf{R}) \tag{10}$$

$$L(\mathbf{R}) = H_{e;ik}(\mathbf{R}) = \int d\mathbf{r}\, \varphi_i^*(\mathbf{r},\mathbf{R}) H_e(\mathbf{r},\mathbf{R})]\varphi_k(\mathbf{r},\mathbf{R}) \tag{11}$$

$i,k \equiv \{\alpha, \beta\}$; H_e is defined by Eq. (5). It is essential that in the diabatic representation the matrix elements $H_{e;ik}$ (i=k) differ from zero. We neglect small terms such as $T_{ii} = \int d\mathbf{R}\, dr\, \phi_i^2 \varphi_i^* T_R \varphi_i$, etc, see [14].

One can calculate the energy splitting ΔE in the presence of the JT crossing. We consider the case of symmetric position of the electronic terms ($\varepsilon_\alpha = \varepsilon_\beta$; a more general case will be described elsewhere). With use of Eqs.(7),(7') we obtain:

$$\Delta E = \int L(\mathbf{R})\, \phi_\alpha(\mathbf{R})\, \phi_\beta(\mathbf{R})\, d\mathbf{R} \tag{12}$$

$L(\mathbf{R})$ is the electronic factor defined by Eq. (11).

The vibrational wavefunctions are different from zero in the region a, where a is the amplitude of the oxygen vibrations; therefore, the integrand in (12) differs noticeably from zero in this region. Since $a \ll L$, where L is the length of the bond, one can put $L(\mathbf{R}) \cong L_0 \equiv L(\mathbf{R}_0)$, and we obtain:

$$\Delta E \cong L_0 F \tag{13}$$

where L_0 is the electronic factor (at $\mathbf{R} = \mathbf{R}_0$), determined by Eq. (11), and

$$F = \int \phi_\alpha(\mathbf{R})\, \phi_\beta(\mathbf{R})\, d\mathbf{R} \tag{14}$$

is the Franck-Condon (FC) factor.

One can see from Eq. (13) that the JT dynamic instability leads to an appearance of the additional FC factor. The inequality, F<1, leads to a decrease in the energy splitting, and, correspondingly, in the inequlity $\Delta E < \Omega$ being realistic.

PAIR CORRELATION AND ITS MANIFESTATION

For finite Fermi-systems, such as small metallic clusters, or π-electrons in aromatic molecules, the pair correlation is manifested, first of all, in strong dependence of their properties on the number of electrons being odd or even (odd-even effect). The finite size of the system leads to a crutial difference in spectra for odd and even number of particles. Similar effect has been observed for atomic nuclei and is a manifestation of pairing correlation. The disproportion of the levels (energy gap) has been observed only for nuclei with even number of protons (neutrons). The presence of the unpaired particle (odd number) leads to disappearance of the energy gap, that is, to the opportunity to absorb the quantum of a smaller frequency. The odd-even effect has been observed for small clusters [4].

Another manifestation of the pair correlation is the anti-Hund structure of the spectra. This feature was indicated and studied by Jolicier in [16]. Indeed, the Cooper pair consists of two electrons with opposite spins, and, therefore, the pairing (with even orbital momenta: s, d, ...), indeed, corresponds to the anti-Hund rule.

Another effect which is directly related to pair correlation is an anomalous diamagnetism. Consider the molecule in an external weak magnetic field; the field is perpendicular to the plane of the molecule. Based on the theory of finite Fermi systems [7], one can evaluate the density matrix, and we obtain [2]:

$$\chi_L = \chi_p + \chi_D \tag{15}$$

where

$$\chi_p = \left(\frac{e^2}{8m^2}\right) \sum_{\lambda \lambda'} F(\lambda, \lambda', \Delta) \left|\hat{M}_{z\lambda\lambda'}\right|^2$$

$$F(\lambda, \lambda', \Delta) = \left(E_\lambda E_{\lambda'} - \xi_\lambda \xi_{\lambda'} - \Delta^2\right)\left[2 E_\lambda E_{\lambda'} (E_\lambda + E_{\lambda'})\right]^{-1}; \quad E_\lambda = \left(\xi_\lambda^2 + \Delta^2\right)^{1/2}$$

and χ_D is the diamagnetic term. If $\chi_D=0$, we obtain the expression for the usual Van Vleck paramagnetism. For a normal metal the terms χ_p and χ_D cancel each other, and the metal display only small resudial Landau diamagnetism. Speaking of molecules, one should note that the cancellation does not take place, but the quantities are of the same order of magnitude. Nevertheless, experimentally one can observe the depression of the paramagnetic term for the aromatic molecules and, as a result, anomalous diamagnetism.

One should note that χ_p vanishes for the molecules with axial symmetry. Indeed, the operator M_z is diagonal in this case and χ_p, regardless of the presence of the pair correlation. Then the magnetic susceptibility of such molecules and, consequently, the π-electron current, are determined exclusively by the diamagnetic part of the Hamiltonian. This occurs, for example, for the benzene molecule whose symmetry is frequently assumed to correspond to the D_{6h} group. However, the anomalous diamagnetism has been observed for molecules which do not have an axial symmetry (e.g. for molecule such as ovalen).

The anomalous diamagnetism has an origin similar to the Meissner effect. The presence of the pair correlation, and therefore, finite value of the gap parameter Δ, leads to depression of χ_P. One can show that $F(\lambda, \lambda', \Delta) \ll F(\lambda, \lambda', \Delta=0)$; it is due to inequality $\Delta\xi \ll \Delta$, Δ is a spacing in the absence of the correlation. As a result, $\chi \cong \chi_D$ in agreement with the experimental data.

MECHANISMS OF PAIR CORRELATION

Pairing in π-electron systems or in small mechanic clusters is due to interactions similar to those in usual superconductors. The major mechanism of the correlation is a polarization of the ionic system and localized electrons. For aromatic molecules we are talking about polarization of the σ-core. This polarization is caused by a number of interactions.

At first, let us discuss the π-electron-vibrational interaction. We discussed this mechanism above; it requires a presence of the degenerate electronic levels. The pairing leads to appearance of the energy gap. In connection with contribution of the electron-vibrational interaction in the pairing of the π-electrons, one should note that in the usual superconductors the value of the energy gap is smaller than the characteristic phonon energy $h\Omega$ ($\Omega \approx \Omega_D$, Ω_D is a Debay frequency). However, the strong couling theory allows the opposite case [18,19]. For example, if $\lambda>3$, the value of the gap can exceed that of the characteristic vibrational frequency. Probably, this is the case for the high Tc oxides [10,20]. A similar situation may occur in the molecules or small clusters. Then the spacing between the electronic levels, that is, the energy gap (0-0' transition) exceeds the vibrational spacing. As was noted above, such situation may occur, if we are dealing with very strong coupling between electrons and lattice. Note that in the adiabatic theory the electron-vibrational interaction is directly related to the non-adiabaticity which is described by the parameter $h\Omega/\Delta E_{el}$. Increase in π-electron system leads to an decrease in ΔE_{el}, whereas the characteristic vibrational energy does not depend strongly on the size of the molecules. As a result, for large molecules $\hbar\Omega$ and $\Delta E_{el.}$ become comparable, and it corresponds to strong electron-vibrational coupling.

Besides of the electron-vibrational interaction, the pairing can be also caused by the electronic mechanism, namely, by the $\sigma-\pi$ virtual excitations (see [3]). In addition, if the molecule contains several different π-electron groups , the Little's mechahism [21], based on the presence of spacially separated groups of electrons, can be also very efficient.

If we study, for example, small metallic clusters of Al, or Pb, etc., we are dealing with a material which is superconducting in the bulk, macroscopic form. There is a very interesting question about the critical size of a small cluster, which corresponds to an appearance of the state with pair correlation (see discussion in [4, 6]).

CHARGE TRANSFER

The electron transfer in the biological membrains is an example of the tunneling phenomenon. Consider the row of aromatic molecules with a fixed distance between them. The pair correlation inside of each molecule causes the phase coherence in the π-electron system. Such a coherence leads to the Josephson tunneling in the membrain and to the possibility of a simultaneous transfer of two π-electrons. This means the existence of dissipationless transfer in the large system. Note that the exidation process which plays a key role for any biologically active system, is a charge transfer of two π-electrons.

One should stress that any living matter necessarily contains conjugated systems, and, therefore, mobile π-electrons, that is, the sets of finite Fermi systems (see e.g. [22]). It is quite possible that the extraordinary stability of such systems as well as the long-range correlations are due to the presence of bound electron pairs.

CONCLUSION

In Summary, we discussed the effect of pair correlation for finite Fermi systems. The pairing is manifested in difference in spectra between systems with odd and even number of particles, anomalous diamagnetism, anti-Hund rule. Non-adiabacity is playing a key role. First of all, the pairing, even in bulk systems, is a giant non-adiabatic effect. In addition, the presence of degenarate unfilled states leads to Jahn-Teller instability which is also a non-adiabatic phenomenon.

REFERENCES

1. V.Z. Kresin, *Phys.Lett. A* **24**, 749 (1967); *Sov. Phys.-Doklady* **12**,1147 (1968).
2. V.Z. Kresin, V. Litovchenko, A. Panasenko, *J. of Chem. Phys.* **63**, 3613 (1975); *J. de Phys.*, **C6**-479 (1978).
3. V.Z. Kresin, in *Organic Superconductivity*, V. Kresin and W. A. Little, Eds., p. 285, Plenum, NY (1990).
4. W.D. Knight, in *Novel Superconductivity*, S. Wolf and V. Kresin, Eds. Plenum, NY (1987), p.47.
5. V.V. Kresin, *Physics Reports*, **220**, 1 (1992).
6. V.V. Kresin and W.D. Knight, this volume.
7. A. Migdal, *Theory of Finite Fermi System*; *Application to Atomic Nuclei* (Interscience, NY, 1967).
8. A. Bohr, B. Mottelson, *Nuclear Structure*, v.1, W.A. Benjamin Inc, NY (1969).
9. P. Ring and P. Schuck, *The Nuclear Many-body Problem*, Springer-Verlag, NY (1980).
10. V. Kresin, H. Morawitz, S. Wolf, *Mechanisms of Conventional and High Tc Superconductivity*, Oxford, NY (1993).
11. B. Geilikman, *J. of Low Temp. Phys.*, **4**, 189 (1971).
12. V. Kresin and W. Lester, *Int. J. of Quan. Chem., Symp.* **23**, 17 (1989).
13. L. Salem, *The Molecular Orbital Theory of Conjugated Systems*, ch. 8, W.A. Benjamin, NY (1966).
14. O'Malley, *Phys. Rev.* **152**, 98 (1967); *Adr. Atomic Molec. Phys.* **7**, 223 (1971).
15. L. Landau and E. Lifshitz, *Quantum Mechanics*, Oxford, NY (1977).
16. L. Bergomi and T. Jolicoeur, *C.R. Acad. Sci. Paris,* t. 318, s. II, p. 283 (1994); T. Jolicoeur, this volume.
17. L. Pauling, *J. Chem. Phys.* **4**, 673 (1936); F.London, J. Phys. Rad. **8**, 397 (1937).
18. P. Allen, R.Dynes, *Phys. Rev. B* **12**, 905 (1975).
19. V.Z. Kresin, H. Gutfreund, W.A. Little, *Solid State Comm.* **51**, 339 (1984).

20. V. Kresin and S. Wolf, *Phys. Rev. B* **41**, 4278 (1990); M. Reeves et al., *Phys. Rev. B* **47**, 6065 (1993).
21. W.A. Little, *Phys. Rev*, **134**, A1416 (1964).
22. B. Pullman, A. Pullman, *Quantum Biochemistry*, Interscience, NY, 1963.

FULLERENE ANIONS AND PAIRING IN FINITE SYSTEMS

Th. Jolicœur

Service de Physique Théorique
Orme des Merisiers
CEA Saclay
91191 Gif-s-Yvette, France

INTRODUCTION

The discovery of superconductivity in alkali-metal-doped fullerenes[1] K_3C_{60} and Rb_3C_{60} has raised interesting questions about the electron-phonon coupling in such compounds and its interplay with Coulomb repulsion. C_{60} is a highly symmetrical molecule i.e. it is a truncated icosahedron and its electronic lowest unoccupied molecular orbitals (LUMO) are threefold degenerate[2,3,4]. They form a T_{1u} representation of the icosahedral group I_h. Filling the LUMO in C_{60}^{n-} anions leads in a naive picture to narrow, partially filled bands in the bulk fullerides. The bandwidth W is determined by the hopping between the C_{60} molecules which are quite far apart and $W \approx 0.5$ eV. The coupling of some H_g phonons with electrons residing in the T_{1u} orbital has been suggested to be responsible for the superconductivity[5,6,7]. The Coulomb repulsion also may be important on the ball[8]. Several authors[9–12] have undertaken the study of the Jahn-Teller distortion that is expected in the fullerene anions. In such calculations one considers the electrons as fast degrees of freedom and the phonon normal coordinates are treated as static[13]. Here we will investigate the interplay between the electronic and phononic degrees of freedom on an isolated fullerene anion. We study an effect that goes beyond the Born-Oppenheimer approximation which is the modification of electronic levels due to phonon exchange. The lifting of degeneracy can be obtained by a perturbation calculation in the case of an undistorted anion. The ordering of levels can be described as "anti-Hund"' rule. This calculation is very close in spirit to the standard treatment of the electron-phonon coupling in superconducting metals. Here the energy scale of this effect may be comparable to that of the Jahn-Teller effect. This is because the phonons have high frequencies as well as medium to strong coupling to electrons. The effect we observe may be sought by spectroscopy of solutions of fullerides in liquid ammonia, for example. We discuss the opposite effect of Coulomb interaction, leading to Hund's rule in ordinary situations. Finally we point out that experimentally observed spectra may be at least partially explained by our crude calculation[23].

THE ON-BALL ELECTRON-PHONON INTERACTION

The electronic structure of π electrons in the C_{60} molecule is well known to be given by a simple Hückel calculation. The levels are labeled[4] by the irreducible representations (irreps) of the icosahedron group I_h. One important property has to be noted: three of the I_h irreps are the $l=0,1,2$ spherical harmonics of $SO(3)$ which do not split under the I_h group. They are commonly named A_g, T_{1u}, H_g. In addition there is also the twofold spin degeneracy.

In the ground-state of the neutral C_{60} molecule all levels up to H_u included are completely filled thus building a singlet state $|\Psi_0\rangle$. The LUMO are the six T_{1u} states. These are occupied upon doping with extra electrons and the ground-state becomes then degenerate. One then expects the Jahn-Teller effect to distort the anion and lift this orbital degeneracy[13,14]. We focus on another effect which goes beyond the Born-Oppenheimer approximation in the sense that nuclear motions are crucial for its very existence: the coupling of the T_{1u} electrons to the vibrational modes of the molecule (also referred to as phonons). Phonon exchange between electrons leads to an effective electron-electron interaction that competes with Coulomb repulsion and may lead to anti-Hund ordering of energy-levels.

For simplicity we treat this effect assuming the absence of the Jahn-Teller distortion. The next logical step would be to compute first the distortion pattern of the anion under consideration, then obtain its vibrational spectrum and the electron-phonon coupling in the distorted structure and then compute again the effective electron-electron interaction. As a first investigation of electron-phonon coupling we use a perturbation scheme suited to degenerate levels we will derive an effective electron-electron interaction with the assumption that filled states lying below the T_{1u} level remain frozen so that intermediate states involve only T_{1u}-T_{1u} excitations. Indeed the H_u-T_{1u} gap is $\approx$ 2eV whereas maximum phonon energies are $\approx$ 0.2eV.

A typical electron-phonon interaction term reads:

$$W = \sum_{\alpha, m_1, m_2, \sigma} f_{\alpha m_1 m_2} X_\alpha c^\dagger_{m_1 \sigma} c_{m_2 \sigma}.$$

Here X_α are normal coordinates, the subscript referring both to the irrep and to the row in the irrep they belong to, $c^\dagger_{m_1 \sigma}$ is the creation operator for an electron with spin σ in the T_{1u} ($l=1$) level, m_1 taking one of the $m=-1,0,1$ values, and $f_{\alpha m_1 m_2}$ are complex coefficients. The $c^\dagger_{m\sigma}$ operators transform as $l=1$ $|l, m\rangle$ vectors under I_h symmetries, and their conjugates $c_{m\sigma}$ transform as $(-1)^{m+1}|l, -m\rangle$ vectors. The $(-1)^{m_2+1} c^\dagger_{m_1 \sigma} c_{-m_2 \sigma}$ products transform then as members of the $T_{1u} \times T_{1u}$ representation, which in the I_h group splits as:

$$T_{1u} \times T_{1u} = A_g + T_{1g} + H_g.$$

This selects the possible vibrational modes T_{1u} electrons can couple to. In fact, only H_g modes split the degeneracy[7].

Let us consider a particular fivefold degenerate multiplet of H_g modes. Their normal coordinates will be labelled X_m, m ranging from -2 to +2. Since H_g appears only once in the product $T_{1u} \times T_{1u}$, the interaction is determined up to one coupling constant g by the usual formula for the coupling of two equal angular momenta to zero total angular momentum:

$$W = g \sum_m (-1)^m X_m \Phi_{-m}. \qquad (1)$$

The X_m may be chosen such that $X_m^\dagger = (-1)^m X_{-m}$ and have the following expression in terms of phonon operators:

$$X_m = \frac{1}{\sqrt{2}} \left(a_m + (-1)^m a_{-m}^\dagger \right) \qquad (2)$$

whereas the Φ_m are the irreducible $l=2$ tensor operators built from the $c^\dagger c$ products according to:

$$\Phi_m = \sum_{m_1} (1,1,2|m_1, m-m_1, m)(-1)^{(m-m_1+1)} c_{m_1\sigma}^\dagger c_{-m+m_1\sigma}, \qquad (3)$$

where $(l_1, l_2, l|m_1, m_2, m)$ are Clebsch-Gordan coefficients.

We now consider a doped C_{60}^{n-}, molecule, $0 \leq n \leq 6$. Its unperturbed degenerate ground-states consist of $|\Psi_0\rangle$ to which n T_{1u} electrons have been added times a zero-phonon state. They span a subspace denoted by $\mathcal{E}_0$. In $\mathcal{E}_0$ the unperturbed Hamiltonian H_0 reads:

$$H_0 = \epsilon_{t_{1u}} \sum_{m,\sigma} c_{m\sigma}^\dagger c_{m\sigma} + \hbar\omega \sum_m a_m^\dagger a_m,$$

where $\epsilon_{t_{1u}}$ is the energy of the T_{1u} level, $\hbar\omega$ is the phonon energy of the H_g multiplet under consideration. Within $\mathcal{E}_0$ the effective Hamiltonian up to second order perturbation theory is given by:

$$H_{eff} = E_0 P_0 + P_0 W P_0 + P_0 W (1-P_0) \frac{1}{E_0 - H_0} (1-P_0) W P_0,$$

where P_0 is the projector onto $\mathcal{E}_0$, E_0 is the unperturbed energy in this subspace which is just the number of doping electrons times $\epsilon_{t_{1u}}$. The linear term in W gives no contribution. Using expressions (1) and (2) for W and X_m one finds:

$$H_{eff} = H_0 - \frac{g^2}{2\hbar\omega} \sum_{m,\sigma_1\sigma_2} (-1)^m \Phi_{m\sigma_1} \Phi_{-m\sigma_2}, \qquad (4)$$

where we have now included spin indices. We can now use equation (3) to express H_{eff} as a function of c and $c^\dagger$ operators and put it in normal ordered form using fermion anticommutation rules. In this process there appears a one-body interaction term which is a self-energy term. We will henceforth omit the H_0 term which is a constant at fixed number of doping electrons.

Let us now define pair creation operators $A_{lm}^{s\sigma\,\dagger}$ which when operating on the vacuum $|0\rangle$ create pair states of T_{1u} electrons that are eigenfunctions of $\mathbf{L},\mathbf{S}, L_z, S_z$, where $\mathbf{L},\mathbf{S}$ are total angular momentum and spin, and L_z, S_z their z-projections. l and s can take the values $0,1,2$ and $0,1$ respectively. This holds also if $|0\rangle$ is taken to be the singlet state $|\Psi_0\rangle$.

$$A_{lm}^{s\sigma\,\dagger} = \sum_{m_1,\sigma_1}(1,1,l|m_1,m-m_1,m)(\tfrac{1}{2},\tfrac{1}{2},s|\sigma_1,\sigma-\sigma_1,\sigma)c^\dagger_{m_1\sigma_1}c^\dagger_{m-m_1\sigma-\sigma_1}. \quad (5)$$

The quantity $A_{lm}^{s\sigma\,\dagger}$ is non-zero only if $(l+s)$ is even and the norm of $A_{lm}^{s\sigma\,\dagger}|0\rangle$ is then equal to $\sqrt{2}$. The inverse formula expressing $c^\dagger c^\dagger$ products as $A^\dagger$ operators is:

$$c^\dagger_{m_1\sigma_1}c^\dagger_{m_2\sigma_2} = \sum_{l,s}(1,1,l|m_1,m_2,m_1+m_2)(\tfrac{1}{2},\tfrac{1}{2},s|\sigma_1,\sigma_2,\sigma_1+\sigma_2)A_{lm_1+m_2}^{s\sigma_1+\sigma_2\,\dagger}. \quad (6)$$

As H_{eff} is a scalar, its two-body part may be written as a linear combination of diagonal $A_{lm}^{s\sigma\,\dagger}A_{lm}^{s\sigma}$ products whose coefficients depend only on l and s:

$$\sum_{ls,m\sigma} F(l,s)A_{lm}^{s\sigma\,\dagger}A_{lm}^{s\sigma}.$$

The $F(l,s)$ coefficients are calculated using expressions (4), (3), (6). We then get H_{eff} in final form:

$$H_{eff} = -\frac{5g^2}{6\hbar\omega}\left(\hat{N} + A_{00}^{00\,\dagger}A_{00}^{00} - \frac{1}{2}\sum_{m,\sigma}A_{1m}^{1\sigma\,\dagger}A_{1m}^{1\sigma} + \frac{1}{10}\sum_m A_{2m}^{00\,\dagger}A_{2m}^{00}\right). \quad (7)$$

In this formula $\hat{N}$ is the electron number operator for the T_{1u} level; the $\hat{N}$ term appears when bringing H_{eff} of expression (4) in normal ordered form. In our Hamiltonian formulation the effective interaction is instantaneous.

There are actually eight H_g multiplets in the vibrational spectrum of the C_{60} molecule. To take all of them into account we only have to add up their respective coefficients $5g^2/6\hbar\omega$, their sum will be called Δ.

THE ELECTRONIC STATES OF FULLERENE ANIONS

We shall now, for each value of n between 1 and 6, find the n-particle states and diagonalize H_{eff}. The Hamiltonian to be diagonalized is that of equation (7) where the prefactor is replaced by $-\Delta$. The invariance group of H_{eff} is $I_h \times SU(2)$. The n-particle states may be chosen to be eigenstates of $\mathbf{L}, \mathbf{S}, L_z, S_z$ and we shall label the multiplets by (l,s) couples, in standard spectroscopic notation (^{2s+1}L stands for (l,s)). The pair (l,s) label $\mathbf{SO(3)} \times \mathbf{SU(2)}$ irreps which, as previously mentioned, remain irreducible under $\mathbf{I_h} \times \mathbf{SU(2)}$ as long as l doesn't exceed 2; for larger values of l $\mathbf{SO(3)}$ irreps split under $\mathbf{I_h}$. Fortunately enough, the relevant values of l never exceed 2. Moreover given any value of n, (l,s) multiplets appear at most once so that the energies are straightforwardly found by taking the expectation value of the Hamiltonian in one of the multiplet states. The degeneracies of the levels will then be $(2l+1)(2s+1)$. We now proceed to the construction of the states.

- n=1: There are six degenerate ^{2}P states $c_{m\sigma}^\dagger|\Psi_0\rangle$ whose energy is $-\Delta$.

- n=2: There are 15 states, generated by applying $A_{lm}^{s\sigma\,\dagger}$ operators on $|\Psi_0\rangle$. There is one ^{1}S state, five ^{1}D states and nine ^{3}P states with energies -4Δ, $-11\Delta/5$, $-\Delta$.

- n=3: There are 20 states. States of given l, m, s, σ can be built by taking linear combinations of $A^\dagger c^\dagger|\Psi_0\rangle$ states according to:

$$\sum_{m_1,\sigma_1} (l_1,1,l|m_1,m-m_1,m)(s_1,\frac{1}{2},s|\sigma_1,\sigma-\sigma_1,\sigma) A_{l_1 m_1}^{s_1 \sigma_1\,\dagger} c_{m-m_1\sigma-\sigma_1}^\dagger |\Psi_0\rangle.$$

These states belong to the following multiplets: ^{2}P (E=-3Δ), ^{2}D (E= -9Δ/5), ^{4}S (E=0).

- n=4: There are 15 states, which are obtained by applying $A_{lm}^{s\sigma\,\dagger}$ operators on $A_{00}^{00\,\dagger}|\Psi_0\rangle$. They are ^{1}S (E=-4$\Delta$), ^{1}D (E= -11$\Delta$/15), ^{3}P (E=-$\Delta$)

- n=5: There are six ^{2}P states which are $c_{m\sigma}^\dagger A_{00}^{00\,\dagger} A_{00}^{00\,\dagger}|\Psi_0\rangle$ and whose energy is $-\Delta$.

- n=6: There is one ^{1}S state whose energy is 0.

It is interesting to note that the above treatment of electron-phonon interaction parallels that of pairing forces in atomic nuclei[15,16]. Of course in the case of finite fermionic systems there is no breakdown of electron number but there are well-known "odd-even" effects that appear in the spectrum. In our case pairing shows up in the ^{1}S ground state for C_{60}^{2-} rather than ^{3}P as would be preferred by Coulomb repulsion i.e. Hund's rule. The construction of the states above is that of the seniority scheme in nuclear physics[16]. We note that similar ideas have been put forward by V. Kresin some time ago, also in a molecular context[17]. The effective interaction that he considered was induced by σ core polarization.

THE EFFECT OF COULOMB REPULSION

We now consider the Coulomb electron–electron interaction and assume it to be small enough so that it may be treated in perturbation theory. To get some feeling of the order of magnitude of this repulsion we use the limiting case of on-site interaction i.e. the Hubbard model. This Hamiltonian is not specially realistic but should contain some of the Hund's rule physics. The two-body interaction now reads:

$$\frac{U}{2} \sum_{i,\sigma} c^\dagger_{i\sigma} c^\dagger_{i-\sigma} c_{i-\sigma} c_{i\sigma},$$

where the i subscript now labels the π orbitals on the C_{60} molecule. The quantity U is $\approx$ 2-3 eV from quantum chemistry calculations[18] Since level degeneracies are split at first order in perturbation theory we confine our calculation to this order and have thus to diagonalize the perturbation within the same subspace $\mathcal{E}_0$ as before. In this subspace it reads:

$$W_H = U \sum_{i,\alpha\beta\gamma\delta} \langle\alpha|i\rangle\langle\beta|i\rangle\langle i|\gamma\rangle\langle i|\delta\rangle \; c^\dagger_{\alpha\uparrow} c^\dagger_{\beta\downarrow} c_{\gamma\downarrow} c_{\delta\uparrow},$$

where greek indices label one–particle states belonging either to $|\Psi_0\rangle$ or to the T_{1u} level. Let us review the different parts of W_H. Note that since the $|\Psi_0\rangle$ singlet remains frozen we have the identity: $c^\dagger_\alpha c_\beta = \delta_{\alpha\beta}$ if α, β label states belonging to $|\Psi_0\rangle$.

–A part involving states belonging to $|\Psi_0\rangle$ only:

$$W_{H_1} = U \sum_{i,\alpha\beta} |\langle\alpha|i\rangle|^2 |\langle\beta|i\rangle|^2 \; c^\dagger_{\alpha\uparrow} c_{\alpha\uparrow} \; c^\dagger_{\beta\downarrow} c_{\beta\downarrow}.$$

α,β belong to $|\Psi_0\rangle$. This term is thus diagonal within $\mathcal{E}_0$ and merely shifts the total energy by a constant that does not depend on the number of doping electrons. It won't be considered in the following.

–A part involving both states belonging to $|\Psi_0\rangle$ and to the T_{1u} level:

$$W_{H_2} = U \sum_{i,\alpha\delta,\beta,\sigma} \langle\alpha|i\rangle\langle i|\delta\rangle |\langle\beta|i\rangle|^2 \; c^\dagger_{\alpha\sigma} c_{\delta\sigma} \; c^\dagger_{\beta-\sigma} c_{\beta-\sigma},$$

where α, δ belong to the T_{1u} level whereas β belongs to $|\Psi_0\rangle$. It reduces to:

$$W_{H_2} = U \sum_{\alpha\delta,\sigma} c^\dagger_{\alpha\sigma} c_{\delta\sigma} \left(\sum_i \langle\alpha|i\rangle\langle i|\delta\rangle \sum_\beta |\langle\beta|i\rangle|^2 \right).$$

The sum over β is just the density on site i for a given spin direction of all states belonging to $|\Psi_0\rangle$ which is built out of completely filled irreps. As a result this density is uniform and since $|\Psi_0\rangle$ contains 30 electrons for each spin direction it is equal to 1/2. W_{H_2} then becomes diagonal and reads:

$$W_{H_2} = \frac{U}{2} \sum_{\alpha,\sigma} c^\dagger_{\alpha\sigma} c_{\alpha\sigma}.$$

Its contribution is thus proportional to the number of T_{1u} electrons. It represents the interaction of the latter with those of the singlet and we won't consider it in the following.

–A part involving only states belonging to the T_{1u} level:
W_{H_3} has the same form as W_H with all indices now belonging to the T_{1u} level. Whereas the interaction has a simple expression in the basis of $|i\rangle$ states, we need its matrix elements in the basis of the T_{1u} states. There are in fact two T_{1u} triplets in the one–particle spectrum of the C_{60} molecule, the one under consideration having higher energy. To construct the latter we have first constructed two independent sets of states which transform as x, y, z under $\mathbf{I_h}$. These are given by:

$$|\alpha\rangle = \sum_i \vec{e}_\alpha . \vec{r}_i \, |i\rangle \quad \text{and} \quad |\alpha\rangle' = \sum_i \vec{e}_\alpha . \vec{k}_i \, |i\rangle,$$

where $\vec{e}_\alpha$ are three orthonormal vectors, i labels sites on the molecule, the $\vec{r}_i$ are the vectors joining the center of the molecule to the sites while the $\vec{k}_i$ join the centre of the pentagonal face of the molecule the site i belongs to to the site i. We assume that the bonds all have the same length. These states span the space of the two T_{1u} triplets. The diagonalization of the tight–binding Hamiltonian in the subspace of these six vectors yields then the right linear combination of the $|\alpha\rangle$ and $|\alpha\rangle'$ states for the upper lying triplet. From the x, y, z states one constructs $l=1$ spherical harmonics. We then get the matrix elements of W_{H_3} in the basis of T_{1u} states. As $\mathcal{E}_0$ is invariant under $\mathbf{I_h}$ operations and spin rotations, W_{H_3} which is the restriction of W_H to $\mathcal{E}_0$ is invariant too. It may thus be expressed using the A, $A^\dagger$ operators by using formula (5) in the same way as the phonon–driven interaction and we finally get:

$$W_{H_3} = \left(\frac{U}{40} A_{00}^{00\,\dagger} A_{00}^{00} + \frac{U}{100} \sum_m A_{2m}^{00\,\dagger} A_{2m}^{00} \right), \tag{8}$$

which is the only part in W_H that we will keep. Note that there is no contribution from $l=1$, $s=1$ $A^\dagger A$ products. Indeed the Hubbard interaction is invariant under spin rotation and couples electrons having zero total S_z. As the coefficients of $A^\dagger A$ products depend solely on l and s they must be zero for $s \neq 0$. The spectrum for any number of T_{1u} electrons is now easily found. Of course the order of the multiplet is now reversed: for n=2, we have ^3P, then ^1D, then ^1S. For n=3, we have ^4S, then ^2D, then ^2P. For n=4, we have ^3P, then ^1D, then ^1S.

CONCLUSION

The ordering of energy levels in the electron-phonon scheme are clearly opposite to those of Hund's rule. The clear signature of what we can call "on-ball" pairing is the ground state ^1S of C_{60}^{2-}: the two extra electrons are paired by the electron-phonon coupling. We note that the U of the Hubbard model appears divided by large factors: this is simply due to the fact that the C_{60} molecule is large. As a consequence, if $U \approx 2$ eV, Coulomb repulsion may be overwhelmed by phonon exchange. With a H_g phonon of typical energy 100 meV and coupling $O(1)$ as suggested by numerous calculations[6,7,10], the quantity Δ may be tens of meV.

It seems to us that the cleanest way to probe this intramolecular pairing would be to look at solutions of fullerides leading to free anions such as liquid ammonia solutions or organic solvents[19-22]. EPR or IR spectroscopy should be able to discriminate between the two types of spectra. Measurements by EPR should determine whether or not the two extra electrons in C_{60}^{2-} are paired, for example. In near-IR spectroscopy the lowest allowed transition for C_{60}^{2-} should be at higher energy than that of C_{60}^{-} due to the pairing energy while in the Coulomb-Hubbard case it is at lower energy.

Present experiments[19,20] have studied the near-IR spectra of solutions of fulleride anions prepared by electrochemical reduction. There are several peaks that do not fit a simple Hückel scheme of levels. They do not have an immediate interpretation in terms of vibrational structure[19,20]. With our energy levels in table I, a tentative fit would lead to $\Delta \approx 80$ meV assuming $U = 0$. Such a value leads to intriguing agreement with the major peaks seen for C_{60}^{2-} and C_{60}^{3-} while this is no longer the case for C_{60}^{4-} and C_{60}^{5-}.

Finally we mention that recent EPR experiments[22] have given some evidence for non-Hund behaviour of the fulleride anions. While one may observe some trends similar to the results of the phonon-exchange approximation, it is clear that the model we used is very crude. In a bulk *conducting* solid we do not expect the previous scheme to be valid since the levels are broadened into bands: then phonon exchange leads of course to *superconductivity*.

Acknowledgements

This work has been done in collaboration with Lorenzo Bergomi and is a part of his PhD thesis. We thank K. M. Kadish, M. T. Jones, C. A. Reed and J. W. White for informing us about their recent experimental work. We have also benefited from the help of the GDR "Fullerenes" supported by the Centre National de la Recherche Scientifique (France). Thanks are also due to V. Kresin for his kind invitation to present this work in Erice summer school.

REFERENCES

[1] For a review see A. F. Hebard, Physics Today, November 1992.

[2] S. Satpathy, Chem. Phys. Letters **130**, 545 (1986).

[3] R. C. Haddon and L. T. Scott, Pure Appl. Chem. **58**, 137 (1986); R. C. Haddon, L. E. Brus, and K. Raghavachari, Chem. Phys. Letters **125**, 459 (1986).

[4] G. Dresselhaus, M. S. Dresselhaus, and P. C. Eklund, Phys. Rev. **B45**, 6923 (1992).

[5] M. Lannoo, G. A. Baraff, M. Schlüter and D. Tomanek, Phys. Rev. **B44**, 12106 (1991). See also K. Yabana and G. Bertsch, Phys. Rev. **B46**, 14263 (1992).

[6] V. de Coulon, J. L. Martins, and F. Reuse, Phys. Rev. **B45**, 13671 (1992).

[7] C. M. Varma, J. Zaanen, and K. Raghavachari, Science **254**, 989 (1991).

[8] S. Chakravarty and S. Kivelson, Europhys. Lett. **16**, 751 (1991); Chakravarty, M. P. Gelfand, and S. Kivelson, Science **254**, 970 (1991); S. Chakravarty, S. Kivelson, M. Salkola, and S. Tewari, Science **256**, 1306 (1992).

[9] A. Auerbach, "Vibrations and Berry phases of charged Buckminsterfullerene", preprint (1993).

[10] N. Koga and K. Morokuma, Chem. Phys. Letters **196**, 191 (1992).

[11] J. C. R. Faulhaber, D. Y. K. Ko, and P. R. Briddon, Phys. Rev. **B48**, 661 (1993).

[12] F. Negri, G. Orlandi and F. Zerbetto, Chem. Phys. Lett. **144**, 31 (1988).

[13] M. C. M. O'Brien and C. C. Chancey, Am. J. Phys. **61**, 688 (1993).

[14] J. B. Bersuker, The Jahn-Teller Effect and Vibronic Interactions in Modern Chemistry, Plenum Press, (1984).

[15] P. Ring and P. Schuck, The Nuclear Many-Body Problem, Springer-Verlag, (1980).

[16] J. M. Eisenberg and W. Greiner, "Nuclear Theory, Microscopic Theory of the Nucleus", North Holland, Amsterdam (1972), Vol. III, see pp. 287-317.

[17] V. Z. Kresin, J. Supercond. **5**, 297 (1992); V. Z. Kresin, V. A. Litovchenko and A. G. Panasenko, J. Chem. Phys. **63**, 3613 (1975).

[18] V. P. Antropov, O. Gunnarsson and O. Jepsen, Phys. Rev. **B46**, 13647 (1992).

[19] G. A. Heath, J. E. McGrady, and R. L. Martin, J. Chem. Soc., Chem. Commun. 1272 (1992).

[20] W. K. Fullagar, I. R. Gentle, G. A. Heath, and J. W. White, J. Chem. Soc., Chem. Commun. 525 (1993).

[21] D. Dubois, K. M. Kadish, S. Flanagan and L. J. Wilson, J. Am. Chem. Soc. **113**, 7773 (1991); ibid. **113**, 4364 (1991).

[22] P. Bhyrappa, P. Paul, J. Stinchcombe, P. D. W. Boyd and C. A. Reed, "Synthesis and Electronic Characterization of Discrete Buckminsterfulleride salts", JACS, November 1993.

[23] L. Bergomi and Th. Jolicoeur, Compte-Rendus de l'Académie des Sciences (Paris), **318** II, 283 (1994), cond-mat/9311011.

QUANTIZED ELECTRONIC STATES IN METAL MICROCLUSTERS: ELECTRONIC SHELLS, STRUCTURAL EFFECTS, AND CORRELATIONS

Vitaly V. Kresin[1] and Walter D. Knight[2]

[1]Department of Physics and Astronomy
University of Southern California
Los Angeles, CA 90089-0484, USA

[2]Department of Physics
University of California
Berkeley, CA 94720, USA

INTRODUCTION

We review some properties of metal microclusters in which one observes an interplay between the quantum shell structure of delocalized valence electrons and the ionic framework.[1] In particular, we point out that even in systems which are strongly disordered by conventional solid-state criteria (mixed, or alloyed, clusters, clusters with impurities) the existence of quantized shells keeps the spectral features ordered and understood. We discuss the transition from electronic to lattice-based periodicities in cluster growth, and the splittings of giant dipole photoabsorption peaks of clusters. In both cases a reduction in conventional structural order, e.g., by heating, does not lead to a disorganized situation. On the contrary, it allows electronic shell effects to emerge more clearly. Thus metallic clusters are nanoscale systems in which both a distinct shell-based ordering principle and its interplay with the geometric degrees of freedom can be observed. We also comment on recent work addressing the possibility of electron pairing in size-quantized particles.

ELECTRONIC SHELL STRUCTURE IN METAL CLUSTERS

The basic ordering principle of solids is their crystalline structure. It determines the geometry of the solid and governs its phonon and electronic band spectra. But what happens as the crystal size is reduced until the remaining piece is made up of only a finite number of atoms, say a thousand, a hundred or a dozen? The study of such agglomerates, occupying the range between small molecules and bulk matter, is the subject of cluster science. Clusters can be generated out of a variety of materials; our focus here is on metallic clusters, especially those of so-called "simple" or "nearly-free-electron metals." Discussions of some other cluster families can be found in recent reviews, monographs, and conference proceedings.[2-5] Further information about simple metal clusters is also available in the above references and in other reviews.[6-12]

When the metal particle becomes so small that it no longer makes sense to talk of a periodic crystal lattice, is there any hope for finding a replacement ordering principle? It would be boring indeed if each small cluster had to be dealt with separately as an individual molecule. Fortunately, this turns out not to be the case. Measurements on free alkali clusters in a molecular beam revealed that their mass spectra displayed a well-defined structure with some specific cluster sizes being especially prominent,[6,13,14] see Fig. 1. By analogy with similar periodicities in atomic and nuclear structure these "magic numbers" were identified with the closing of *electronic shells* occupied by cluster valence electrons. Further experimental and theoretical evidence has confirmed this interpretation (see the aforementioned reviews). Electronic shell effects have now been observed in the alkali and many other metals,[9,15] and seen to persist in clusters with hundreds and thousands of atoms (see, e.g., Fig. 2).

The picture that characterizes metal microclusters is thus as follows: when atoms condense into clusters, the weakly bound outer valence electrons leave their parents and form a delocalized cloud extending throughout the cluster. This cloud is the ancestor of the Fermi sea of conduction electrons in nearly-free electron metals.[16] What is special about clusters is that they are small (e.g., the radius of the Na_{40} cluster is $\approx 7Å$), resulting in size quantization: the energy levels (shells) of the valence electrons are separated by sizable gaps. As illustrated in Figs. 3 and 4, the magnitude of these gaps can be an appreciable fraction of an eV. This means that shell structure in small clusters will not be washed out by ionic pseudopotentials, surface or temperature effects, etc. We are thus dealing with a distinct type of ordering: electronic shell structure in clusters, as opposed to crystal lattice structure in solids.

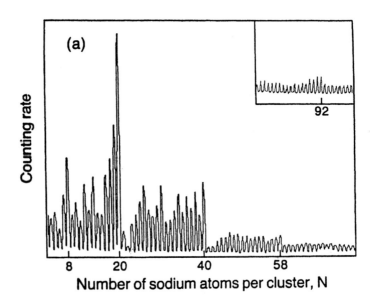

Figure 1. An abundance mass spectrum for Na clusters.[13] Note the prominent steps associated with electronic shells.

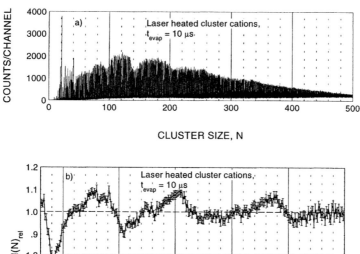

Figure 2. Examples of shell structure in large clusters.[17] Top (a): a directly measured abundance spectrum of Na_n^+ clusters. Bottom (b,c,d): The relative abundances from this spectrum are compared with other spectra obtained under quite different experimental conditions. The abundance variations are remarkably similar.

The manifestations of shell effects are manifold and have now been explored in photoionization and photoelectron spectroscopy of clusters, in polarizability and photoabsorption measurements, as well as chemical reactivity, stability, and evaporation/fission studies. Details of this and related work can be found in the aforementioned reviews and compendia.* In the following three sections we will restrict ourselves to one particular issue: to emphasize that size quantization introduces a novel type of order and to highlight its interplay with the more familiar concepts of order and disorder as encountered in the physics of macroscopic solids. In particular, we will discuss mixed clusters, the transition to crystalline order, and the behavior of cluster photoabsorption resonances.

---

* An interesting variant of shell structure has been observed in alkali metal layers coating a fullerene core.[18] Mass spectra of $C_{60}Cs_n$, $n \leq 500$, have been interpreted in terms of successive electronic shell filling by the Cs valence electrons confined to a metal layer surrounding the C_{60} core.

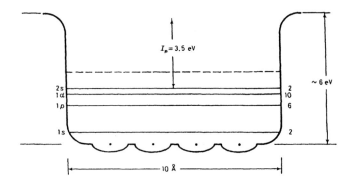

Figure 3. A schematic potential well for electrons in the Na_{20} cluster.[19] The filled shells, indicated by solid horizontal lines, accommodate 20 valence electrons. The positive ionic cores are screened, producing a rather smooth well. The effect of the cores can be taken into account by introducing pseudopotentials, see, e.g., [10].

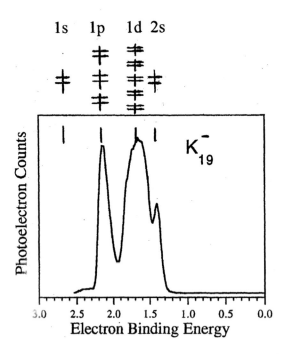

Figure 4. Photoelectron spectrum of a closed-shell potassium cluster anion with 20 valence electrons, showing electronic shell levels.[20] Designations and electron occupations for individual levels are shown at the top. This spectrum is part of a series taken for K^-_{2-19}, showing the evolution of the energy levels.

MIXED CLUSTERS AND CLUSTERS WITH IMPURITIES

A classic way to introduce disorder into a crystalline lattice is to bring in foreign atoms as impurities. What happens with the electronic structure of an alkali cluster if different atoms are admixed to the latter? Several variants of this situation have been explored, e.g.: (a) mixing different alkalis or other simple metals, (b) introducing oxygen, hydrogen or halogen impurities into the cluster, and (c) introducing a divalent metal impurity into the cluster.

(a) Mixed alkali clusters. This can be said to represent a microparticle version of alloying. It is found (see Fig. 5) that although compositional order is destroyed, the electronic shell structure remains unchanged. The valence electrons of both types of alkalis become delocalized and fill up the shells together. Thus the observed pattern of magic numbers remains unchanged.

Similarly, ionization potentials of mixed bimetallic clusters such as Al_nNa_m, Al_nCs_m and Au_nNa_m show discontinuous changes when the *total* number of delocalized electrons fills an electronic shell.[21]

(b) Oxides, hydrides, and halides. A beautiful pattern emerges when oxygen[22,23] or hydrogen[24] impurities are introduced into the cluster. As can be seen from Fig. 6, the mass spectra behave precisely as if each O atom binds two valence electrons, and the rest remain in the delocalized cloud and form shells with the same filling factors as in the pure metal case. The ordering principle remains; each added oxygen atom simply shifts the mass spectrum in a well-defined way.

Similarly, the effect of adding a hydrogen atom is to remove one valence electron from the delocalized group, once more leaving the remainder to organize into shells and thereby govern the electronic properties. This is well illustrated by the ionization potential data in Fig. 7. Evaporation rates of $Li_nH_m^+$ clusters[25] show a similar behavior.

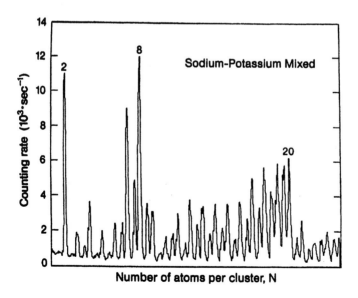

Figure 5. A mass spectrum containing pure K_n and mixed $K_{n-1}Na$ (intermediate peaks) clusters,[26] showing that both display electron shell closings.

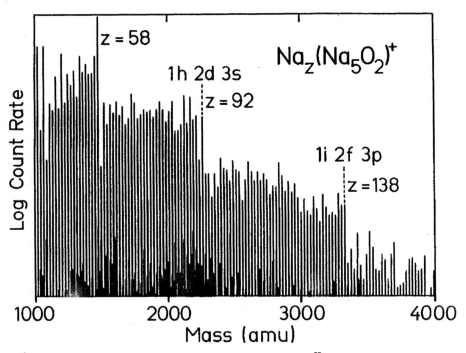

Figure 6. Laser photoionization spectrum of $Na_z(Na_5O_2)$ clusters.[23] The number of delocalized electrons z controls the closing of electronic shells.

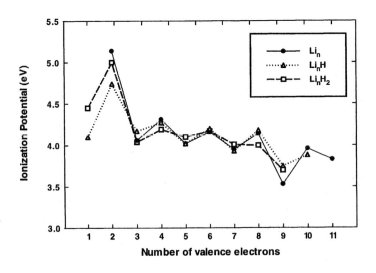

Figure 7. Ionization potentials of Li_n (•), Li_nH (Δ), and Li_nH_2 (□) clusters displayed as a function of the number of delocalized electrons: n, $n-1$, and $n-2$, respectively (after [24]).

250

Photoionization and photoelectron spectra of Na_nCl_m and Na_nI_m clusters[27,28] likewise show remarkable similarities with those of Na_{n-m}.*

(c) A divalent impurity. An interesting change appears in the mass spectrum of mixed K_nMg_m clusters[30] (Fig. 8). Strong abundance steps are seen both at $n=8, m=0$ and at $n=8, m=1$. At first glance this appears difficult to reconcile with the shell structure picture. Further reflection shows, however, that there is a natural explanation for the pattern. The explanation is based on the following two considerations: *1*. The two valence electrons of Mg join the cluster's electron cloud, and *2*. The resulting Mg^{2+} ion settles in the middle of the cluster and thus introduces a dip at the center of the potential well. The effect of the dip is to lower the energies of the *s*-shells (only these electrons have a large probability of sampling the potential near the origin). Considering the energy level scheme in Fig. 3, one recognizes that a likely consequence will be that the *2s* level will drop below the *1d* and a noticeable gap will arise; this reasoning is supported by computations.[31] Counting the shell filling factors, we now see that the new "magic numbers" of electrons will include both 8 and 10, in agreement with the mass spectrum in Fig. 8. A similar picture is observed for clusters K_8Zn, $K_{18}Mg$, and $K_{18}Zn$.[30]

Recent photoelectron spectra of K_nZn^- and K_nOH^- clusters[32] provided a direct verification of such level shifts. In Zn-doped clusters the *s*-levels were observed to be pulled down in agreement with the preceding argument, while in OH-doped clusters these levels were pushed upward as befits an additional repulsive potential introduced into the middle of the cluster. This effect serves as an excellent illustration of the power and adaptability of the electronic shell picture for clusters.

We emphasize that in macroscopic crystals the foregoing examples would all imply significant degrees of disorder, whereas in microscopic metallic clusters the electron spectrum is strongly quantized and is not disrupted by changes in the positive ion structure.

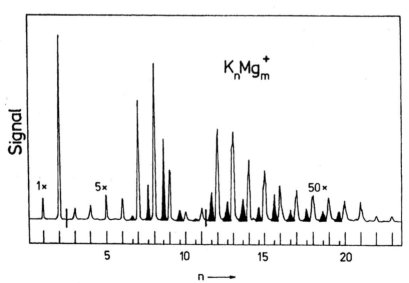

Figure 8. Mass spectrum for a mixed K/Mg expansion.[30] Both pure K_n and mixed $K_nMg_{m=1}$ (shaded peaks) clusters are observed. Note the relative abundance maxima at K_8 as well as at K_8Mg_1.

---

* However, this is not the case with Cs_nCl_m and Cs_nI_m clusters, suggesting that here the metallic and salt phases do not segregate.[28,29]

TRANSITION FROM SHELL-BASED TO STRUCTURAL ORDERING

As the cluster size increases, the shell spacing becomes very small, and at some point the organizational principle switches to that of crystalline packing. In other words, at this point the cluster is more appropriately described as a microcrystal than an electron droplet (a detailed review of clusters with geometrical packing can be found in [33]). The advent of high-range mass spectrometers made it possible to locate this cross-over.[34] As shown in the Na_n mass spectrum of Fig. 9, there is a periodicity change that occurs when cluster size reaches n~1500. This change was shown to be consistent with a switch from shell-filling to crystal-building.

A thorough and elegant analysis of this transition was given by Clemenger.[35] The central point of the argument is that the crystal field will become important when it is able to disturb the electronic shells and break their symmetry. The latter are classified by the angular momentum l and the radial quantum number n, while the crystal field has octahedral or icosahedral symmetry. As the cluster size increases, higher shell indices come into play, and it is possible to estimate from perturbation theory at what point the spherical shell ordering will become disrupted by the protocrystalline structure. Such a calculation in [35] results in the critical size estimate $n \approx 1600$, in excellent agreement with experiment. This work is valuable in that it clarifies the physics of the transition between the two ordering principles.

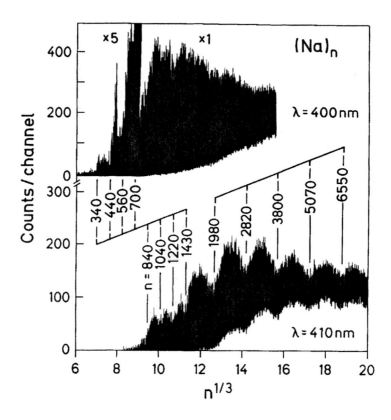

Figure 9. High-range mass spectra of Na_n clusters photoionized with 400 nm and 410 nm light, plotted on the $n^{1/3}$ scale.[34] The change in periodicity pattern at n~1500 is interpreted as a sign of the transition from electronic shell ordering to a protocrystalline one.

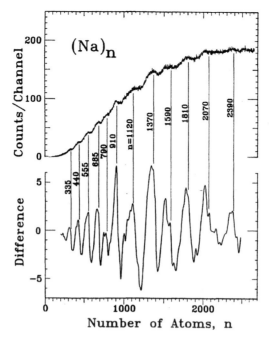

Figure 10. Top curve: mass spectrum of spectra of Na_n clusters photoionized by 4.0 eV photons and warmed with 2.71 eV continuous laser light.[36] Bottom curve: difference between full mass spectrum and a smoothed envelope, highlighting the appearance of steps. The pattern shows the presence of electronic shell structure effects up to n~2500.

In a follow-up experiment to [34], the large sodium clusters in transition to crystalline packing, as described above, were "warmed up" with a laser beam.[36] This melted the lattice structure; as a result, shell regularities returned, no longer inhibited, and were seen for up to n~2500 (Fig. 10). Later such transitions were observed for sizes up to ~10 000 atoms.[37] In the context of our discussion, this is quite a remarkable result. Indeed, melting of a bulk solid is unequivocally associated with a drastic loss of structural order. In the present case, though, this loss leads to the recovery of another kind of order: that based on electronic shells.*

SPLITTINGS OF GIANT DIPOLE RESONANCES

Much insight into the dynamics of delocalized electrons in metal clusters has been gained through the study of their giant optical resonances. The discovery of these collective oscillations (plasmons) in alkali microclusters[41] (see, e.g., Fig. 11) has been followed by a lot of work, both experimental and theoretical, devoted to the spectroscopy of such excitations and what they tell us about the electronic and vibrational states of clusters.

---

* Shell ordering in large clusters has led to extensive recent investigations of "supershells," characteristic shell bunchings or eigenvalue density oscillations. In semiclassical language, these may be interpreted as beats, or Moiré patterns,[38] of the density of closed particle orbits. Further reviews and details can be found, e.g., in [2,10,35,39]. An illustration of the correspondence between classical and wave-mechanical orbits in a potential well has been given in [40].

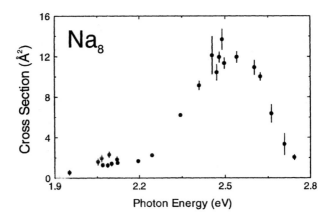

Figure 11. Experimental photoabsorption cross sections of Na_8.[41] A collective resonance is seen at 2.51 eV, taking up 70% of the valence electrons' dipole oscillator strength. The classical surface plasma resonance of a sodium sphere is at 3.41 eV. Error bars denote absolute uncertainty in the values of cross sections.

The giant dipole resonances in microclusters are due to highly correlated electronic states; they can be viewed as the precursors of plasmon oscillations on the surfaces of macroscopic conducting particles, thus earning the conventional label of "surface plasmons." Their deviations from the macroscopic limit and implications of these deviations are currently subjects of active research in the field. For more complete details and references, the reader is referred, e.g., to the reviews [2,8-10,39,42-44].

Here we would like to address one particular aspect of the subject, namely the splittings of the photoabsorption lines. Most observed resonances are in fact not contained in single lines as in Fig. 11, but are fragmented. It turns out that this fragmentation is caused not by a single mechanism but by a rich variety of effects, not all of them fully understood.

First of all, it is necessary to bring in a fact that was sidestepped in the preceding sections. Whereas clusters with closed electronic shells (total $L=0$) are spherical in shape, those with incompletely filled shells find it energetically favorable to deform (a manifestation of the Jahn-Teller effect). As originally shown in [45], the optimal deformations can be found by minimizing the total energy of electronic states in a potential well (see also the reviews [2,5-10,14]).

A clean example of a deformed cluster is one with 10 valence electrons. It is predicted to acquire a spheroidal shape with an axial ratio of ≈ 0.64.[41] Intuition based on the classical electrodynamics of a spheroidal conductor leads one to expect that the two axes will have different restoring forces acting on the electron cloud, producing two resonance frequencies. This expectation (which also can be justified quantum-mechanically[42]) is indeed borne out, as can be seen from Fig. 12.

However, it turns out that a cluster does not need to be deformed in order to display a split resonance. Consider the photoabsorption spectrum of Na_{20} in Fig. 13. This is a closed-

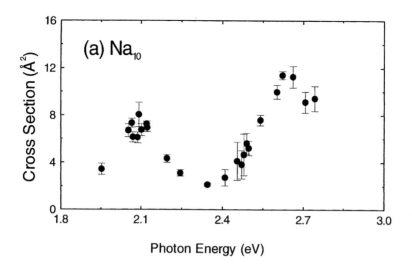

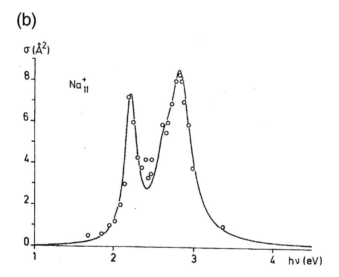

Figure 12. Experimental photoabsorption cross sections of spheroidally deformed clusters: (a) Na$_{10}$ (from [41], error bars denote absolute uncertainty in the values of cross sections); (b) Na$_{11}^+$ (from [46], absolute values of the cross sections are accurate to within ±25%). Note that the ratios of energies of the two resonance peaks are essentially the same in both cases. This illustrates the fact that the deformation of a metal cluster is determined by the total number of delocalized electrons, here 10 in both cases.

shell spherical particle, and yet the resonance is clearly fragmented. This effect has been explained as being due to the interference of the collective state with weak single-particle transitions.[47,48,42] Degeneracy lifting is manifested as a splitting of the giant resonance line. Put another way, this picture is a precursor of the famous Landau damping in solid-state and plasma physics, which is the decay of a plasmon wave into single-particle excitations. Since microclusters have a quantized rather than a continuous electron-hole spectrum, we observe

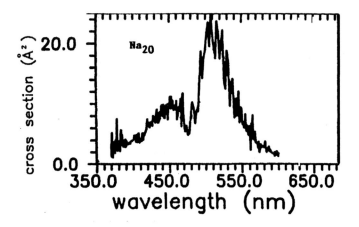

Figure 13. Photoabsorption profile of Na_{20} (from [49], absolute cross section values are accurate to within ±50%). The peak is split even though Na_{20} is a closed-shell spherical cluster.

discrete fragmentation rather than continuous broadening of the collective state. This fragmentation channel represents yet another manifestation of the discrete electronic spectrum of a metallic microparticle.

It turns out that this is still not the end of the story. The measurements described above were done on cluster beams prepared in expansion ovens; the smaller clusters synthesized in this manner are likely to be rather hot, that is, highly vibrationally excited or even liquid.* It is not unreasonable that the optical response of such a cluster should be describable as that of a quantum-sized metallic droplet. However, what happens if the cluster is cooled down?

Such an experiment has been carried out on clusters generated in a specially-designed cell in which cluster ions grow by aggregation in a flow of helium gas of adjustable temperature.[51]. This development brought about the long-sought control over the cluster temperature. Thus optical absorption measurements could now be performed on the same cluster as a function of its thermal state. Lo and behold, it was observed that the smooth resonance spectra started to break up as the condensation temperature decreased. This is illustrated in Fig. 14, where the clusters in 14(a) are probably liquid and in 14(c) solid.

Clearly, we once again observe the interplay between the purely electronic description of the cluster particle and its structural features. Evidently, the spectra in Fig. 14 reflect the emerging influence of the ionic backbone.[52]

---

* The melting point and heat capacity of an individual free cluster in a beam was recently determined using an adjustable-temperature cluster source and laser fragmentation spectroscopy.[50] For the spherical Na^+_{139} cluster the melting point lies at 267K, which is 104K below that of bulk sodium. The latent heat of fusion is also significantly reduced (by 46%), while the width of the solid-to-liquid phase transition is increased (by 59%) with respect to a bulk-like system.

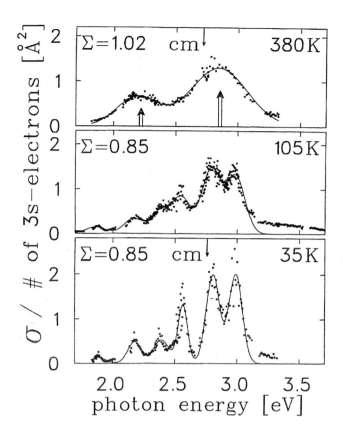

Figure 14. Photoabsorption cross sections of Na_{11}^+ at three different nominal temperatures.[51]

In fact, the line positions for the smaller clusters can be described in detail by means of the quantum-chemical approach.[53] By specifying the molecular geometry of a particular cluster and carrying out an *ab initio* computation of its excitation spectrum, it has been possible to reproduce the experimental data quite closely. From this reference point, the smoothing-out of the spectra with increasing cluster temperature or size is the result of a thorough averaging out of the molecular spectral features. No coherent understanding has yet emerged, though, of how to bridge the two modes of description and to establish the essential physical factors *rigorously defining* the conversion between two well-established limits: one strictly molecular and the other that of a quantized metallic droplet.

On the other hand, we can return to a dominant theme of this paper by following the evolution of the spectra in Fig. 14 in the direction of increasing temperature and noting that it once again reflects the interesting behavior we remarked on in earlier sections: as the temperature (and thereby the degree of disorder in the conventional sense), increases, the spectrum simplifies and approaches that of an electron droplet. Thus once again, the loss of one measure of order translates into the emergence of another: that of delocalized electrons in a finite potential.

SUPERCONDUCTING CORRELATIONS

One of the most intriguing questions in the microcluster field is the possibility of superconductivity in size-quantized particles. One would like to understand: how large does a piece of metal have to become in order to support electron pairing; how would such pairing correlations manifest themselves experimentally; how does the electron-phonon coupling evolve from the electron-vibrational interaction in molecules; can alternative pairing channels (e.g., surface oscillations) be present in microclusters; etc. At present, most of these questions remain unanswered.

No experimental search for superconductivity has yet been undertaken on free size-selected clusters in beams. First of all, as indicated earlier, most such cluster beams are born at rather high temperatures, so they are likely to be too hot to display any superconducting features. Recent progress on the growth of metal microclusters inside ultracold helium droplets[54] may offer a solution. Secondly, it is not yet clear how one would go about detecting paired states in free particles experimentally.

An alternative approach is to study nanoparticles deposited or fabricated on surfaces. So far, this approach does not allow precise size selectivity and is limited to much larger particles with correspondingly smaller electron energy level spacings δ. On the other hand, it makes it possible to employ cryogenic temperatures and very sensitive single-electron tunneling techniques. Black, Ralph, and Tinkham[55] have performed such an experiment on individual nanometer-sized Al grains connected to external leads via high-resistance tunnel junctions. The size of the grains ranged from ~5-13 nm, corresponding to ~10^5-10^6 atoms, thus still significantly exceeding the microcluster size regime. Nevertheless, the electron level spacings in these Al grains (δ~0.3-0.02 meV) were already comparable to the superconducting energy gap in bulk Al (Δ_{bulk}=0.18 meV). The tunneling measurements of electronic energy levels at T≤50mK showed a superconducting excitation gap in the spectrum of the Al grains. The excitation spectrum revealed a clear difference between particles with odd and even numbers of electrons, and the gap could be extinguished by an applied magnetic field.

Subsequent theoretical work investigated solutions of the BCS Hamiltonian

$$\hat{H} = \sum_j \varepsilon_j^0 (\hat{c}_j^\dagger \hat{c}_j) - G \sum_{ij}{}' (\hat{c}_i^\dagger \hat{c}_{\bar{i}}^\dagger \hat{c}_{\bar{j}} \hat{c}_j) \tag{1}$$

for finite particles. Here ε_j^0 are the energies of the discrete single-electron states j, $\bar{j}$ are their time-reversed states, and G is the phenomenological coupling constant. The possibility of mean-field solutions for the superconducting order parameter was considered under the assumptions of an approximately equal electronic level spacing[56] or of a random-matrix distribution.[57] Quantum fluctuations of the order parameter were also investigated.[58] A significant difference between odd- and even-numbered cases was predicted, and an approximate limit of δ<5-10Δ_{bulk} was estimated for the existence of a mean-field solution.

It is interesting to speculate whether a pair-correlated state can exist in the much smaller microclusters we have discussed above. The distribution of electronic states in such a cluster does not need to be approximated by either a constant-spacing or a random-matrix model; instead, it is given by the well-defined shell ordering. Mean-field estimates cited in the preceding paragraph are not encouraging as regards the probability of finding pairing effects in a closed-shell cluster with a large shell spacing (although the model BCS Hamiltonian and the mean-field approach itself may not be reliable guides in this situation).* Note, however, that deformed clusters with unfilled shells can possess dramatically smaller electronic level spacings and may be promising systems for the appearance of electron correlations. There is significant overlap between this problem and the corresponding questions arising in the spectroscopy of isolated fullerene anions and conjugated organic molecules addressed elsewhere in this volume.

CONCLUSIONS

Our aim has been to highlight the variety of effects which can be understood on the basis of the electronic shell structure of finite metallic clusters. The picture of delocalized electrons occupying a set of size-quantized orbits is clearly manifested not only in pure clusters, but also in those which in the solid-state limit would be considered quite disordered: clusters with impurities, mixed clusters, or hot clusters. In a number of cases (crystallization, photoabsorption spectroscopy) it is observed that a decrease in the structural ("lattice") order reduces its competition with the shell structure and allows the latter to emerge more clearly. Metallic microclusters turn out to be situated at the interface between two types of order, electronic and structural. They provide an interesting and quite unique laboratory for better understanding of the interplay between these components.

ACKNOWLEDGEMENTS

This work was supported by the U.S. National Science Foundation under Grant No. PHY-9600039 (V.V.K.) and by the Berkeley Faculty Committee on Research (W.D.K.).

REFERENCES

1. The present article is based upon an earlier review: V.V.Kresin and W.D.Knight, Z. Phys. Chem. **203** (1998).
2. *Clusters of Atoms and Molecules*, ed. by H.Haberland (Springer, Berlin, 1994).
3. *Advances in Metal and Semiconductor Clusters*, ed. by M.A.Duncan (JAI Press, Greenwich, 1993-1997).
4. *Proceedings of the International Symposia on Small Particles and Inorganic Clusters (ISSPIC)*: Surf. Sci. **156** (1985); Z. Phys. D **12** (1989); Z. Phys. D **19-20** (1991); Z. Phys. D **26** (1993); Surf. Rev. Lett. **3** (1996); Z. Phys. D. **40** (1997).
5. *Large Clusters of Atoms and Molecules*, ed. by T.P.Martin (Kluwer Academic, Dordrecht, 1996).
6. W.A. de Heer, W.D.Knight, M.Y.Chou and M.L.Cohen, in *Solid State Physics*, vol. 40, ed. by H.Ehrenreich and D.Turnbull (Academic, New York, 1987).
7. M.Moskowitz, Annu. Rev. Phys. Chem. **42**, 465 (1991).
8. V.O.Nesterenko, Sov. J. Part. Nucl. **23**, 726 (1992).
9. W.A. de Heer, Rev. Mod. Phys. **65**, 611 (1993).
10. Comments At. Mol. Phys. **31**, No. 3-6 (1995), *Special issue: Nuclear Aspects of Simple Metal Clusters*, ed. by C.Bréchignac and Ph.Cahuzac.
11. K.D.Bonin and V.V.Kresin, *Electric-Dipole Polarizabilities of Atoms, Molecules and Clusters* (World Scientific, Singapore, 1997).

---

* There have been several attempts[59-61] to apply the BCS Hamiltonian (1) to shell-filling electrons in a simple metal cluster and to consider whether it can lead to any significant consequences for the cluster ionization potentials. No conclusive results were obtained, and the degree to which this model is appropriate to the microcluster size regime is in fact unclear.[61]

12. U.Näher, S.Bjørnholm, S.Frauendorf, F.Garcias, and C.Guet, Phys. Rep. **285**, 245 (1997).
13. W.D.Knight, K.Clemenger, W.A. de Heer, W.A.Saunders, M.Y.Chou, and M.L.Cohen, Phys. Rev. Lett. **52**, 2141 (1984).
14. M.L.Cohen and W.D.Knight, Phys. Today (December 1990, p.42).
15. M.Ruppel and K.Rademann, Z. Phys. Chem. **184**, 265 (1994).
16. N.W.Ashcroft and N.D.Mermin, *Solid State Physics* (Holt, Rinehart and Winston, New York, 1976).
17. S.Bjørnholm, J.Borggreen, H.Busch, and F.Chandezon, in Ref. 5, p. 111.
18. M.Springborg, S.Satpathy, N.Malinowski, U.Zimmermann, and T.P.Martin, Phys. Rev. Lett. **77**, 1127 (1996).
19. W.D.Knight, in *The Chemical Physics of Atomic and Molecular Clusters*, ed. by G.Scoles (North-Holland, Amsterdam, 1990), p. 413.
20. J.G.Eaton, L.H.Kidder, H.W.Sarkas, K.M.McHugh, and K.H.Bowen, in *Nuclear Physics Concepts in the Study of Atomic Cluster Physics*, ed. by R.Schmidt, H.O.Lutz, and R.Dreizler (Springer, Berlin, 1992), p. 291.
21. A.Nakajima, K.Hoshino, T.Naganuma, Y.Sone, and K.Kaya, J. Chem. Phys. **95**, 7061 (1991); K.Hoshino, T.Naganuma, K.Watanabe, A.Nakajima, and K.Kaya, Chem. Phys. Lett. **211**, 571 (1993); K.Hoshino, K.Watanabe, Y.Konishi, T.Taguwa, A.Nakajima, and K.Kaya, Chem. Phys. Lett. **231**, 499 (1994).
22. T.Bergmann and T.P.Martin, J. Chem. Phys. **90**, 2848 (1989).
23. N.Malinowski, H.Schaber, T.Bergmann, and T.P.Martin, Solid State Commun. **69**, 733 (1989).
24. B.Vezin, Ph.Dugourd, D.Rayane, P.Labastie, J.Chevaleyre, and M.Broyer, Chem. Phys. Lett. **206**, 521 (1993); B.Vezin, P.Rambaldi, Ph. Dugourd, and M.Broyer, J. Physique IV, C4-651 (1994).
25. R.Antoine, Ph.Dugourd, D.Rayane, E.Benichou, B.Vezin, and M.Broyer, Z. Phys. D **40**, 436 (1997).
26. W.D.Knight, W.A. de Heer, K.Clemenger, and W.A.Saunders, Solid State Commun. **53**, 445 (1985).
27. S.Pollack, C.R.C.Wang, and M.M.Kappes, Z. Phys. D **12**, 241 (1989).
28. D.J.Fatemi, F.K.Fatemi, and L.A.Bloomfield, Phys. Rev. B **55**, 10094 (1997).
29. S.Frank, N.Malinowski, F.Tast, M.Heinebrodt, I.M.L.Billas, and T.P.Martin, J. Chem. Phys. **106**, 6217 (1997).
30. M.M.Kappes, P.Radi, M.Schär, and E.Schumacher, Chem. Phys. Lett. **119**, 11 (1985).
31. S.B.Zhang, M.L.Cohen, and M.Y.Chou, Phys. Rev. B **36** 3455 (1987).
32. H.L. de Clercq, Ph.D. Thesis, Johns Hopkins University, 1997; H.L. de Clercq, C.A.Fancher, and K.H.Bowen, to be published.
33. T.P.Martin, Phys. Rep. **273**, 199 (1996).
34. T.P.Martin, T.Bergmann, H.Göhlich, and T.Lange, Chem. Phys. Lett. **172**, 209 (1990).
35. K.Clemenger, Phys. Rev. B **44**, 12991 (1991).
36. T.P.Martin, S.Bjørnholm, J.Borggreen, C.Bréchignac, Ph.Cahuzac, K.Hansen, and J.Pedersen, Chem. Phys. Lett. **186**, 53 (1991).
37. T.P.Martin, U.Näher, H.Schaber, and U.Zimmermann, J. Chem. Phys. **100**, 2322 (1994).
38. G.Oster, *The Science of Moiré Patterns,* 2d ed. (Edmund Scientific, Barrington, N.J.,1969).
39. M.Brack, Rev. Mod. Phys. **65**, 677 (1993).
40. R.W.Robinett, Am. J. Phys. **64**, 440 (1996).
41. W.A. de Heer, K.Selby, V.Kresin, J.Masui, M.Vollmer, A.Châtelain, and W.D.Knight, Phys. Rev. Lett. **59**, 1805 (1987); K.Selby, V.Kresin, J.Masui, M.Vollmer, W.A. de Heer, A.Scheidemann, and W.D.Knight, Phys. Rev. B **43**, 4565 (1991).
42. V.V.Kresin, Phys. Rep. **220**, 1 (1992).
43. U.Kreibig and M.Vollmer, *Optical Properties of Metal Clusters* (Springer, Berlin, 1995).
44. V.Bonačić-Koutecký, P.Fantucci, and J.Koutecký, Chem. Rev. **91**, 1035 (1991).
45. K.Clemenger, Phys. Rev B **32**, 1359 (1985).
46. C.Bréchignac, Ph.Cahuzac, F.Carlier, M. de Frutos, and J.Leygnier, Chem. Phys. Lett. **189**, 28 (1992).

47. C.Yannouleas, R.A.Broglia, M.Brack, and P.F.Bortignon, Phys. Rev. Lett. **63**, 255 (1989); C.Yannouleas and R.A.Broglia, Ann. Phys. **217**, 105 (1992).
48. G.F.Bertsch and R.A.Broglia, *Oscillations in Finite Quantum Systems* (Cambridge, New York, 1994).
49. C.R.C.Wang, S.Pollack, J.Hunter, G.Alameddin, T.Hoover, D.Cameron, S.Liu, and M.M.Kappes, Z. Phys. D **19**, 13 (1991).
50. M.Schmidt, R.Kusche, W.Kronmüller, B. von Issendorff, and H.Haberland, Phys. Rev. Lett. **79**, 99 (1997).
51. C.Ellert, M.Schmidt, C.Schmidt, T.Reiners, and H.Haberland, Phys. Rev. Lett. **75**, 1731 (1995).
52. Indications of similar resonance-curve fragmentation in a cold cluster beam also have been observed in small silver clusters: B.A.Collings, K.Athanassenas, D.M.Rayner, and P.A.Hackett, Chem. Phys. Lett. **227**, 490 (1994).
53. V.Bonačić-Koutecký, J.Pittner, C.Fuchs, P.Fantucci, M.F.Guest, and J.Koutecký, J. Chem. Phys. **104**, 1427 (1996).
54. A.Bartelt, J.D.Close, F.Federmann, F.Quaas, and J.P.Toennies, Phys. Rev. Lett. **77**, 3525 (1996).
55. C.T.Black, D.C.Ralph, and M.Tinkham, Phys. Rev. Lett. **76**, 688 (1996).
56. J. von Delft, A.D.Zaikin, D.S.Golubev, and W.Tichy, Phys. Rev. Lett. **77**, 3189 (1996).
57. R.A.Smith and V.Ambegaokar, Phys. Rev. Lett. **77**, 4962 (1996).
58. K.A.Matveev and A.I.Larkin, Phys. Rev. Lett. **78**, 3749 (1997).
59. M.Barranco, E.S.Hernández, R.J.Lombard, and Ll.Serra, Z. Phys. D **22**, 659 (1992).
60. F.Iachello, E.Lipparini, and A.Ventura, in *Nuclear Physics Concepts in the Study of Atomic Cluster Physics*, ed. by R.Schmidt, H.O.Lutz, and R.Dreizler (Springer, Berlin, 1992), p. 318.
61. N.N.Kuzmenko, V.O.Nesterenko, S.Frauendorf, and V.V.Pashkevich, Nuovo Cimento **18D**, 645 (1996).

V. FINITE SYSTEMS. NUCLEI.

Role of Pairing Correlations in Cluster Decay Processes

D. S. Delion[1], A. Insolia[2], R.J. Liotta[3]

[1]Institute of Atomic Physics, Bucharest
Măgurele, POB MG-6, Romania
[2]Department of Physics, University of Catania and INFN, I-95129 Catania, Italy
[3]KTH, Physics at Frescati, Frescativägen 24, S-10405 Stockholm, Sweden

Introduction: BCS approach to cluster decay

One of the most obvious features of nuclear spectra is the energy gap seen in even-even nuclei, as stressed in the pioneering work of Bohr, Mottelson and Pines [1]. The even nuclei nearly all have just one and much more often no excited state less than 0.5 MeV in energy, except for deformed nuclei which may have a couple of excited states below that energy. The odd nuclei usually have a number of excited levels in this energy range, in fact more than ten in several cases.

If obviously collective states are discounted (the vibrational and rotational bands), the remaining intrinsic spectrum shows a gap. The lowest intrinsic excitation is at least 900 keV in the even nuclei. On the other hand, the first intrinsic excitation in odd nuclei is often at less than 100 keV and rarely at more than 300 keV, except for nearly magic nuclei (where the corresponding even nucleus is a spherical normal core) [2]. All these demonstrate the suitability of a superconducting description of nuclei [1]. In the absense of any pairing, the spherical or deformed single-particle spectrum would predict the same typical lowest intrinsic excitation energy and level density for even and odd nuclei since the excitation would simply require the shift of a valence particle to a new single-particle orbit in either case. A diagonal pairing energy for each $(j^2)_{J=0}$ pair would raise in the even nuclei the excitation energy of those states in which a particle has shifted its orbit and thus broke a pair, but those excitations in which a pair shifted orbits would be unaffected. This would lead to the lowest even intrinsic state at about twice the energy of the typical odd excitation in marked contrast with experiment, which shows this ratio to be at least five. It was thus early recognized that the pairing interaction acts also upon non-diagonal configurations giving rise to collective excitations, the so-called pairing vibrations [3, 4]. This was important to understand features related to two body transfer reactions. But the most important consequence of these studies in relation to the present paper was the realization that the pairing interaction plays a fundamental role in clustering the nucleons that eventually becomes the alpha-particle in alpha-decay processes [5, 6, 7, 8]. We will limit ourself to discuss only what we think is of interest for the discussion of the main point in the following. In spite of the large amount of work on pairing correlations the interplay between pairing and surface clustering was fully understood when it was shown the role played by the mixing of levels of different parity in the nuclear wave functions [7, 9].

To clarify this important point, and in order to select the appropriate measure of two-particle surface clustering in spherical nuclei, in ref. [8] were used the coordinates $\vec{R} = \frac{(\vec{r}_1+\vec{r}_2)}{\sqrt{2}}$, $\vec{r} = \frac{(\vec{r}_1-\vec{r}_2)}{\sqrt{2}}$, associated with the center of mass and relative motion, respectively, and considered the probability distribution

$$P(r, R) = r^2 R^2 \int d\hat{r} d\hat{R} d\chi_1 d\chi_2 |\Psi(\vec{r}_1, \chi_1; \vec{r}_2, \chi_2)|^2 \tag{1}$$

The coordinate transformation, from the individual nucleon coordinates r_1 and r_2, is done by means of the Talmi-Moshinsky transformation brackets. The probability turns out to be mainly localized in (n+1) circular shells with a dominance of the outer shell. In this region it has a number of maxima equal to the orbital angular momentum quantum number l, in the case of antiparallel spin ($j = l-1/2$) and equal to $l+1$, in the case of parallel spin ($j = l+1/2$). The effect due to the interference between two levels of opposite parity is such that if they interfere with the *right* phase the probability distribution is enhanced on the surface and strongly suppressed for larger values of the relative coordinate. This happens to be the case for the pairing interaction. This fact has also been understood in terms of a simple analytical model using perturbation theory [9].

It is expected that in a deformed nucleus the mixing of levels with different parity is quite large around the Fermi surface. This fact is going to produce a remarkable enhancement of the cluster probability distribution. In the case of deformed systems, one can define [10] (assuming K=0^+ pairs – namely axially symmetric two-particle system –) the probability distribution as

$$P(R, \theta_R) = R^2 |\int d^3\vec{r} \Phi_{00}(\nu_0 \vec{r}) \Psi(\vec{r}_1, \vec{r}_2)|^2 \tag{2}$$

In this equation $\Psi(\vec{r}_1, \vec{r}_2)$ is the two-particle wave function in the intrinsic frame. The overlap has now been considered with a 0s state of the relative motion with a frequency ν_0 characterizing the size of the cluster.

In the four-particle case (α-cluster probability distribution) one can, similarly, calculate

$$P_\alpha(R_\alpha, \theta_{R_\alpha}) = R_\alpha^2 |\int d^3\vec{r}_\pi d^3\vec{r}_\nu d^3\vec{r}_{\pi\nu} \Phi_{00}(\nu_\alpha \vec{r}_\pi) \Phi_{00}(\nu_\alpha \vec{r}_\nu) \Phi_{00}(\nu_\alpha \vec{r}_{\nu\pi}) \cdot$$
$$\Psi(\vec{r}_{1\pi}, \vec{r}_{2\pi}, \vec{r}_{1\nu}, \vec{r}_{2\nu})|^2 \tag{3}$$

In ref. [10] Nilsson levels are generated in a deformed harmonic oscillator potential. Particles are correlated by the pairing interaction within the BCS approximation (see discussion below). It was found [10] that different pure (protons$\otimes$neutrons) configurations are localized in different nuclear region according to the quantum number K of the orbits.

Summarizing the previous results one can conclude that the pairing interaction strongly enhances the probability of finding a pair or a quartet of nucleons on the suface, and, for the spin singlet ($S = 0$) component, the probability of finding a pair shows a sharp peak around relative distance zero. Furthermore, while pure Nilsson configurations produce pair and quartet densities peaked at particular points of the deformed nuclear surface, configuration mixing smears these peaks into a more or less uniform density ridge around the nucleus. At the same time, the formation amplitude is very much enhanced in comparison with pure Nilsson orbits [10].

Therefore, to investigate clustering properties on nuclei, one has to look for a simple frame where complex spherical and deformed nuclear systems could be conveniently and realistically described.

The BCS theory, introduced for the description of system with very large number of particles, was quite soon used as the convenient frame in which to exploit the importance of a pairing force [5]. It provided a nearly satisfactory model for the description of some cluster-decaying, especially α-decaying, nuclei. This model is appealing because it is realistic just in between two closed shells, where the ordinary shell model gets very involved, and because it lends itself to a consistent description of the parent and the daughter nuclei. In the BCS model the nuclei are hardly "individualized", and thus the results change rather smoothly from nucleus A to nucleus $A+2$, which is consistent with the systematics of the empirical reduced α-decay widths of heavy nuclei far from magic numbers (see e.g. [11]). Recently phenomenological evidence has been found [12] for an even more striking uniformity in even–even actinide nuclei. Experimental branching ratios of α-transitions to different final states were used in a coupled-channel barrier-penetration model of the α-decay of these deformed nuclei. The results seem to imply that the amplitudes belonging to different α–core orbital momenta L are essentially the same for all these nuclei. Such an observation reaffirms that it is promising to apply the BCS theory to the cluster decay of these nuclei. It is to be emphasized, however, that the pairing force only acts between like nucleons, and the feasibility of large-scale BCS calculations hinges on the neglect of the proton–neutron force.

The technical simplicity of the BCS approach comes from the fact that it transforms the wave function of a correlated system into a form of a pure configuration. The particles that form such a pure configuration are, however, not real particles but quasiparticles, which are defined by their creation operators $\alpha_k^\dagger$. These are linear combinations of creation and annihilation operators for real particles:

$$\alpha_k^\dagger = U_k a_k^\dagger - V_k a_{\bar{k}}, \qquad \alpha_{\bar{k}}^\dagger = U_k a_{\bar{k}}^\dagger + V_k a_k, \tag{4}$$

where $\bar{k}$ labels the time-reversed of the state labelled by k. This definition makes it obvious that in a BCS state the number of particles is not a good quantum number.

Both for the parent nucleus, B, and for the daughter, A, the BCS wave function is a product of the proton and of the neutron parts:

$$\Phi^{(i)}(x_i) \equiv \langle x_i | \text{BCS} \rangle_i = \Phi_\pi^{(i)}(x_i^\pi) \Phi_\nu^{(i)}(x_i^\nu) \equiv \langle x_i^\pi | \text{BCS} \rangle_i^\pi \langle x_i^\nu | \text{BCS} \rangle_i^\nu \quad (i = B, A), \tag{5}$$

where x_i stands for the set of all s.p. coordinates involved. (We disregard the centre-of-mass motion.) For even particle numbers, any of the state vectors $|\text{BCS}\rangle$ can be written in occupation-number representation as

$$|\text{BCS}\rangle = \prod_{k>0}(U_k + V_k a_k^\dagger a_{\bar{k}}^\dagger)|0\rangle, \tag{6}$$

where $|0\rangle$ stands for the particle vacuum and k denotes all s.p. quantum numbers. The restriction $k > 0$ expresses that, to avoid double counting, k and $\bar{k}$ should run over complementary domains. For spherical nuclei rotational symmetry in eq. (6) is brought about by the angular-momentum coupling implicit in V_k, while for non-spherical nuclei eq. (6) is just the wave function in the intrinsic (body-fixed) coordinate system, and rotational symmetry is restored by a transformation to the laboratory frame. A BCS calculation consists in a minimization of the expectation value of the total Hamiltonian

$$H = \sum_{k>0} \epsilon_k (a_k^\dagger a_k + a_{\bar{k}}^\dagger a_{\bar{k}}) - G \sum_{kk'>0} a_k^\dagger a_{\bar{k}}^\dagger a_{\bar{k}'} a_{k'} \quad (7)$$

with respect to the parameters U_k, V_k, subject to the condition that the expectation value of the particle-number operator $N = \sum_k a_k^\dagger a_k$ is equal to the number of protons or neutrons in the nucleus i. In eq. (7) ϵ_k are the s.p. energies and G is the strength parameter of the pairing force.

The BCS approach was introduced long ago in studying α-decay by Mang and Rasmussen [6, 13] and by Soloviev [5] and was subsequently applied by Poggenburg, Mang and Rasmussen [14]. This approach lends itself to treating deformed nuclei, and it was applied to the decay of such nuclei. The collective motions were described in the standard Bohr–Mottelson rotational model [3], while the intrinsic motions were represented by BCS wave functions built up from Nilsson s.p. orbits expressed in terms of harmonic-oscillator wave functions. The penetration through the deformed potential barrier was described in the WKB aproximation in the formalism of Fröman [23].

These pioneering calculations necessarily suffered from serious basis limitations: they used single-shell valence spaces which imply basis sizes very small in comparison with what is possible now with the present-day computers. Nevertheless, their results were very promising. Hundreds of transitions were calculated, and the empirical trends were mostly reproduced. In particular, the phenomenological notions of "favoured" α-decay transitions were clearly identified with cases that involve neither rearrangements of unpaired nucleons nor quantum number mismatch between the Nilsson orbits involved in the transition. The "hindrance factors", which characterize unfavoured transitions quantitatively, were reasonably reproduced. All these calculations, however, were only successful in reproducing relative transition rates. The absolute decay rates were undershot by orders of magnitude, and the degree of disagreement depends on parameters that are uncertain enough to diminish seriously the predictive power of the model for the absolute decay rates [13].

Recently the BCS treatment was applied to studying the α-decay of spherical systems with many particles outside the core [15]. In that work a large number of s.p. states were used. Later the approach was used again for deformed nuclei [16, 17, 18, 19, 21, 20] and, even more recently, it was generalized to heavy-cluster decays [22, 24, 25].

The next section will review these recent works.

A short outline of BCS calculations for α decay

The BCS approach is built upon a (truncated) set of orthonormal single particle (s.p.) states, The s.p. states should be Woods–Saxon eigenfunctions; for spherical nuclei ordinary orbits, whereas for deformed nuclei Nilsson orbits. In both cases they may be expanded in terms of the eigenfunctions of a spherical harmonic oscillator belonging to good orbital and total angular momenta. For a realistic calculation many major shells need to be considered.

The BCS calculations on such large s.p. bases are feasible though challenging computationally. The challenge is again that the formation amplitude, i.e. the projection of the mother state onto the (antisymmetrized) product of the states of the two fragments, should be reliable up to far enough from the centre, and that requires a great number of s.p. states. With the wave function given in terms of an oscillator expansion, the formation amplitude is analytically calculable, and that is much simpler if all oscillator parameters are the same.

The BCS approach to α-decay was tested on the lead isotopes ^{186,188,190,192}Pb [15]. The ingredients of the model were chosen, in a fairly standard way, so as to describe the known properties of the structures of the ground states. The proton and neutron harmonic-oscillator states were chosen to span all shells up to $7\hbar\omega$ and $8\hbar\omega$, respectively. It was confirmed that configuration mixing enhances surface clustering.

A fully logical treatment of the proton and neutron gaps would be to fit them to the ^{206}Hg and ^{206}Pb energies, respectively, since these involve a pair of holes over the doubly magic core. However, when applied to another nucleus, these parameters should produce a discrepancy in energy because, when there are both proton and neutron holes, there must be a residual interaction between them. The neglected proton(-hole)–neutron(-hole) interaction can be mocked up to some extent by adjusting the gaps to each particular daughter nucleus. Since in this way extra correlation is induced between like nucleons, rather than between unlike nucleons, this is not an undisputable prescription. Nevertheless, it works qualitatively as expected.

Table 1: Calculated widths in comparison with the experimental ones. The calculated half-lifes and penetrabilities are also reported. The channel radius is 8.2 fm. This corresponds to a region of values in which the α-decay width Γ is almost constant. Reduced widths are given in MeV and half-lifes in seconds.

A	$T_{1/2}^{th}$	P_{th}	γ_{th}^2	$T_{1/2}^{exp}$	P	γ_{exp}^2
186	5.5	$.11 \times 10^{-20}$	.1086	4.7	$.68 \times 10^{-20}$	.09
188	17.9	$.34 \times 10^{-22}$	.1123	22.	$.11 \times 10^{-20}$	.114
190	72.	$.45 \times 10^{-24}$	.1311	72.	$.42 \times 10^{-21}$	.094
192	356.	$.16 \times 10^{-26}$	.1363	210.	$.28 \times 10^{-21}$	.049

The r_c-dependence of the width has a plateau between 8 and 8.5 fm. With such a channel radius and with the prescription simulating the proton–neutron interaction, the agreement of the results with experiment exceeds any expectations as seen in table 1.

Following the classical works by Mang et al. [6, 13, 14], the recent BCS calculations for α-decay for deformed nuclei were combined with the semiclassical description of barrier penetration [23].

The aim of the calculations for the deformed even–even Ra, Rn and Th isotopes [16, 18] was to reproduce the transitions from the g.s.'s to the g.s.'s and to the first 2^+ excited states. Calculations for the decay of some odd-mass nuclei (^{241}Am, ^{221}Fr, and a range of At and Rn isotopes) [17, 19] were made to describe the anisotropy of the α-particle emission. These have non-zero spins in their g.s.'s, thus can be oriented, and the angular distribution of the emitted particles with respect to the orientation of the nucleus has been measured for quite a few nuclei (see, e.g., [26, 27, 28]).

Throughout the calculations for deformed nuclei all s.p. orbits (Woods–Saxon or Nilsson s.p. states) up to the barrier top were included, and the expansion over the spherical s.p. states were cut beyond $18\hbar\omega$. The gap constants were again fitted to experiment.

For even–even nuclei the g.s.-to-g.s. transitions involve the $L = 0$ formation amplitudes. These were found to be substantially larger than those with $L \neq 0$. Furthermore, these amplitudes depend on the deformation parameters rather weakly. The resulting widths are collected in table 2. The agreement with the experimental data is quite remarkable. For the transitions to the 2^+ states the agreement [18] is comparable with those in table 2. For odd - mass nuclei (non-zero spin), the anisotropy of α-emission

Table 2: Alpha-decay widths Γ_0 produced by the BCS theory for even–even deformed nuclei compared with experiment (in MeV).

Mother nucleus	Experiment [33, 34, 58]	Theory
^{222}Rn	1.38×10^{-27}	0.70×10^{-27}
^{220}Rn	8.20×10^{-24}	2.7×10^{-24}
^{218}Rn	1.30×10^{-20}	0.25×10^{-20}
^{226}Ra	9.0×10^{-33}	6.6×10^{-33}
^{224}Ra	1.44×10^{-27}	0.83×10^{-27}
^{222}Ra	1.20×10^{-23}	0.52×10^{-23}
^{232}Th	1.01×10^{-39}	0.60×10^{-39}

may arise from two reasons: both the penetrability of the barrier and the nuclear wave functions are anisotropic in the body-fixed frame. Earlier, the observed preference for emission in the direction of the spin vector was attributed to the penetration through the deformed barrier [3], but recently the deformed intrinsic nuclear structure has also been proposed to be important [29, 30]. The BCS approach made it possible, for the first time, to describe this anisotropy microscopically, by taking into account the deformation of the nucleonic orbits.

A BCS state $|BCS\rangle_{A+1}$ in an odd-mass system can be generated by

$$|BCS\rangle_{A+1} = a^\dagger_{k_0}|BCS\rangle_A, \qquad (8)$$

where $|BCS\rangle_A$ is similar to the state $|BCS\rangle$ given by eq. (6). The anisotropy arises since the angular momentum carried away by the α-particle may be different from zero. The emission probability will be the modulus square of a coherent sum of amplitudes; each amplitude contains a matrix product of the penetrability matrix, whose elements belong to penetration transitions between states of particular angular momenta, with a one-column matrix of the multipoles of the formation amplitude defined in the body-fixed frame [17].

Calculations were performed for the favoured g.s.-to-g.s. transition ^{241}Am$\rightarrow$ ^{237}Np$+\alpha$, for which data are available. Both states have spin-parities $\frac{5}{2}^-$, with spin projections onto the symmetry axis $\frac{5}{2}$. The parameters, including the deformation parameters, were taken from independent sources. Both the absolute width and the anisotropy were reproduced very well (see table 3). The dominant contribution was found to arise from the quadrupole deformation. By suppressing the effect of the non-zero multipoles of the formation amplitude, the absolute strength as well as the angular distribution of the emission probability changed just about 10%, which shows that the bulk of the anisotropy comes from the penetrability effect. A systematic survey of a number of At

Table 3: Width and ratio $\mathcal{R}$ of emission probability in directions parallel with respect to that in directions perpendicular to the symmetry axis for the transition ^{241}Am→^{237}Np+α, with a partial alignment corresponding to the experimental situation.

Quantity	Theory	Experiment [27]
Γ (MeV)	2.09×10^{-34}	3.34×10^{-34}
$\mathcal{R}$	2.038	2.255

and Rn isotopes has corroborated these trends [19]. It was shown that emission in polar directions tends to be favoured mainly by prolate nuclei, whereas oblate nuclei prefer equatorial emission. Furthermore, the relation between the deformation parameter β_2 and the anisotropy of α-decay was found monotonic, which enables one to determine β_2 from experimental data for the anisotropy.

This is illustrated by table 4, taken from Ref. [19], which contains, for a set of nuclei spanning a monotonic sequence of β_2, the calculated ratios $\mathcal{R}$ of the probability of emission in the direction parallel to the symmetry axis with respect to that in the perpendicular direction. A detailed and critic discussion on the BCS approach for the anisotropic α decay can be found in ref. [31]. These considerations as well as many

Table 4: Deformations, experimental and theoretical widths and emission probabilities in the direction of the symmetry axis relative to those in the perpendicular direction ($\mathcal{R}$) for odd Rn isotopes.

Mother nucleus	β_2	Γ_{exp} (MeV)[32]	Γ_{th} (MeV)	$\mathcal{R}$
$^{205}_{86}$Rn	0.005	2.68×10^{-24}	6.76×10^{-25}	1.034
$^{207}_{86}$Rn	0.016	8.17×10^{-25}	8.50×10^{-26}	1.118
$^{209}_{86}$Rn	0.023	2.62×10^{-25}	4.27×10^{-26}	1.177
$^{219}_{86}$Rn	0.081	1.15×10^{-22}	2.05×10^{-22}	1.749

additional microscopic calculations have shown that the continuum part of the nuclear spectrum plays an important role in α-decay processes [35, 36, 37, 7]. In particular, one needs to include the continuum, or high lying configurations in bound representations, to ensure a proper asymptotic behaviour of the α-particle formation amplitude at large distances.

This is in sharp contrast to usual spectroscopic calculations, where shell-model representations that include only a few bound states of a realistic Woods-Saxon potential is enough to describe well low energy properties [38, 39, 40, 41]. Spectroscopic properties of well deformed nuclei could also be conveniently studied assuming that the core is a realistic Woods-Saxon potential. However, the diagonalization of such a potential is usually performed within spherical harmonic oscillator basis, and this also requires the introduction of high lying configurations [42, 43, 44, 45]. In the calculation of absolute decay widths this is even more important, in spherical [46] as well as in deformed nuclei [18]. For instance, in ref. [35] it was found that the inclusion of 13 major harmonic oscillator shells was not enough to explain the ground-state to ground-state α-decay width of ^{212}Po.

In deformed nuclei it was possible to reproduce the total α-decay width within a factor of three, but only after including 18 major shells [18]. In heavy cluster decay the use of very large number of shells in the basis improves the calculation but still the calculated absolute decay widths are too small by 2–3 orders of magnitude with respect to the corresponding experimental data [18, 25].

All these indicate that the description of cluster-decay processes (including α - particles) within standard harmonic oscillator representations, i. e. in terms of the eigenstates of an harmonic oscillator potential with parameters just suited to the nucleus under study, is inadequate to guide experimental searches, or even reproduce experimental data [47]. Yet this is important particularly in relation to present experimental facilities and methods that would allow one to detect even tiny signals which may correspond to the decay of highly hindered processes, as the decay of heavy clusters or very unstable states [32, 48], where very high lying configurations may play an important role.

The aim of the next section is to present a single-particle representation which is rather small but at the same time is adequate to describe the quasi-continuum. The basis consists of the eigenstates of two different sets of harmonic oscillators states. One is suited to describe the discrete part of the spectrum and the other one the quasi-continuum. Since the problem outlined above lies in the poor description of the decay process at large distances, i. e. deficient or incomplete high lying configurations in the basis used in the calculations, we hope that the introduction of such a representation would cure those shortcomings while the calculation itself remains feasible.

The basis becomes non-orthogonal, but the problem of dealing with a hermitean Hamiltonian can be solved rather easily by using an orthogonalization procedure .

Formalism: The basis

The many-body problem in nuclear physics can conveniently be treated using harmonic oscillator representations. With this in mind, we write the stationary Schrödinger equation describing the single-particle motion of a particle of mass M_0 in a spherical mean field $V(r)$ as

$$H\psi(\xi) \equiv [-\frac{\hbar\omega}{2\lambda_0}\vec{\nabla}^2 + V(r)]\psi(\xi) = E\psi(\xi) \qquad (9)$$

where $\xi = (\vec{r}, s)$ is the set of spatial and spin coordinates, and

$$\lambda_0 = \frac{M_0\omega}{\hbar} \qquad (10)$$

is the harmonic oscillator parameter that best suits the potential $V(r)$ within the nuclear volume, i. e. with $\hbar\omega = 41A^{-1/3}$.

The wave function that solves the eigenvalue problem (9) have a separable form, i.e.

$$\psi_{Elj\Omega}(\xi) = u_{Elj}(r)[i^l Y_l(\hat{r})\chi_{\frac{1}{2}}(s)]_{j\Omega} \qquad (11)$$

where l is the angular momentum of the particle, j its total spin and Ω the corresponding projection on the z-axis. The radial part of the wave function satisfies the equation

$$-\frac{\hbar\omega}{2\lambda_0}[\frac{1}{r}\frac{d^2}{r^2}r - \frac{l(l+1)}{r^2}]u_{Elj}(r) \equiv H_{lj}^{(\lambda_0)}u_{Elj}(r) = [E - V(r)]u_{Elj}(r) \qquad (12)$$

The radial wave functions u_α, where α labels the set of quantum numbers $\{E,l,j\}$, is usually expanded in an harmonic oscillator basis, i. e.

$$u_\alpha(r) = \sum_{n=0}^{\infty} c_{\alpha n} R_{nl}^{(\lambda)}(r) \qquad (13)$$

where R_{nl}^λ is the h. o. radial wave function with radial quantum number n and angular momentum l. The h.o. parameter λ determines the representation. It may be convenient to choose it according to the number of shells that one includes in the basis and, therefore, it might differ from the value of λ_0 given in (10). We define λ as

$$\lambda = f\lambda_0 = f\frac{M_0\omega}{\hbar} \qquad (14)$$

and the constant f is chosen such that one obtains an h. o. spectrum that best fits the discrete part of $V(r)$ (this usually is a realistic Woods-Saxon potential). In figure 1 we show schematically the Woods-Saxon potential (dark line) and the h. o. potential determined by λ (full line).

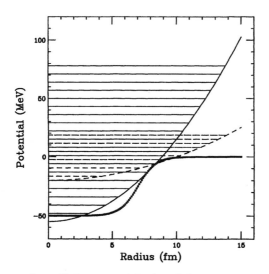

Figure 1: The harmonic oscillator potentials that define our representation. The full line is the potential that provides the low lying shells of the basis and the dashed line the one that provides the high lying shells. The dark line is the Woods-Saxon potential.

The solution given by (13) is exact if one takes all the terms in the expansion. For spectroscopic calculations (discrete part of the spectrum and transition probabilities) in heavy nuclei it is sufficient to take in the expansion of the radial wave function terms up to $N_0 = 2n_0 + l = 6$ major shells. One thus gets a good description of the wave functions up to the geometrical nuclear radius $R_0 = 1.2\ A^{1/3}$.

As discussed above, this approach is not enough for a microscopic description of the alpha or cluster decay processes, for which it is very important to reproduce the wave function at distances larger than the nuclear radius. It is known that in order to have a proper asymptotic behaviour of the wave functions at large distances it is necessary to include up to $N = 18$ major shells in heavy nuclei [18]. But this can also be achieved by using a mixed non-orthogonal h.o. basis in the diagonalisation procedure, thus reducing the number of shells needed in the expansion and, at the same time, obtaining a better description of the decay process. That is, instead of the expansion (13) we will use the following representation [49]

$$u_\alpha(r) = \sum_{2n_1+l=N_1 \leq N_0} c^{(1)}_{\alpha n_1} R^{(\lambda_1)}_{n_1 l}(r) + \sum_{2n_2+l=N_2 > N_0} c^{(2)}_{\alpha n_2} R^{(\lambda_2)}_{n_2 l}(r) \qquad (15)$$

where λ_1 is the h.o. parameter coresponding to the h.o. potential which fits the Woods-Saxon interaction in the region of the discrete spectrum (the full line of figure 1), while λ_2 corresponds to an h. o. potential that describes better the continuum part of the spectrum (dashed line in figure 1). Notice that these two different h. o. potentials are both centered at the origin of coordinates.

Below we will describe how to evaluate the amplitudes c in eq. (15). But it is worthwhile to point out here that although the representation used in (15) is discrete, the density of states is larger at high energies. In fact it can be made as large as one wishes by properly decreasing the parameter λ_2, as indicated in figure 1. This is just what is needed to describe processes occurrying at distances larger than the nuclear surface.

For each of the h. o. potentials of figure 1 the radial wave function R satisfies

$$H^{(\lambda_i)}_{lj} R^{(\lambda_i)}_{nl}(r) = \hbar\omega(2n + l + \frac{3}{2} - \frac{\lambda_i r^2}{2}) R^{(\lambda_i)}_{nl}(r) \qquad (16)$$

where

$$H^{(\lambda_i)}_{lj} = -\frac{\hbar\omega}{2\lambda_i}[\frac{1}{r}\frac{d^2}{r^2}r - \frac{l(l+1)}{r^2}] R^{(\lambda_i)}_{nl}(r) \qquad (17)$$

and λ_i is the h. o. parameter of the potential i.

The diagonalization of the potential V in terms of the two h.o. representations can be performed by inserting in eq. (9) the expanded wave function (15). One then obtains the following set of equations

$$\begin{pmatrix} \mathcal{H}^{(11)}_{n_1 n'_1} & \mathcal{H}^{(12)}_{n_1 n'_2} \\ \mathcal{H}^{(21)}_{n_2 n'_1} & \mathcal{H}^{(22)}_{n_2 n'_2} \end{pmatrix} \begin{pmatrix} c^{(1)}_{\alpha n'_1} \\ c^{(2)}_{\alpha n'_2} \end{pmatrix} = E \begin{pmatrix} \mathcal{I}^{(11)}_{n_1 n'_1} & \mathcal{I}^{(12)}_{n_1 n'_2} \\ \mathcal{I}^{(21)}_{n_2 n'_1} & \mathcal{I}^{(22)}_{n_2 n'_2} \end{pmatrix} \begin{pmatrix} c^{(1)}_{\alpha n'_1} \\ c^{(2)}_{\alpha n'_2} \end{pmatrix} \qquad (18)$$

where the amplitudes c are as in eq. (15) and the overlap integrals are defined by

$$\mathcal{I}^{(ik)}_{n_i n'_k} \equiv < \lambda_i n_i l | \lambda_k n'_k l > = < R^{(\lambda_i)}_{n_i l} | R^{(\lambda_k)}_{n'_k l} > \qquad (19)$$

while the Hamiltonian kernels are given by

$$\mathcal{H}^{(ik)}_{n_i n'_k} = \hbar\omega f_k [(2n'_k + l + \frac{3}{2}) \mathcal{I}^{(ik)}_{n_i n'_k} -$$
$$\frac{1}{2} < \lambda_i n_i l | \lambda_k r^2 | \lambda_k n'_k l >] + < \lambda_i n_i l | V | \lambda_k n'_k l > \qquad (20)$$

where, as in eq. (14), it is

$$f_k = \frac{\lambda_k}{\lambda_0}; \qquad k = 1, 2 \qquad (21)$$

The overlap integrals (19) can be written in terms of gamma functions [50].

A method to diagonalise the hermitean system (18), corresponding to the representation of the Schrödinger operator in a non-orthogonal basis, is described in Appendix A.

Formation amplitude

Let us consider the ground state to ground state α-decay process for an even-even nucleus

$$B \to A + \alpha \tag{22}$$

We will assume that the mother nucleus (B) as well as the daugther nucleus (A) are spherically symmetric. Within the R-matrix approach [51, 13] the total α-decay width can be written as

$$\Gamma(R) = \sum_{L_\alpha=0}^{\infty} \gamma_{L_\alpha}^2(R) P_{L_\alpha}(R) \tag{23}$$

where the reduced width $\gamma_{L_\alpha}^2(R)$ is proportional to the formation amplitude $F_{L_\alpha}(R)$ squared, i.e.

$$\gamma_{L_\alpha}^2(R) = \frac{\hbar^2}{2M_\alpha} R |F_{L_\alpha}(R)|^2 \tag{24}$$

The L_α component of the formation amplitude is the overlap integral between the entrance and exit α-decay channels, i. e.

$$F_{L_\alpha}(R) = \int d\hat{R} d\xi_\alpha d\xi_A d\xi_B [Y_{L_\alpha}(\hat{R}) \Psi_\alpha(\xi_\alpha) \Psi_A(\xi_A)]^* \Psi_B(\xi_B) \tag{25}$$

where $\hat{R} = (\theta, \phi)$ are the angular center of mass coordinates and $\Psi_X(\xi_X)$ is the internal wave function of the nucleus X with internal coordinates ξ_X.

As usual [13] we write the internal α-particle wave function as a product of $n = l = 0$ h. o. states, i.e.

$$\Psi_\alpha(\xi_\alpha) = \Phi_{00}^{(\lambda_\alpha)}(\vec{r}_\pi) \Phi_{00}^{(\lambda_\alpha)}(\vec{r}_\nu) \Phi_{00}^{(\lambda_\alpha)}(\vec{r}_{\pi\nu}) \chi_{00}^{(\pi)}(s_1, s_2) \chi_{00}^{(\nu)}(s_3, s_4) \tag{26}$$

where Φ is the radial h. o. wave function with parameter $\lambda_\alpha = 0.513 fm^{-2}$ [13], χ the corresponding spin wave function and

$$\vec{r}_\pi = \frac{\vec{r}_1 - \vec{r}_2}{\sqrt{2}}, \quad \vec{r}_\nu = \frac{\vec{r}_3 - \vec{r}_4}{\sqrt{2}}, \quad \vec{r}_{\pi\nu} = \frac{\vec{r}_\pi - \vec{r}_\nu}{\sqrt{2}} \tag{27}$$

where we have labelled by 1, 2 the proton and by 3, 4 the neutron coordinates.

We will describe the mother and daughter nuclei within the BCS formalism [13]. That is,

$$\Psi_X(\xi_X) = \psi_{X\pi}(\xi_\pi) \psi_{X\nu}(\xi_\nu), \quad X = A, B \tag{28}$$

where

$$\psi_{X\tau}(\xi_\tau) = <\xi_\tau|BCS>_{X\tau}, \quad \tau = \pi, \nu \tag{29}$$

One can write the mother nucleus wave function assuming the daughter to be the core. Thus, in configuration space one gets, for protons

$$\psi_{B\pi} = \sum_{\alpha_\pi} \chi_{\alpha_\pi} \sum_{\Omega>0} \mathcal{A}[\psi_{\alpha_\pi \Omega}(1) \psi_{\alpha_\pi \overline{\Omega}}(2)] \psi_{A\pi} \tag{30}$$

and a similar expresion for the neutrons. Here

$$\psi_{\alpha_\pi \overline{\Omega}} = (-)^{j-\Omega} \psi_{\alpha_\pi -\Omega} \tag{31}$$

is the time reversed wave function. In the Fock space this corresponds to the following expansion

$$|BCS>_{B\tau} = \sum_{\alpha_\tau} \chi_{\alpha_\tau} \sum_{\Omega>0} a^\dagger_{\alpha\Omega} a^\dagger_{\alpha\overline{\Omega}} |BCS>_{A\tau} \qquad (32)$$

where $a^\dagger_{\alpha_\tau\Omega}$ are the creation operators corresponding to normal particles. The coefficients χ_{α_τ} can be written in terms of the BCS occupation amplitudes as

$$\chi_{\alpha_\tau} = {}_{B\tau}<BCS|a^\dagger_{\alpha_\tau\Omega} a^\dagger_{\alpha_\tau\overline{\Omega}}|BCS>_{A\tau} \cong U^{(A)}_{\alpha_\tau} V^{(B)}_{\alpha_\tau} \qquad (33)$$

By using eq. (15) one obtains for the 2π expansion

$$\psi_{B\pi} = \sum_{\alpha_\pi} \sum_{n_i n_k} B_\pi(\alpha_\pi n_i n_k) [\phi^{(\lambda_i)}_{n_i(lj)_\pi}(1) \phi^{(\lambda_k)}_{n_k(lj)_\pi}(2)]_0 \psi_{A\pi} \qquad (34)$$

where

$$B_\pi(\alpha_\pi n_i n_k) \equiv \frac{\hat{j}}{\sqrt{2}} \chi_{\alpha_\pi} c^{(i)}_{\alpha_\pi n_i} c^{(k)}_{\alpha_\pi n_k} \qquad (35)$$

A similar expansion is obtained for the neutron part

$$\psi_{B\nu} = \sum_{\beta_\nu} \sum_{n'_i n'_k} B_\nu(\beta_\nu n'_i n'_k) [\phi^{(\lambda'_i)}_{n'_i(lj)_\nu}(3) \phi^{(\lambda'_k)}_{n'_k(lj)_\nu}(4)]_0 \psi_{A\nu} \qquad (36)$$

where

$$B_\nu(\beta_\nu n'_i n'_k) \equiv \frac{\hat{j}}{\sqrt{2}} \chi_{\beta_\nu} c^{(i')}_{\beta_\nu n'_i} c^{(k')}_{\beta_\nu n'_k} \qquad (37)$$

One can perform all integrals over the integral coordinates in eq. (25) analitically by transforming to relative coordinates. Using the generalized Talmi-Moshinsky transformation with different masses [52] (i. e. with the different size parameters λ_1 and λ_2 in our case) one obtains for $L_\alpha = 0$

$$F_0(R) = \sum_{Niki'k'} \Phi^{(\Lambda^{i'k'}_{ik})}_{N0}(R) \sum_{nN_\pi N_\nu} <n0N0;0|N_\pi 0 N_\nu 0;0>_{D^{i'k'}_{ik}} \mathcal{I}^{(\lambda^{i'k'}_{ik},\lambda_\alpha)}_{n0} G^{(ik)}_{N_\pi} G^{(i'k')}_{N_\nu} \qquad (38)$$

where $G^{(ik)}_\pi (G^{(i'k')}_\nu)$ are geometrical coefficients depending on the proton (neutron) single particle parameters. Thus, for proton it is [16]

$$G^{(ik)}_{N_\pi} = \sum_{\alpha_\pi n_\pi} \sum_{n_i n_k} B_\pi(\alpha_\pi n_i n_k) <(l_\pi l_\pi)0(\frac{1}{2}\frac{1}{2})0;0|(l_\pi \frac{1}{2})j_\pi (l_\pi \frac{1}{2})j_\pi;0> \times$$

$$<n_\pi 0 N_\pi 0;0|n_i l_\pi n_k l_\pi;0>_{D_{ik}} \mathcal{I}^{(\lambda_{ik},\lambda_\alpha)}_{n_\pi 0} \qquad (39)$$

where B_π are the coefficients given by eq. (35). The first bracket denote $jj - LS$ recoupling coefficients while the second one the generalised Talmi-Moshinsky symbol depending on the ratio $D_{ik} = \lambda_i/\lambda_k$. The quantities $\mathcal{I}^{(\lambda_i,\lambda_\alpha)}_{n_\pi 0}$ are the overlap integrals of eq. (19), which now reads

$$\mathcal{I}^{(\lambda_{ik},\lambda_\alpha)}_{n_\pi 0} = <R^{(\lambda_{ik})}_{n_\pi 0}|R^{(\lambda_\alpha)}_{00}> \qquad (40)$$

with the relative h.o. parameter corresponding to one particle moving in the potential i and the other in the potential k, i. e.

$$\lambda_{ik} = \frac{\lambda_i \lambda_k}{\Lambda_{ik}} \qquad (41)$$

and the corresponding center of mass parameter

$$\Lambda_{ik} = \frac{1}{2}(\lambda_i + \lambda_k) \tag{42}$$

The radial c.m. wave function $\Phi_{N_\alpha L_\alpha=0}^{(\Lambda_{ik}^{i'k'})}(R)$ corresponds to an h.o. potential with the relative and c.m. parameters given by

$$\lambda_{ik}^{i'k'} = \frac{\Lambda_{ik}\Lambda_{i'k'}}{\Lambda_{ik}^{i'k'}} \tag{43}$$

$$\Lambda_{ik}^{i'k'} = \frac{1}{2}(\Lambda_{ik} + \Lambda_{i'k'}) \tag{44}$$

The overlap integrals $\mathcal{I}_{n_\alpha 0}^{\lambda_{ik}^{i'k'}}$ are defined in a similar way as above with the generalised Talmi-Moshinsky coefficients depending on the mass ratio

$$D_{ik}^{i'k'} = \Lambda_{i'k'}/\Lambda_{ik} \tag{45}$$

Since in the applications it may be interesting to know in which potential a given particle is moving, we will clearly specify all the possible combinations in the summation (38). There are five terms, they are

1) $i = k = 1$, $i' = k' = 1$;
2) $i = k = 2$, $i' = k' = 2$;
3) $i = k = 1$, $i' = k' = 2$;
   $i = k = 2$, $i' = k' = 1$;
   $i \neq k$, $i' \neq k'$;
4) $i = k = 1$, $i' \neq k'$;
   $i \neq k$, $i' = k' = 1$;
5) $i = k = 2$, $i' \neq k'$;
   $i \neq k$, $i' = k' = 2$;

The formation amplitude can then be written as

$$F_0(R) = \sum_{k=1}^{5} F_0^{(k)}(R) \tag{46}$$

We will refer to this summation in the applications below, but from the outset one may expect that properties mainly induced by the motion of the particles inside the nuclear volume would be determined by the therm $F_0^{(k=1)}(R)$.

Diagonalisation procedure and numerical results

In this section we will apply the method developed above to describe the alpha decay of ^{220}Ra. This nucleus, as well as the daughter nucleus ^{216}Rn, has a quadrupole deformation $\beta_2 \cong 0.1$. As it was shown in ref. [18] for this value of the deformation the inclusion of the non-spherical terms in the estimation of the microscopic formation amplitude change the final result (total decay width) only by a few percent. It is therefore important to analyse the extent to which a description of this case in terms of a spherical mean field is valid.

For the Woods-Saxon mean field we take the so called universal parametrisation [44].

We use the value $N_0 = 6$ for the principal quantum number defining the limit between the two kinds of terms in the expansion (15). We will perform the calculation by using a total of 11 shells. For the h. o. parameter that is supposed to describe the discrete part of the spectrum we choose the value $f_1=1.2$ (see eq. 14). This part of the spectrum should be insensitive to the value of the h. o. parameter defining the quasicontinuum part, i. e. to f_2. One indeed sees in figure 2 that this is the case.

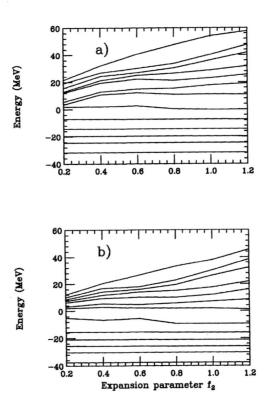

Figure 2: Single particle levels in ^{220}Ra as a function of the harmonic oscillator parameter f_2 for $f_1 = 1.2$. The basis consists of the shells $N_1 = 1-6$ (corresponding to f_1) and $N_2 = 7-11$ (corresponding to f_2). We call this "minimal basis". a) Protons, b) Neutrons.

This is an important conclusion because it implies that the orthogonalization procedure, which mixes all states, will not affect the discrete part of the spectrum. That is, one knows well the h. o. parameters that provide a good representation to describe bound properties in the nucleus. Our procedure, which is intended to allow the calculation of processes in the continuum, increases the dimension of the representation, but without affecting the calculated bound properties.

One also sees in figure 2 that the high energy part of the spectrum is strongly dependent upon f_2. For $f_2 = f_1=1.2$ the spectrum corresponds to the one used in standard calculations, i. e. by using one h. o. potential only. As f_2 decreases the level density increases and the representation is, therefore, more suited to describe the quasi -

continuum spectrum. That is, by choosing a small h.o. parameter λ_2 the representation becomes more adequate to describe the wave functions at large distances. But the drawback of using a small value for λ_2 is that the dimension of the representation may become big and the calculation cumbersome, a feature that we want to avoid. That would happen if e.g. one wants to include all possible shells up to a given energy. Yet one can find a value of f_2 that greatly improves the calculation while the number of shells needed is less than the one used within only one h. o. potential, as will be shown below.

The width (23) consists of two very distinct parts, namely the formation probability and the penetration through the Coulomb barrier. Deformations are very important regarding the penetration through the Coulomb barrier, but the formation amplitude corresponding to ground state to ground state decays is not dependent on the deformation parameters [18]. We therefore will calculate in this section the formation amplitude neglecting deformations. The effects of deformations on the penetration will be shown at the end.

In order to evaluate the formation amplitude we have first to perform the BCS calculation that defines the intrinsic wave functions of the nuclei involved in the decay. We estimated the BCS parameters by using the third order mass difference relation [53], which gives for ^{220}Ra: $\Delta_p = 1.091\ MeV$, $\Delta_n = 0.914\ MeV$ and for ^{216}Rn: $\Delta_p = 1.066\ MeV$, $\Delta_n = 0.926\ MeV$. We then computed the occupation amplitudes U and V for each system as a function of the parameter f_2.

It is worthwhile to stress here that a large scale shell model calculation would be the best proper way to include a large configuration mixing. Anyway this is almost impossible when many active particles are considered. That's why the BCS method has been widely used since the pionering work by Soloviev [5]) in the alpha decay calculations. In addition we have shown [8, 10] how the mixing induced by pairing interaction at the level of BCS approximation, for spherical as well as for deformed nuclei, is able to induce a large enhancement of clustering properties in heavy sperical or deformed nuclei. In particular we have arrived to the conclusion that the effective inclusion of the nuclear correlations (particularly the proton-neutron interaction) through the procedure of fitting the strength to reproduce the experimental pairing gap is somehow a proper way of including the very complicated correlations that induce clustering in nuclei [31].

The idea of this calculation was to find a minimal basis that describes well the calculated width. As mentioned above, f_1 was fixed to have the value $f_1=1.2$ and, therefore, the possible different bases are determined by the parameter f_2. But let us stress once more that the results do not (and cannot) depend on the value that one chooses for f_2. If f_2 is big then one needs many shells to reach the final values of the calculated quantities, as expected [18]. We found that the minimal basis is given by the choice $f_2 = 0.7$. The basis consists then of the h. o. shells $N_1 = 0 - 6$ corresponding to the parameter $f_1 = 1.2$ and the shells $N_2 = 7 - 11$ corresponding to $f_2 = 0.7$. Considering that the basis provided by only one harmonic oscillator potential requires at least 18 major shells to reproduce the experimental results [18] one can say that our basis is very small.

We performed the calculation within the minimal basis around the touching radius which is defined by

$$R_c = 1.2\ (A_{^{216}Rn}^{1/3} + A_{^4He}^{1/3}) = 9.15\ fm \tag{47}$$

We found that, among the different terms entering the formation amplitude (46) as a function of the c.m. radius R, the contribution corresponding to the discrete part of the spectrum (k=1) is peaked on the nuclear radius while the mixed contribution between discrete and quasicontinuum parts (k=3) is centered on the touching radius.

An important test for the calculation is that it should be independent upon the distance R outside the nuclear surface. This is actually the case. The main contribution to the total width is provided by the term in eq. (46) that contributes most to the formation amplitude around the touching radius, i. e. the term $k = 3$. The contribution from the inner part of the nucleus (k=1) is very small. The total contribution is indeed practically independent upon the distance for $R \geq 9 fm$, i.e. beyond the touching radius. For smaller radius the width is very small. This is because we have neglected antisymmetrisation effects between the daughter nucleus and the α-cluster [54, 55, 56] but this does not affect our results because we are able to compute the formation amplitude well beyond the touching point, where those effects are negligible [46]. This is the main reason why it is important to have a reliable theory capable of describing the decay process at large distances.

It is interesting to analyse the reason why the minimal basis reproduces so well the experimental data, although it is relativelly a very small basis. We thus show in figure 3.a the ratio between the calculated width and the corresponding experimental value as a function of the parameter f_2. One sees that the calculated width reaches the experimental value just in a region of f_2 values around $f_2=0.7$. Outside this region the theory does not reproduce the experimental value within the shells included in the minimal basis. That is, outside this region the basis does not span the Hilbert subspace that contains the functions determining the alpha decay process around the touching radius.

In figure 3.b one see that increasing the basis dimension one gets saturation.

One also observes in figure 3 that the main contribution to the formation amplitude is provided by the term $F_0^{(k=3)}$ in the expansion (46) (plotted by dots), while the contribution from the discrete part (k=1, plotted by a dashed line) is very small.

One concludes that it is not possible to reproduce the experimental value in a region around the touching point with a less number of major shells for any value of the expansion parameter because then the maximum would lie under the experimental value.

Another quantity that can be calculated with our formalism is the spectroscopic factor

$$S = \int_0^\infty |R_0(R)|^2 R^2 dR \qquad (48)$$

The result of the calculation for this decay is $S = 7.0 \; 10^{-2}$, which is in agreement with a calculation where the Pauli principle acting among the particles in the core and in the α-particle was taken into account properly [55].

It is also interesting to analyse the influence of the pairing correlations on the decay process. We found that the calculated results are very sensitive to the BCS parameters, as shown in figure 4. In this figure the dependence of the ratio between the calculated and experimental widths as a function of the gap parameter is presented (full line). One sees that the total width can change by several orders of magnitude by changing Δ within "reasonable" limits. It is thus remarkable that choosing the Δ value prescribed by spectroscopic calculations, in agreement with experimental data one gets just the right width.

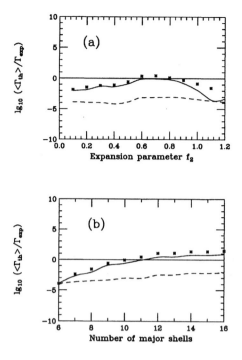

Figure 3: Ratio between the theoretical and experimental decay width calculated as in figure 3 as a function of (a) the parameter f_2 and (b) the number of shells N_2.

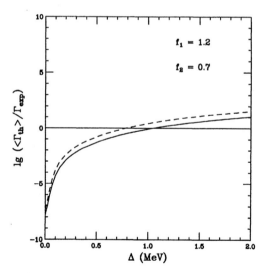

Figure 4: Dependence of the averaged width on the BCS gap parameter corresponding to an spherical barrier (solid line) and a deformed barrier (dashed line) for the ground state to ground state decay of ^{220}Ra.

The calculation of the formation amplitude was performed by neglecting deformations. These are only important regarding the penetration through the Coulomb barrier. This is also shown in figure 4 (dashed line), where one sees that the calculated decay width, for a quadrupole deformation $\beta_2 = 0.3$, increases by a factor of 5, practically independently of the BCS parameters. Again in this case, it is remarkable that one gets the experimental width by just taken the deformation parameters extracted from spectroscopic studies.

Alpha decay systematics

In this section we will analyse ground state to ground state alpha decays from even-even nuclei by using the minimal basis discussed in the previous section, i.e. by choosing the h.o. shells according to

$$f_1 = 1.2, \quad N_1 = 0-6; \quad f_2 = 0.7, \quad N_2 = 6-11 \tag{49}$$

for the Woods-Saxon potential we adopted again the universal set of values [44]. As in the previous section, the BCS gap parameters were calculated according to the third order mass formula [53]. Even if the dependence of the width upon the distance around the touching point is weak, we will present the calculated values averaged on the interval $R = 8 - 10 fm$, i.e. around the touching point.

We have calculated the total α-decay width corresponding to 26 cases of even-even heavy nuclei, i. e. above the ^{208}Pb in the periodic table. The aim of this calculation is to probe the validity of the conclusions reached in the previous section for a large number of nuclei as well as to analyse the effect of deformations on the decay process. For this we computed the total width by using a spherical as well as a deformed Coulomb barrier for the penetration problem. In both cases we applied the semiclassical (WKB) approximation, which is known to be very good in alpha decay [18, 23]. The deformation parameters were taken from ref. [57] and the experimental values of the widths from refs. [32, 58]

In table 5 are given the results of the calculation. Considering that this is a complete microscopic calculation without any free parameter, one can say that the agreement between theory and experiment is excellent.

One important feature that can be seen in this table is the contributions of deformations to the decay widths. As expected, when the deformation increases the difference between the calculated spherical and deformed widths also increases, but the contribution of deformations always improves the calculation, sometimes by large factors, as in the decay of ^{244}Pu.

The effects of deformations shown in table 5 can perhaps better be seen in figure 5, where the ratio between theoretical and experimental width is presented for the different nuclei of table 5. Open circles correspond to calculations performed within a spherical framework while the stars are the values calculated with deformations. One again sees in this figure that the agreement between theory and experiment is excellent if all the ingredients, including deformations, are properly included in the calculation. The mean value of the ratio between the theoretical and experimental widths is 0.79, with a standard deviation of 0.21. For comparison, one can mention that neglecting deformations (i. e. the "spherical" values in table 5) the average value of the ratio is 0.41 with the same standard deviation, i. e. 0.21.

Table 5: Absolute α-decay widths calculated in a spherical BCS model with a two-size oscillator basis and with barriers assumed spherical (Γ_{sph}) and deformed (Γ_{def}) compared with the experimental data (Γ_{exp}).

Nucleus	β_2 [57]	Γ_{sph} (MeV)	Γ_{def} (MeV)	Γ_{exp} (MeV) [58, 32]	$\Gamma_{sph}/\Gamma_{exp}$	$\Gamma_{def}/\Gamma_{exp}$
^{212}Po	0.00	1.7×10^{-15}	1.7×10^{-15}	1.5×10^{-15}	1.14	1.14
^{214}Po	0.00	1.5×10^{-18}	1.5×10^{-18}	2.9×10^{-18}	0.53	0.53
^{216}Po	0.00	2.0×10^{-21}	2.0×10^{-21}	3.0×10^{-21}	0.68	0.68
^{218}Po	0.00	1.3×10^{-24}	1.3×10^{-24}	2.4×10^{-24}	0.56	0.56
^{216}Rn	0.07	4.8×10^{-18}	5.0×10^{-18}	1.0×10^{-17}	0.47	0.50
^{218}Rn	0.09	6.7×10^{-21}	7.3×10^{-21}	1.3×10^{-20}	0.51	0.56
^{220}Rn	0.13	4.5×10^{-24}	5.6×10^{-24}	8.1×10^{-24}	0.55	0.69
^{222}Rn	0.14	6.1×10^{-28}	8.3×10^{-28}	1.4×10^{-27}	0.44	0.60
^{220}Ra	0.11	1.5×10^{-20}	1.8×10^{-20}	2.0×10^{-20}	0.79	0.91
^{222}Ra	0.19	5.4×10^{-24}	9.1×10^{-24}	1.2×10^{-23}	0.46	0.77
^{224}Ra	0.18	5.2×10^{-28}	8.4×10^{-28}	1.4×10^{-27}	0.37	0.61
^{226}Ra	0.20	2.6×10^{-33}	5.0×10^{-33}	8.6×10^{-33}	0.30	0.58
^{224}Th	0.21	1.4×10^{-22}	2.5×10^{-22}	3.5×10^{-22}	0.39	0.71
^{226}Th	0.23	4.7×10^{-26}	1.0×10^{-25}	1.8×10^{-25}	0.26	0.56
^{228}Th	0.23	1.7×10^{-30}	3.9×10^{-30}	5.5×10^{-30}	0.30	0.71
^{230}Th	0.24	5.0×10^{-35}	1.4×10^{-34}	1.5×10^{-34}	0.34	0.94
^{232}Th	0.26	2.2×10^{-40}	7.6×10^{-40}	8.0×10^{-40}	0.27	0.94
^{230}U	0.26	7.1×10^{-29}	2.1×10^{-28}	1.7×10^{-28}	0.42	1.24
^{232}U	0.26	4.6×10^{-32}	1.5×10^{-31}	1.4×10^{-31}	0.32	1.04
^{234}U	0.27	1.1×10^{-35}	4.0×10^{-35}	4.2×10^{-35}	0.26	0.96
^{236}U	0.28	9.8×10^{-38}	4.1×10^{-37}	4.6×10^{-37}	0.22	0.89
^{238}U	0.28	3.7×10^{-40}	1.6×10^{-39}	2.5×10^{-39}	0.14	0.65
^{238}Pu	0.28	3.5×10^{-32}	1.4×10^{-31}	1.2×10^{-31}	0.30	1.21
^{240}Pu	0.29	3.1×10^{-34}	1.3×10^{-33}	1.6×10^{-33}	0.19	0.82
^{242}Pu	0.29	5.4×10^{-36}	2.5×10^{-35}	3.0×10^{-35}	0.18	0.82
^{244}Pu	0.29	2.9×10^{-38}	1.4×10^{-37}	1.4×10^{-37}	0.20	0.97

The good agreement between theory and experiment shown in table 5 and figure 5 has to be consider in the context not only of the dependence of the width upon deformations (shown there) but also on its strong dependence upon the Q-value [13] and the BCS parameters, as seen in figure 4. All these quantities can be very different in the rather disparate collection of nuclei of table 5.

Applications to exotic cluster decay

The discovery of heavy-cluster decay [59] triggered a febrile theoretical activity. Many models, both phenomenological and microscopical, were proposed trying to understand the mechanisms that induce the clustering of the nucleons that eventually become the cluster as well as to predict likely candidates for experimental searches. Phenomenological models were rather successful in this [60, 61]. One of the simplest

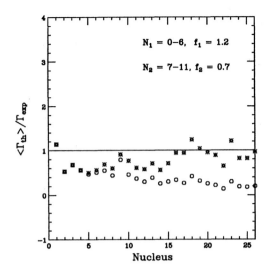

Figure 5: Ratio between the theoretical and experimental decay widths for spherical (circles) and deformed (stars) barriers corresponding to the decays in table 1. The parameters N and f defining the minimal basis are also given.

phenomenological models describes the penetration of a preformed cluster in a square well through the Coulomb barrier [62]. It reproduces all ground state to ground state partial half-lives of even-even nuclei within a factor of three. The effect of the nuclear ground state deformation on cluster decay was for the first time discussed in ref. [63]. Together with shell attenuation the effect is estimated to modify the absolute half-lives up to an order of magnitude. First attempts for a microscopic approach to the exotic cluster decay are already available. We mention, for instance, the approach of ref. [64] for the calculation of the hindrance factors for ^{34}Si emission in the framework of the R-matrix theory for axially deformed nuclei. As a new ingredient, in ref. [64], it was used a four-body interaction term connected with alpha-like staggering in binding energies. The total decay width of ^{14}C was also recently microscopically calculated [19, 25]. The theoretical results disagree with experiment by $1-3$ orders of magnitude even after including up to 18 major shells in the basis [19, 25].

In the ground state to ground state decay of α-particles from deformed nuclei it was found that deformations influence mostly the penetration through the Coulomb barrier. The formation amplitude can be evaluated considering the nuclei involved as spherically symmetric. This approximation, which will be used here, affects the results within $10-30\%$ only [18].

In order to calculate the wave function of the mother nucleus we will use the multistep shell-model method (MSM) [65]. In this method one calculates a many-body system in several steps. In each step one includes all the interactions and Pauli effects properly. For instance, to evaluate a nucleus consisting of a core plus an equal number of pairs of neutrons and protons, one calculates first the core plus two-neutrons and two-protons (core plus an "α-particle"). In the second step one takes the core as the system calculated previously and adds a second "α-particle" and apply the formalism again to evaluate the original core plus two "α-particles". One thus continues in successive steps up to the final system. Within this approximation, and including the pairing correlations through the BCS formalism, we calculated the first step as in the case of α-decay [18]. For the second step, i. e. daughter nucleus plus two alpha-like excitations, we proceeded in the same fashion. Finally, we repeated this procedure for the third step thus reaching the mother nucleus corresponding to the decay of ^{12}C. For the case of the decay of ^{14}C we added a neutron pair.

The internal wave functions of the $^{12,14}C$ clusters were taken as in ref. [19].

For the mean field we adopted, again, the so-called universal parametrization [44] of the Woods-Saxon potential. We then expanded the single particle wave functions in terms of the corresponding single-particle wave functions in the minimal basis, as in eq. (15). With standard transformations to center of mass (c.m.) and relative motion and performing the integrals analytically one gets for the formation amplitude the expression

$$F_{^{14}C}(\vec{R}) = \sum_{Nik} \Phi_{N0}^{(\lambda_{ik})}(\vec{R}) F_N(i,k) \tag{50}$$

where

$$F_N(i,k) = \sum_{nn_{12}n_3 N_{\alpha_1} N_{\alpha_2} N_{\alpha_3} N_{\nu_4}} <n0N0;0|N_{12}0N_30;0> \mathcal{I}_{n2}^{(\lambda_{ik},\lambda_{^{14}C})}$$

$$\times <n_{12}0N_{12}0;0|N_{\alpha_1}0N_{\alpha_2}0;0> \mathcal{I}_{n_{12},2}^{(\lambda_{ik},\lambda_{^{14}C})} <n_30N_30;0|N_{\alpha_3}0N_{\nu_4}0;0> \mathcal{I}_{n_3,1}^{(\lambda_{ik},\lambda_{^{14}C})}$$

$$\times F_{N_{\alpha_1}}(i,k,i,k) F_{N_{\alpha_2}}(i,k,i,k) F_{N_{\alpha_3}}(i,k,i,k) G_{N_{\nu_4}}^{(ik)} \tag{51}$$

In this equation all angular momenta are zero because the different MSM steps proceed through ground states. This is also consistent with the internal wave function of the cluster [66]. The principal quantum numbers in the summations denote the motion of the different blocks used in the MSM. Thus, e.g., n_{12} and N_{12} refer to the relative and c. m. motion of the alpha-like blocks 1 and 2 while the numbers n_3 and N_3 refer to the relative and c. m. motion of the neutron pair with respect to the alpha-like block 3. This notation was also used in refs. [19, 66]. The indices i and k label the h.o. potentials that define the minimal basis in which the particles move. For instance, $\lambda_{i=1,k=2}$ is the size parameter of the h. o. potential corresponding to the relative motion of one particle moving in the potential defined by $f_1 = 1.2$ and the other particle in the h. o. potential defined by $f_2 = 0.7$, where f_i is defined by $\lambda_i = f_i M_0 \omega/\hbar$, with $\hbar\omega = 41A^{-1/3}$. One has to use generalized Talmi-Moshinsky brackets [52].

Since the minimal basis is not an orthogonal basis, different components of the basis elements may overlap with each other. This overlap is expressed by the quantities $\mathcal{I}$. Thus $\mathcal{I}_{n_3,1}^{(\lambda_{ik},\lambda_{^{14}C})}$ indicates the overlap between the wave function of a pair of nucleons moving in the h. o. given by λ_{ik} with principal quantum number $n = n_3$ and the wave function of the same pair of nucleons moving in the cluster, i.e. in the h. o. with size parameter $\lambda_{^{14}C}$ and $n = 1$ [66]. The same is valid for the other overlap integrals. The coefficients G are defined as in ref. [18]. They contain geometrical factors and the BCS amplitudes, which we evaluated by using the mass difference formula for the gap parameters [53]. Finally, $\Phi_{N0}^{(\lambda_{ik})}(\vec{R})$ is the h.o. wavefunction (with radial part as in eq. (15)) with size parameter λ_{ik}.

We calculated the penetration through the deformed Coulomb barrier semiclassically within the WKB approximation, as usual (see e.g. [18]). The quadrupole deformations were taken from ref. [57].

Integrating over the angles we calculated the absolute decay width and the spectroscopic factor, defined as before.

We found in all cases of table 6 that the widths are only weakly dependent upon distance in the range $R = 10fm - 13fm$, i. e. beyond the touching point, confirming the soundness of this approach. One sees that the influence of deformations can be very important. With the exception of ^{222}Ra, the theoretical and experimental absolute

Table 6: Absolute ^{14}C-decay widths calculated in a spherical BCS model combined with a multistep shell model and a two-size oscillator basis. The widths Γ_{sph} and Γ_{def}, respectively, were calculated with spherical and deformed barriers. For comparison included are also the absolute experimental ^{14}C-widths (Γ_{exp}) and the theoretical ^{14}C-widths ($\Gamma_{^{14}C} \equiv \Gamma_{def}$) divided by the respective α-widths. The relative spectroscopic factors $S_{^{14}C}/S_\alpha$ are of the order of 10^{-7}.

Mother nucleus	β_2	Γ_{sph} (MeV)	Γ_{def} (MeV)	Γ_{exp} (MeV)	$(\Gamma_{^{14}C}/\Gamma_\alpha)_{exp}$	$(\Gamma_{^{14}C}/\Gamma_\alpha)_{th}$
^{222}Ra	0.19	1.9×10^{-35}	2.1×10^{-34}	4.1×10^{-33}	3.4×10^{-10}	2.3×10^{-11}
^{224}Ra	0.18	1.2×10^{-39}	1.4×10^{-38}	6.3×10^{-38}	4.5×10^{-11}	1.7×10^{-11}
^{226}Ra	0.20	3.8×10^{-44}	1.1×10^{-42}	2.9×10^{-43}	3.3×10^{-11}	2.2×10^{-10}

decay widths agree within a factor of five. The corresponding values relative to α-decay are also reasonable.

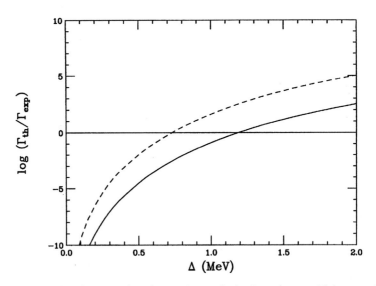

Figure 6: Ratio of the theoretical and experimental absolute decay widths as a function of the gap parameter corresponding to a spherical Coulomb barrier (solid line) and a deformed barrier with a quadrupole deformation $\beta_2 = 0.3$ (dashed line) for $^{224}Ra \longrightarrow ^{210}Pb + ^{14}C$.

The spectroscopic factors are within the accepted order of magnitude [55]. It is worthwhile to mention that the main contribution to this spectroscopic factor is provided by the first N_0 major shells, i.e. from the bound states of the minimal representation ($i = 1$ in eq. (50)), while in the absolute decay width the main contribution is provided by shells with $N > N_0$ ($i = 2$), i. e. from high-lying configurations in the representation. Although expected, this is an important feature because the spectroscopic properties (e. g. energies and electromagnetic transition rates) are practically not affected by the high lying configurations.

It is also interesting to analyse the dependence of the results on the nucleon-nucleon interaction. In fig. 6 we show the ratio of the theoretical and experimental absolute decay width as a function of the pairing gap parameter. For each value of Δ we have calculated the penetrability assuming a spherical barrier (solid line) and a deformed one (dashed line) with a quadrupole deformation $\beta_2 = 0.3$. for the ^{14}C-decay from ^{224}Ra.

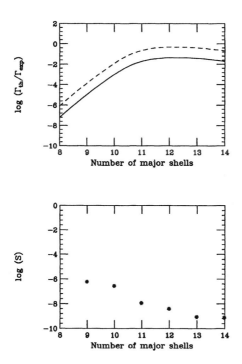

Figure 7: (a) Ratio between the calculated and experimental decay widths as a function of the number of major shells included in the calculation for the decay of ^{14}C from ^{224}Ra in the spherical case (solid line) and in the deformed case (dashed line); (b) As figure 7.a) for the spectroscopic factor.

One sees that the influence of the deformation is nearly independent of the gap parameter, but strongly dependent on the cluster charge, i. e. two orders of magnitude larger for the decay of ^{14}C-decay as compared to α-decay [49]. Even the dependence of the width upon the gap parameter is stronger for the case of the decay of ^{14}C. It is therefore remarkable that one gets results that agree reasonably well with experiment by just taking the gap and deformation parameters from independent spectroscopic studies.

Our theoretical widths are very smooth functions of the distance R between the cores [18]. It is also important to attest to that that there is a "saturation" value for the width as the number of shells taken into account in the calculation increases. In figure (7.a) the ratio between the calculated width and the corresponding experimental value for the decay of ^{14}C from ^{224}Ra is presented as a function of the total number of shells (i. e. $N_1 + N_2$, see Eq. 15). The solid line corresponds to a spherical barrier while the dashed line corresponds to a deformed one. Figure (7.a) shows that in both cases there is a saturation around 12 major shells. The results of fig. 7 were obtained using a fixed pairing gap. One also sees in this figure that the deformation changes the width by about one order of magnitude. This is a very interesting feature since it shows the suitability of cluster decay processes to determine nuclear deformations.

It is important to mention that the saturation seen in figure (7.a) takes place at about the same value of the numbers of shells as in the case of alpha decay [49].

In figure (7.b), where the spectroscopic factor S is plotted versus the number of major shells, one also sees a clear convergence of S at about 13 shells.

Conclusions

The calculation of absolute α-decay widths is a very difficult undertaking because either one calculates it at a short distance, as for spectroscopic quantities, where the Pauli principle acting among the particles in the core and in the α-particle is important, or one calculates it at large distances, where standard shell model representations are inadequate to describe the decay process. In this paper we have attacked this problem by choosing as representation a basis provided by two different harmonic oscillator potentials. The lowest shells of the representation are taken from an h.o. potential that describes well spectroscopic properties, i. e. distances up to around the nuclear surface. This is the h.o. potential used generally in spectroscopic calculations, i.e. the one that fits best the corresponding Woods-Saxon potential (see figure 1). The highest shells are taken from an h.o. potential that is rather shallow, thus with a greater density of states in the quasicontinuum, i. e. suited to describe distances outside the nuclear surface. The corresponding basis is formed by the shells $N_1 = 0 - 6$ obtained from the h.o. potential with parameter $f_1 = 1.2$ (see eq. (21) and the shells $N_2 = 7 - 11$ obtained from the h.o. potential with $f_2 = 0.7$. In other words, one can conclude that we have used a realistic single particle basis. The good asymptotic behavior of the single particle wave function means a non exponential decreasing in the tail (for radial distances larger than the nuclear radius). The location of the last maximum for the s.p. radial wave function is found at a point where

$$N\hbar\omega = V(r) \qquad (52)$$

so that if the h. o. parameter λ is smaller this radius is pushed far away. If this is the case, the previous equation is satisfied for a larger radius and smaller N for a shallower h.o. potential. The coherent superposition coming from the large mixing of different contributions in the region behind the last maximum produces a non exponential decay of the final α formation amplitude [35, 17].

Within this representation one is able to calculate the formation amplitude of the α-particle at large distances, thus avoiding the formidable task of considering exactly the Pauli principle among the core and the α-particle. Besides, one achieves this with a rather small basis, since it is just constructed to fit the decay process. This basis is non-orthogonal, but one can use any of the methods available to treat such basis. One of such methods, which is the one used in this paper, is presented in appendix A.

In our calculations we have used for the mean field standard Woods-Saxon potentials with parameters that we have taken from independent spectroscopic calculations. We included the pairing correlations among the particles as prescribed by standard BCS calculations and, therefore, there is not any free parameter in the calculated absolute decay widths.

With these prescriptions we calculated a rather large number of α-decay widths in heavy nuclei. To probe the validity of the calculations, we have checked in all cases that the widths are only weakly dependent upon the distances between the cores and the α-particle outside the nuclear surfaces, which in the calculated nuclei is between 8 fm and 10 fm. The calculated values of the widths agree well with the corresponding experimental values. The mean value of the ratio between the theoretical and experimental decay widths for all calculated cases is 0.79 with a standard deviation of 0.21.

We also calculated spectroscopic factors, which are in good agreement with other calculations, where the Pauli principle was taking into account properly [55]. In this context, it is important to notice that the basis used in this paper (11 major h.o. shells) is rather small. For comparison one may mention that similar calculations using only one h.o. potential required the inclusion of 18 major shells [18]. This small basis may make microscopic calculations of more complicated decays (as the decay of heavy clusters) feasible.

We found that the main contribution to the total width is provided by that part of the formation amplitude centered around the touching point, i.e. the term connecting the discrete and quasicontinuum parts of the s.p. spectrum. We also found that deformations are important to describe properly the decay process.

We have also shown that, within the framework of the R-matrix theory, quantities corresponding to the decay of ^{14}C from Ra isotopes can be easily calculated using the new single particle basis. The single-particle representation consists of the eigenvectors of two different harmonic oscillator potentials adjusted to reproduce the bound states and the quasi-continuum part of a realistic Woods-Saxon potential. The calculation of the mother-nucleus wave functions was performed within the multistep shell-model method, including pairing correlations through the BCS approximation. The calculated absolute decay widths agree reasonable well with the corresponding experimental data. The observed discrepancies are explained by the peculiar behavior of the penetration factor in the WKB approximation. We found that both the pairing correlations and the quadrupole deformation play an important role in exotic cluster decay. As an important conclusion we stressed that the calculated quantities converge to a definite value as the number of major shells is increased.

APPENDIX A: Diagonalisation procedure in a non-orthogonal basis

In this Appendix we will remind the method to solve the eigenvalue problem for an hermitean matrix in a non-orthogonal basis(18) Let us consider the general eigenvalue problem

$$H\Psi_k = E_k \Psi_k \qquad (A.1)$$

where the eigenfunction Ψ_k, cooresponding to the eigenvalue E_k, is expanded in a non-orthogonal basis ϕ_n

$$\Psi_k = \sum_n C_{kn}^T \phi_n = \sum_n C_{nk} \phi_n \qquad (A.2)$$

T denoting the matrix transposition. Insertig (A.2) into eq. (A.1), multiplying to left by ϕ_m and integrating one obtains

$$\sum_n <\phi_m|H|\phi_n> C_{kn}^T = E_k \sum_n C_{kn}^T <\phi_m|\phi_n> \qquad (A.3)$$

This is just the general eigenvalue problem to be solved given by eq. (2.13), with the metric matrix

$$I_{mn} \equiv <\phi_m|\phi_n> \qquad (A.4)$$

Let us first find the eigenvalues F_j and eigenvectors Y_{lj} of the symmetric metric matrix

$$\sum_l I_{ml} Y_{lj} = F_j Y_{mj} \qquad (A.5)$$

where the orthonormality conditions for the eigenvectors holds

$$\sum_i Y_{mi} Y_{in}^T = \delta_{mn} \qquad (A.6)$$

The system of functions

$$\psi_i = \sum_k \frac{Y_{ik}^T}{\sqrt{F_i}} \phi_k \qquad (A.7)$$

is orthonormal, because from (A.5) and (A.6) one obtains

$$<\psi_i|\psi_j> = \sum_k \frac{Y_{jk}^T}{\sqrt{F_i}} \sum_l I_{kl} \frac{Y_{lj}}{\sqrt{F_j}} = \sum_k \frac{Y_{ik}^T}{\sqrt{F_i}} \sqrt{F_j} Y_{kj} = \delta_{ij} \qquad (A.8)$$

Expanding the initial eigenfunctions Ψ_k (A.2) in terms of this basis

$$\Psi_k = \sum_i X_{ki}^T \psi_i \qquad (A.9)$$

one obtains an eigenvalue problem

$$\sum_i <\psi_l|H|\psi_i> X_{ki}^T = E_k X_{ki}^T \qquad (A.10)$$

for the symmetric matrix

$$<\psi_l|H|\psi_i> = \sum_{mn} \frac{Y_{lm}^T}{\sqrt{F_l}} <\phi_m|H|\phi_n> \frac{Y_{ni}}{\sqrt{F_i}} \qquad (A.11)$$

Using (A.7) and (A.9) one finally obtains the following expression for the expansion coefficients in (A.2)

$$C_{kn}^T = \sum_i X_{ki}^T \frac{1}{\sqrt{F_i}} Y_{in}^T \qquad (A.12)$$

Here the eigenvalues F_i and the eigenvectors Y_{in}^T are solutions of the system (A.5),(A.6) and the eigenvectors X_{ki}^T the solutions of (A.10) and (A.11).

References

[1] A. Bohr, B. R. Mottelson and D. Pines, Phys. Rev. 110 (1958) 936.

[2] A.M. Lane, Nuclear Theory; Benjamin (New York - 1964).

[3] A. Bohr and B. R. Mottelson: Nuclear Structure, Vol. 2 (Benjamin, New York, 1975).

[4] D.R.Bes, R.A.Broglia, O.Hansen and O.Nathan, Phys. Rep. **34C** (1977) 1

[5] V. G. Soloviev, Phys. Lett. **1** (1962) 202

[6] H. J. Mang and J. O. Rasmussen, Kgl. Danske Videnskab. Selskab, Mat.-Fys. Skifter 2 (1962) No. 3.
J. O. Rasmussen, in: Alpha-, Beta- and Gamma-Ray Spectroscopy, ed. K. Siegbahn (North-Holland, Amsterdam, 1965) Vol. I, p. 701.

[7] F. A. Janouch and R. Liotta, Phys. Rev. **C27** (1983) 896

[8] F. Catara, A. Insolia, E. Maglione and A. Vitturi, Phys. Rev. C 29 (1984) 1091.

[9] W. T. Pinkston, Phys. Rev. C29 (1984) 1123.

[10] F. Catara, A. Insolia, E. Maglione and A. Vitturi, Phys. Lett. **149B** (1984) 41

[11] E. Roeckl, Radiochim. Acta 70/71 (1995) 107;
E. Roeckl, in: Nuclear Decay Modes, ed. D. N. Poenaru and W. Greiner (Institute of Physics, Bristol) in press.

[12] N. Rowley, G. D. Jones and M. W. Kermode, J. Phys. G 18 (1992) 165;
T. L. Stewart, M. W. Kermode, D. J. Beachey, N. Rowley, I. S. Grant and A. T. Kruppa, Phys. Rev. Lett. 77 (1996) 36;
T. L. Stewart, M. W. Kermode, D. J. Beachey, N. Rowley, I. S. Grant and A. T. Kruppa, Nucl. Phys. A, in press.

[13] H. J. Mang, Ann. Rev. Nucl. Sci. **14** (1964) 1;
H. J. Mang and J. O. Rasmussen, Mat. Fys. Skr. Dan. Vid. Selsk. **2**, no. 3 (1962);

[14] J. K. Poggenburg, H. J. Mang and J. O. Rasmussen, Phys. Rev. **181** (1969) 1697

[15] A. Insolia, R. J. Liotta and E. Maglione, Europ. Lett. **7** (1988) 209

[16] A. Insolia, P. Curutchet, R. J. Liotta and D. S. Delion, Phys. Rev. C44 (1991) 545.

[17] D. S. Delion, A. Insolia and R. J. Liotta, Phys. Rev. C 46 (1992) 884.

[18] D. S. Delion, A. Insolia and R. J. Liotta, Phys. Rev. C 46 (1992) 1346.

[19] D. S. Delion, A. Insolia and R. J. Liotta, Phys. Rev. C 49 (1994) 3024.

[20] D. S. Delion, A. Insolia and R. J. Liotta, Nucl. Phys. A549 (1992) 407.

[21] D. S. Delion, A. Florescu, M. Huyse, J. Wauters, P. Van Duppen, ISOLDE Collaboration, A. Insolia and R. J. Liotta, Phys. Rev. Lett. 74 (1995) 3939;
D. S. Delion, A. Florescu, M. Huyse, J. Wauters, P. Van Duppen, ISOLDE Collaboration, A. Insolia and R. J. Liotta, Phys. Rev. C 54 (1996) 1169.

[22] D. S. Delion, A. Insolia and R. J. Liotta, J. Phys. G 19 (1993) L189.

[23] P. O. Fröman, Mat. Fys. Skr. Dan. Vid. Selsk. **1**, no. 3 (1957)

[24] D. S. Delion, A. Insolia and R. J. Liotta, J. Phys. G 20 (1994) 1483.

[25] A. Florescu and A. Insolia, Phys. Rev. C 52 (1995) 726.

[26] S. H. Hanauer, J. W. T. Dabbs, L. D. Roberts and G. W. Parker, Phys. Rev. 124 (1961) 1512;
Q. O. Navarro, J. O. Rasmussen and D. A. Shirley, Phys. Lett. 2 (1962) 353.

[27] A. J Soinski and D. A. Shirley, Phys. Rev. C 10 (1974) 1488.

[28] J. Wouters, D. Vandeplassche, E. van Walle, N. Severijns and L. Vanneste, Phys. Rev. Lett. 56 (1986) 1901;
J. Wouters, D. Vandeplassche, E. van Walle, N. Severijns and L. Vanneste, Nucl. Instr. and Meth. B26 (1987) 463.

[29] T. Berggren, Phys. Lett. B 197 (1987) 1.

[30] P. Schuurmans et al., Report No. CERN/ISC 93-3.

[31] R. Lovas, R. J. Liotta, A. Insolia, K. Varga and D. S. Delion, Phys. Rep. (1997) (in press).

[32] B. Buck, A. C. Merchand and S. M. Perez, Atomic data and Nuclear Data Tables **54** (1993) 53

[33] T. Berggren and P. Olanders, Nucl. Phys. A473 (1987) 189.

[34] T. Berggren and P. Olanders, Nucl. Phys. A473 (1987) 221.

[35] A. Arima and I. Tonozuka, Nucl. Phys. **A323** (1979) 45

[36] G. Dodig-Crnkovic, F. A. Janouch and R. J. Liotta, Phys. Scr. **37** (1988) 523; Nucl. Phys. **A501** (1989) 533

[37] A. Insolia, R. J. Liotta and E. Maglione, Europhys. Lett. **7** (1988) 209

[38] R. H. Lemmer and A. E. Green, Phys. Rev. **119** (1960) 1043

[39] A. Faessler and R. Sheline, Phys. Rev **148** (1966) 1003

[40] P. Röper, Z. Phys. **195** (1966) 316

[41] E. Rost, Phys. Rev **154** (1967) 994

[42] V. V. Paskevitsch and V. M. Strutinsky, Nucl. Phys. (russian) **9** (1969) 56

[43] J. Dudek, A. Majhofer, J. Skalski, T. Werner, S. Cwiok and W. Nazarewicz, J. Phys. **G5** (1979) 1359

[44] S. Cwiok, J. Dudek, W. Nazarewicz, J. Skalski and T. Werner, Comp. Phys. Comm. **46** (1987) 379

[45] R. Bengtsson, J. Dudek, W. Nazarewicz and P. Olanders, Phys. Scrip. **39** (1989) 196

[46] K. Varga, R. G. Lovas and R. J. Liotta, Nucl. Phys. **A550** (1992) 421

[47] K. Varga and R.J. Liotta, Phys. Rev **C50** (1994) R1292

[48] I. S. Grant et. al., Proposal to the ISOLDE committee, CERN/ISC 94-8, P61

[49] D.S. Delion, A. Insolia and R.J. Liotta, Phys. Rev. **C54** (1996) 292

[50] J. M. Eisenberg and W. Greiner, Nuclear Theory I, Nuclear Models, (North-Holland, Amsterdam, 1970)

[51] R. G. Thomas, Prog. Theor. Phys. **12** (1954) 253

[52] M. Sotona and M. Gmitro, Comp. Phys. Comm. **3** (1972) 53

[53] A.Bohr and Mottelson, Nuclear structure, vol. 1 (Benjamin, New York, 1975)

[54] T. Fliessbach, H. J. Mang and J. O. Rasmussen, Phys. Rev. **C13** (1976) 1318

[55] R. Blendowske, T. Fliessbach and H. Walliser, Nucl. Phys. **A464** (1987) 75

[56] D. F. Jackson and M. Rhoades-Brown, J. Phys. **G4** (1978) 1441

[57] S. Raman, C. H. Malarkey, W. T. Milner, C. W. Nestor,JR and P. H. Stelton, Atomic Data and Nuclear Data Tables **36** (1987) 1

[58] Table of Isotopes, Ed.s C.M. Lederer and V.S. Shirley, seventh edition (Wiley - Interscience Publication, 1978)

[59] H.J. Rose and G.A. Jones, Nature **307** (1984) 245,
D.V. Aleksandrov. et. al., JETP Lett. **40** (1984) 909

[60] P.B. Price, Nucl. Phys. **A502** (1989) 41c

[61] A. Sandulescu and W. Greiner, Rep. Prog. Phys. **55** (1992) 1423

[62] B. Buck, A.C. Merchand and S.M. Perez, J. Phys. **G17** (1991) L91

[63] Y.-J. Shi and W.J. Swiatecki, Nucl. Phys. **A464** (1987) 205

[64] O. Dumitrescu, Phys. Rev. **C49** (1994) 1466, **C51** (1995) 3264

[65] R. J. Liotta and C. Pomar, Nucl. Phys. **A382** (1982) 1

[66] K. Wildermuth and RTh. Kanellopolus, Nucl. Phys. **7** (1958) 158; Nucl. Phys. **8** (1958) 449

INDEX

A phase, 197
A-decay, 267, 272, 282
A. C. Josephson effect, 77
Alloys, 249
Anderson theorem, 18
Anharmonicity, 33
Anisotropic superconductor, 16
Anisotropic superfluidity, 205, 218
Anti-Hund rule, 235
Anufriev's apparatus, 190
Aslamazov–Larkin theory, 101

B phase, 197
BCS-theory, 3
Bose–Einstein condensation, 7, 207, 208
Broken (gauge) symmetry, 207, 210, 211, 218
Broken symmetry, 4, 8, 16, 17

Clausius–Clapeyron equation, 187
Clebsch–Gordan coefficients, 237
Clogston limit, 148, 152
Cluster-decay process, 272
Clusters, 223, 245
Coherence length, 7, 18, 22
Collective oscillations, 253
Cooper
  instability, 4
  pair, 5
  phenomenon, 3, 4
  vertex part, 6
Cosmic strings, 218
Coulomb
  pseudopotential, 164
  repulsion, 13
Critical exponent, 96
Critical fluctuations, 96
Cross-technic, 18
Cuprate superconductors, 89

d electrons, 58
Daughter nuclei, 274
Double exchange, 68
Double-well, 65
"d-wave" superconductivity, 16, 17

Electron–phonon interaction, 161
Electrosolitons, 173
ET molecule, 150

Ferroelectrics, 65
FFLO state, 150
Fluctuations, 18, 22
Flux-creep, 114
Fluxon dynamics, 112
Formation amplitude, 276, 280
Fredhold equation, 16
Frohlich Hamiltonian, 174
Fullerene anion, 235
Fullerenes, 155

Gap equation, 11
Gapless superconductivity, 21, 22
Gauge invariance, 8
Gaussian fluctuations, 92
Giant resonances, 253
Ginsburg–Landau theory, 22
Gor'kov equations, 10
Gor'kov functions, 8, 9

Half magnetic flux quanta, 82
Heavy fermions, 4, 22, 81
Heavy-cluster decay, 283
Hebel–Slichter peak, 11, 22, 148
High Tc rings, 85
High temperature superconductors, 3, 22
Hopping integral, 69
Hubbard model, 240
Hund's rule, 66
Hybrid donors, 140
Hyperfine interaction, 57

Icosahedral group, 235
Impurity, 18, 249
  destroying effects of, 11, 22
Indirect exchange, 60
Ionic radius, 68
Isotope effect, 4, 25
Isotropic model, 11, 12

Jahn–Teller effect, 43, 68, 225, 235, 254
Josephson penetration length, 78

Knight shift, 57, 148
Kosterlitz–Thouless transition, 145

Landau damping, 255
Landau–Zener effect, 227
London limit, 35
Lorentz force, 116

Magic core, 269
Magnetic oscillatory effects, 141
Maki–Thompson theory, 102
McMillan tunneling model, 40
Melting, 253, 256
Metal–insulator transition, 158
Metallic particles, 60, 245
Migdal theorem, 13
Molecular soliton, 173
Monte-Carlo methods, 160
Mother nuclei, 275
Multistep shell-model, 284

Nesting, 141
Nilsson levels, 266
Non-adiabaticity, 225
Nonlinear response, 117
Nuclear magnetic ordering, 187
Nuclear magnetic resonance, 57

"Odd frequency" pairing, 9
Odd nuclei, 265
"On-ball" pairing, 241
Orbital degeneracy, 158
Order parameter, 4, 8
Oxygen octahedra, 63

π-electrons, 223
Pair condensation, 208
Pair wave function, 210
Pair-breaking, 36
Pairing state symmetry, 81
Paramagnetism,
    Pauli, 58
    van Vleck, 58
Paramagnetic impurities, 19
Phase correlation, 76
Phase diagram
    for Helium3, 206
    for Helium4, 206
Phonon mechanism, 3, 12
Photoemission, 161
Plasmons, 253
Pomeranchuk effect, 187
Power absorption, 112
Probability distribution, 266
Proximity effect, 41
Pseudopotential, 28

Reentrant phase, 143
Retardation effects, 165

Scaling in isotropic model, 11, 12
Screening, 166
SDW ordering, 141
Self-consistence equation, 11
Shell structure, 245
Sine–Gordon equation, 77
Size quantization, 246
Skin depth, 114
Solid helium, 192
Spectroscopic factor, 280
Spin dynamics, 215–217
Spin entropy, 201
Spin-flip, 36
Spin-lattice relaxation rate, 148
Strong vs weak coupling, 13, 22
Structural correlation length, 108
Superconductivity, 167
    and NMR, 60
    in metallic particles, 258
Superfluid Helium3, 205
Supershells, 253
Surface impedance, 125
Symmetric donors, 138
Symmetry approach, 4

T_{1u} electrons, 236
Talmi–Moshinsky symbol, 276
Time-reversal symmetry, 5, 6
TMTSF salts, 136
Topological defects, 214, 218
Touching radius, 279
Transverse mode, 65
Triplet state, 148, 150, 179
Tritium decay, 188
TTF derivative, 135
Tunnel junction, 259
Two-fluid model, 114

Ultrasound excitations, 217
Universality class, 95
Universality principle, 96
Upper critical field H_{C2}, 150

van Hove singularity, 31
Vertex part, 4, 6
Vortex tangle, 218

Woods–Saxon potential, 271, 273

XY model, 96

Zeroes in the gap, 20
Zn doping, 37